Applied Photonics

Applied Photonics

Chai Yeh

Department of Electrical Engineering
and Computer Science
The University of Michigan
Ann Arbor, Michigan

ACADEMIC PRESS

A Division of Harcourt Brace & Company
San Diego   New York   Boston   London   Sydney   Tokyo   Toronto

Cover illustration: A spatial soliton. After T. E. Bell, *Spectrum* **29** (1990) p. 57. © August, 1990 *IEEE*, with permission.

Academic Press, Inc.
525 B Street, Suite 1900, San Diego, California   92101-4495

United Kingdom Edition published by
Academic Press Limited
24–28 Oval Road, London NW1 7DX

Library of Congress Cataloging-in Publication Data

Yeh, Chai.
    Applied photonics / Chai Yeh.
        p.   cm.
    Includes bibliographical references and index.
    ISBN 0-12-770458-2
    1. Photonics.   I.   Title.
  TA1520.Y45   1994
  621.36 --dc20   93-21070
                                                          CIP

PRINTED IN THE UNITED STATES OF AMERICA
94   95   96   97   98   99   EB   9   8   7   6   5   4   3   2   1

This book is dedicated to my wife, Ida Shuyen Chiang

Contents

CHAPTER 3

Recent Advances in Semiconductor Laser Technology

CHAPTER 4

Fiber Lasers

CHAPTER 5

Solid-State Lasers

CHAPTER 6

Other Laser Sources

CHAPTER 7

Photonic Detection

CHAPTER 14

Photonic Image Processing

Preface

In analogy to electronics, *Photonics* was coined to reflect the importance of using the photon nature of light in many scientific and engineering devices. Electronics involves the control of electron-charge flow in a vacuum or in matter, whereas photonics deals with the control of photons in free space or in matter. In the past decade, we were fortunate to have witnessed the beginning of the revolution of photonics technology. We call it a revolution because photonics literally changed the concept of telecommunciations by introducing new methods of transmission and switching, allowing the proliferation of broadband services into homes and businesses. It all started with the development of optical fibers, which carry light-wave signals for transmission. It soon became evident that optical fibers can perform much more efficiently and at much lower cost than the usual copper-wire pairs or coaxial cables. Optical fibers have much lower line loss and smaller dispersion per unit length than coaxial cables. They have a broad bandwidth, which favors higher transmission speed and permits multiplexing to further increase the transmission capacity. Prior to photonics, signals transmitted at optical frequencies had to undergo electronic conversion to be amplified and processed. This conversion not only is costly, but also may introduce losses and even increase the probability of making more errors. Now, optical signals can be amplified and switched directly without conversion. Moreover, as optical amplifiers have a broad bandwidth, many channels in a multiplexing system can be processed simultaneously, thus further simplifying the system and improving its reliability. According to a survey by Kessler Marketing Institute (Newport, Rhode Island), because of world demand for fiberoptic communication, optical fiber installation, which in 1991 was 8.5 million kilometers, will grow 20% annually for many years. The fiberoptic component business is also predicted to grow from the present $4.5 billion to more than $11 billion by 1997. Photonic technology will have far-reaching effects. Long-haul transmission is only one field that has benefited. Photonic technology has been implemented both in the home and in business. On the home front, compact disk players have come on strong. On the business front, laser printers, data memory, and light display demand large volumes of fiberoptic components. The medical field is another rapidly rising field that uses fiberoptics in almost every branch of its services, including diagnostics, surgery, ophthalmology, dentistry, and medical and biological research. Although still evolving, photonic products have become a part of everyday life. With the increased lifetime and output power and reduced threshold operating conditions and lower cost, the future market of photonics is unlimited. Educators should be prepared to meet the demand for more scientists and engineers to handle the rapid growth of photonic technology. The purpose of this book is to help train scientists and engineers to meet these needs.

This book is intended to provide state-of-the-art information in the field of photonics for practicing engineers and scientists with college-level training in fundamental science and mathematics. The topics discussed are those in which progress is most rapid. The sequence of presentation has no relation to the importance of the topics. We present more physical concepts and fewer mathematical derivations. It is hoped that this approach will help readers understand the physics behind the work in optical fiber and photonics technology and allow them to use the up-to-date information to contribute to future advancements in this field.

The author is grateful to professor Yin Yeh at the University of California, Davis, for his many helpful comments and suggestions. He also expresses his appreciation for the cooperation he received from his colleagues and staff of the Department of Electrical Engineering and Computer Science at the University of Michigan. Finally, he thanks his wife, Ida Shuyen Chiang, for her patience and her continuing encouragement during the preparation of this book. She proofread the entire manuscript meticulously.

Chai Yeh

CHAPTER 1

Introduction

From Electronics to Photonics

Photonics is a new word: It parallels electronics. In an electronic circuit, signal transmission is usually carried out by electrons. In a photonic circuit, photons perform this function. As photons travel at the speed of light, it is obvious that the rate of transmission in a photonic circuit can be several orders of magnitude faster than that in an electronic circuit. Also, electrical interconnections using metallic wires often run into trouble when they cross each other, particularly, in electronically integrated circuits where space is limited. Because small optical interconnections can cross each other without interference, it is advantageous to replace these wired interconnections with photonic ones.

Photonics became popular when telecommunication systems adopted optical waves to transmit signals over optical fiber transmission lines, replacing copper wire pairs and coaxial cables. The rapid development of erbium-doped fiber amplifiers has changed the concept of fiber telecommunications significantly. Soon we will discuss long-distance signal transmission systems that can cover thousands of kilometers without using regenerative repeater stations. Erbium-doped fiber amplifiers, when used in conjunction with soliton pulses, can bridge oceans in high-bit-rate transmissions. Optical signal transmission, both detection and processing, without optoelectronic and electrooptic conversions, will be a possibility within the next few years. Interconnections within integrated circuits may also adopt the optical route soon. In fact, the major impact of photonic technology in this information age of the 1990s may be increased optical means of information processing and data transmission. The trend will soon spill over into both the home and business. Local area networks (LANs) have exploited the use of synchronous optical networks (SONETs) to take advantage of the high speed and broad bandwidth of optical signaling services. More and more devices will be designed using photons as signal carriers. This is only the beginning. New developments arise daily.

A Brief History

Scientists and engineerers in this century are fortunate to have witnessed or even participated in the technological revolution in electronics. Although an electron emission phenomenon known as the Edison effect was discovered in the 19th century, the first vacuum triode was built in 1903 by Lee DeForest. Amplifier and oscillator electronics followed immediately. This marked the beginning of electronic engineering. Wireless signal transmission via high frequencies began to dominate the communication field in the 1920s. High-power broadcasting stations were built around the world. Amplitude and frequency modulations were introduced. Short-wave transmission began to boost worldwide radio transmission, followed by microwave transmission. During the Second World War, radar, first introduced by the British Royal Air Force, used microwave frequencies to detect enemy war planes in distant skies. This technological advance was even credited for helping to win the war, which may not be an overstatement. The next breakthrough was the announcement of semiconductor transistors in 1948, which have since replaced vacuum tubes for most amplifying and oscillatory needs. This was followed by the integrated circuit and microelectronics, by which more efficient circuitry could be built on a smaller landscape. A radically new concept emerged from the invention of lasers in the late 1950s and early 1960s. Encouraged by the discovery of low-loss glass fiber in the 1970s, telecommunication technology shifted toward the use of optical fibers as a transmission medium and lasers as an optical source in the 1980s, thus establishing optical telecommunication systems. This switch revolutionized the communication industry. More and more optical components are required to meet this new technology. When the advantages of the optical means of communication were realized, its applications spread into many other fields. Just like electronics in the early 20th century, photonics is now the frontline subject in many research and development projects.

Future Outlook of Photonics

Perhaps a more precise name for photonics is quantum electronics, which describes the relationship between electrons and the quantum nature of matter and radiation. Quantum electronics is the field that expresses the physics involved over and above the broad range of applications. We make an arbitrary distinction between quantum electronics and photonics here for convenience. The former emphasizes theory, whereas the latter applies to engineering practice. So *Applied Photonics* was chosen as the title of this book.

Since the discovery of the maser and laser in 1958–1960, the field of quantum electronics has progressed far. We will try to share our feelings of excitement and vitality in this field with our readers.

The development of photonics is multidirectional. For the purpose of discussion, we subdivide this development into three directions: (1) lasers, (2) optical functional devices, and (3) operational optical components.

With respect to lasers, once the theory of laser operation is understood, the object is to find materials that lase at different frequencies. The first ruby laser operated at a wavelength of 694 nm. New coherent radiation in the X-ray laser has produced lasing action at wavelengths as short as 3 nm. Gases, liquids, semiconductors, solid-state materials, and other materials have been used in the effort to find materials that lase at other wavelengths. One is not confined to pure substances; any mixture of gases, solids, alloys or organic materials can be made to lase. For any special application, one may need to determine a suitable structure of the laser that gives the most efficient operation. For example, specific lasers of single frequency and single mode, operating at a specific wavelength with good spectral characteristics required for long-haul optical communications, have been developed. Vertical quantum-well lasers and surface emitting lasers are suitable for monolithic integration with other optical or electronic components. Other solid-state lasers, designed to cover a wide tunable range of wavelengths, can be modified to replace the dye lasers. Many types of powerful lasers are now commercially available. With the use of external pumping schemes, different wavelengths and power levels can be reached. Double and multiple pumping techniques will further extend the power range of the laser. There has already been further extension of laser technology into high-power devices of various kinds, including ion lasers, free-electron lasers, and even X-ray lasers.

Functional devices include the fiber amplifiers, where erbium-doped silica fiber can be pumped to deliver amplified power for light transmission at 1550 nm; the Nd-doped ZBLAN and the Pr-doped fluoride fiber can do the same for the 1300-nm fibers for optical communication systems. Also available are schemes for soliton transmission of pulses that maintain their shape for long distances and times without distortion.

Nonlinear crystals have been synthesized to fill the need to effect a change in the refractive index of crystals for phase-conjugate mirrors, which can be arranged to undo the distortions introduced into the system. Other crystals may have larger electrooptic and Kerr effects. Harmonic generation using nonlinear crystals can further extend the usefulness of these crystals. Frequency mixing using many pump schemes with these nonlinear crystals can generate a variety of devices for many applications.

Finally, operational optical systems will require many components to complete their connections. Optical switches and interconnects find uses in many fields, particularly in networking and microelectronic integration. Optical signaling will soon spread to the home, where many electronic devices may be replaced by their optical equivalents in the near future.

Each of the items we have mentioned represents many opportunities for research and application. The door is wide open, and work can be found in any of these fields. Photonics is still in the infant stage, just as electronics was at the beginning of the

20th century. With the help of advanced electronics technology, we could progress much more easily than did electronics.

Organization of the Book

The topics selected to be presented in this book are those in which progress is most rapid, as evidenced by the existence of a large number of productive research reports.

In this chapter, the organization of the book is outlined. Chapter 2 introduces photons and their interactions with other matter. Photon–phonon interactions, nonlinear optics, and nonlinear crystals are briefly described. These discussions form the basics of photonics to be described in later chapters. The remaining chapters, Chapters 3 to 14, are devoted to applications of these effects in building photonic devices. To cover a wide range of applications, we divide the presentation into three groups: (1) laser light sources and detectors, (2) nonlinear devices, and (3) components and other applications.

Laser Light Sources and Detectors

Light sources and detectors are the fundamental components of photonic technology. Lasers of all kinds are used as light sources and semiconductor detectors are used for detection.

Recent Advances in Long-Wavelength Semiconductor Lasers for Optical Fiber Communications

Since 1980, semiconductor lasers have assumed a major role as light sources in the rapid development of optical fiber communications. Semiconductor lasers are well suited for this purpose because they are comparable in size to optical fibers, and one can select a wide range of wavelengths from the semiconductor groups to use as carrier waves. To make full use of the low-loss, low-dispersion, and wideband capability of optical fibers, the characteristics of semiconductor lasers are far-reaching and stringent. The trend is to look for semiconductor lasers that have a very low threshold current; are capable of high-speed modulation and single-frequency and stable operation at room temperature; have an adequate power output; produce low noise; are of low cost; and are highly reliable. Frequency tunability is also highly desirable. Chapter 3 is devoted to this objective.

Fiber Lasers

For long-haul telecommunication systems involving distances of thousands of kilometers or greater, optical fibers still require the installation of repeater stations to boost the signals and correct the dispersion effect. Repeater stations are the most expensive equipment in a telecommunication system. Er-doped and other fiber lasers are developed to reduce the number or even eliminate repeater stations. Chapter 4 is devoted to this subject.

Solid-State Lasers

Solid-state lasers have many applications besides telecommunication systems. These lasers are intended to satisfy the demand of higher-power output, and they operate over a widely different frequency range that semiconductor lasers cannot easily fulfill. Chapter 5 briefly describes these lasers.

Other Laser Sources

Practically any matter can be made to lase. Gaseous lasers, dye lasers, excimer lasers, free-electron lasers, and X-ray lasers are all sources with high-power capability and together they span a wide frequency range. A short introduction to these lasers is given in Chapter 6.

Photonic Detection

Methods and devices for photonic detection are described in Chapter 7, which concludes the first part of the book.

Nonlinear Optical Devices

This division is rather arbitrary. Many devices involved in the discussion of laser sources operate on the nonlinear properties of materials. But the following few are special cases worthy of mention as nonlinear devices.

Optical Amplifiers

In Chapter 8, optical amplifiers, which are rapidly replacing electronic regenerative repeaters in telecommunication systems, are the subject of discussion. An optical amplifier uses photons to amplify the optical signal directly. In the 1980s, we experimented with light transmission using a stimulated Raman optical amplifier with some success. Recently developed erbium-doped optical amplifiers, pumped by other lasers, have made rapid advances in optical fiber telecommunications. Repeaterless signal transmission of many thousands of kilometers could become a reality if this type of amplifier is used in conjunction with optical fibers. That an optical amplifier can amplify optical signals without converting them through electronics has major implications for the transmission of voice, video, and high-speed data. An erbium-doped amplifier, with its wideband and low-noise capability, can amplify the signal more efficiently. Erbium-doped optical amplifiers can also be concatenated to increase the amplification. A distributed erbium-doped fiber amplifier can also be built, if needed. Fiberoptic cable with built-in optical amplifiers, including semiconductor laser amplifiers, will never be obsolete. A transoceanic transmission system can be updated to carry more channels or frequencies at shore-based terminals without going through the expensive process of pulling up the cable for routine maintenance. The long-term operating and maintainence costs of such a system will be dramatically reduced. Furthermore, without the process of regenerative repeater and signal conversion, the error rate and reliability of the transmission can be improved. Less noise can also be expected.

Solitons

A special pulse, known as a soliton, is discussed in Chapter 9. The term *soliton* is used to describe a pulselike wave traveling in a nonlinear and dispersive medium such as an optical fiber. Solitons offer the possibility of a pulse with unchanged shape and pulsewidth, that is, distortionless transmission in a certain wavelength range of an optical fiber transmission line. But solitons offer no gain to offset fiber losses. External amplifiers have to be used to make up the losses.

Although a wave solution of solitons was known almost 100 years ago (around 1890), its application to long-haul telecommunications is only recent. If soliton pulses are used in conjunction with erbium-doped optical amplifiers, signal transmission across the Atlantic or Pacific Ocean on a submarine optical fiber cable without regenerative relay stations may be possible. At present, solitons have been tested in laboratories at 2.5 Gbit/s over a distance of 14,000 km, using only optical amplifiers (with recirculating loops), and error rates below 10^{-10} have been achieved. Practical cables for carrying out actual transmission may be only a few years away. It is hoped that the soliton will become the transmission method of choice by the year 2000.

Phase Conjugators

A unique property of phase conjugation is that a light wave reflected from a nonlinear crystal retraces its transmission path back to the origin. In doing so, phase distortions introduced in its first passage can be canceled. Methods for generating phase conjugation and their potential applications are the subject of Chapter 10.

Photonic Components and Other Applications

Photonic Components

As the range of applications extends beyond telecommunications, different types of photonic components are required to supplement optical fibers and to operate the optical system. Chapter 11 introduces some components currently in use, as well as devices for potential applications.

Photonic Switches

In Chapter 12, optical switches and switching are discussed. The development of optical switching schemes may have major implications for both telecommunications and optical computing. All-optical logic gates, the building block of optical computers, and optical processors are only two of many challenging subjects in this field; however, in this chapter, only the general principles underlying some of these subjects are described. Optical computing is not covered in detail. Solitons have been suggested for use in all optical switches.

Photonic Interconnections

A hot subject in photonics today is the development of photonic interconnections in integrated photonic devices, integrated microelectronics, and optical computing. The important problems in high-density photonic integrated devices are crosstalk between channels, excessive power dissipation, and device interconnections. It has been suggested that optical connections can minimize or alleviate the interconnection problem. A complex all-optical switching system may be required for large networks to speed up data transmission. In Chapter 13, we introduce this subject to reveal the many ways in which the interconnection problem may be handled.

Photonic Imaging Processing

In modern telecommunication systems, signal processing is done electronically as digital processing. The advantages of digital processing are accuracy and speed as a result of advances in digital computer technology. In optical communications, however, optical signals have to be converted into electrical signals before the advantages of the system can be realized. In the conversion process, not only is a lot of equipment required, but there is a greater chance of introducing errors. On the other hand, optical processing may be simple to implement, although less accurate. In Chapter 14, we investigate different types of optical imaging processors and suggest that at present, hybrid systems consisting of both optical and digital processing components may be advantageous.

Summary

Photonic technology has advanced very rapidly. It has been found to be broadly applicable to all branches of science and engineering, to supplement or even replace electronic technology. Others have suggested that photonic technology has already invaded the business world as well as the home. Mass production of minilasers by integrated techniques is reported. These new light sources, which cost very little to fabricate, will certainly encourage innovative designs of devices useful business and the home. They can easily flood the market with new applications for years.

We have omitted many other optical signal processing methods in this book. In particular, the application of photonics to the medical field has not been touched for lack of space. Also left untouched are photonic sensors for both scientific and industrial applications. But these fields are wide open. Keen readers will have to keep watching for other developments.

Bibliography

1. Special issue on semiconductor diode lasers. *IEEE J. Quantum Electron.* **QE-25,** No. 6 (1989).
2. T. P. Lee, Recent advances in long-wavelength semiconductor lasers for optical fiber communication. *Proc. IEEE* **79,** 253–276 (1991).
3. Special issue on quantum electronics. *Proc. IEEE* **80,** 337–464 (1992).
4. Nonlinear optical materials and devices for photonic switching. *SPIE* **1216,** 1–299 (1990).
5. C. S. Tsai, *Guided-Wave Acousto-optics,* Springer-Verlag, Berlin, 1990.
6. Optical interconnections and networks. *SPIE* **1281,** 1–289 (1990).
7. H. M. Gibbs, G. Khitrova, and N. Peyghambarian (Eds.), *Nonlinear Photonics,* Springer-Verlag, New York, 1990).
8. Special section on acousto-optic signal processing (six papers). *Proc. IEEE* **69,** 48–118 (1981).
9. H. A. Haus, *Optical Fiber Solitons, Their Properties and Uses, Proc. IEEE,* **81,** 970–983 (1993).

CHAPTER 2

Photons and Interactions

Introduction

The newly coined word *photonics* is used to emphasize the photon nature of light in operating many optical devices. Stimulated by the rapid expansion of optical fibers in communications, optical devices reclaimed their importance in modern communication technology. New developments in laser technology, the rapidly expanding use of optical fibers in many fields besides communications, and the rapid development of semiconductor optical devices for many applications all led us to recognize the need for a new discipline, photonics.

Before we explore photonics in detail, we wish to review the essential properties of photons, the principal participant of photonics, and their relationships in modern optics, quantum optics.

Quantum Optics

The classical electromagnetic theory of light, as developed by James Clark Maxwell in the late 19th century, has been successful in explaining a great many effects in optics. After the turn of this century, however, the failure of classical theory to explain certain optical phenomena, particularly those exhibiting the quantum nature of light, began to be realized in optical experiments. The first experiment that defied classical prediction was the "black body radiation" phenomenon. The use of classical equipartition of energy for radiation field energy led to the ultraviolet catastrophe, where the calculated total field energy becomes infinite. This paradox was resolved when Max Planck introduced the idea that radiant energy comes in energy units of

$h\nu$, where ν is the frequency of light and h is Planck's constant [1]. He derived a valid expression for the spectral distribution of black body radiation by boldly assuming that the radiated energies are quantized, instead of continuous as assumed by classical theory. In this quantum theory, the energies can take on only certain integrally related discrete values: $0, h\nu, 2h\nu, 3h\nu, \ldots$.

The photoelectric effect was another experiment that showed the discrete quantum effect of light. Einstein, in photoelectric effect experiments [2], asserted that (1) there is a threshold for the onset of photoelectron emission, and (2) the amount of photoelectric current is proportional to the intensity of excitation. Whereas the first observation pointed to the quantum nature of matter, the second observation showed that the number of electrons is proportional to the number of "photons" of well-defined energy.

Compton scattering of light by free electrons further showed that these entities of light with energies $h\nu$ have ballistic properties and can be treated like particles bombarding the electron. Conservation of energy and momentum then leads to changes in the resultant energy of the photon on collision [3].

The countering experiment usually cited is Young's interference experiment [4]. In this study, light incident on a screen with two slits resulted in a far-field pattern that was not ballistic at all, but wavelike. The far-field screen exhibited an interference pattern entirely consistent with Huygen's principle of secondary wave propagation.

The photon nature of light has only recently been established in this experiment with the availability of sensitive detectors capable of "seeing" single photon events. As it has now been shown that a single photon can nonetheless yield interference phenomenon, a photon is said to have both particle-like and wavelike characteristics, thus establishing the concept of particle–wave duality for photons. Traditionally, it was thought that all optical media respond linearly to an impressed light field. We were taught that (1) optical properties, such as the absorption coefficient and the refractive index of a dielectric medium, are independent of light intensity; (2) the principle of linear superposition of light waves holds; (3) the frequency of light cannot be changed by passage through the medium; and (4) light does not interfere with light. At present, all these have to be modified. Quantum optics has provided explanations to all these changes in optical effects in light–matter interaction. Many of these irregularities are called nonlinear effects. The new theory is not, however, intended to replace classical theory, but rather regards classical theory as a limiting case.

What are quantitatively observable parameters in quantum optics? The physical state of the photon is not represented by specification of such classical variables as its position or momentum. Instead, the formal description of the state of photon is contained in its wavefunction Φ, which describes the probability amplitude of the photon's presence, analogous to the fields in Maxwell's equations. This wave amplitude contains all the information available about the photon probability distribution, including the polarization state of the photon. But it is the square modulus of Φ

($\Phi * \Phi$), which plays the role of the intensity that gives a measure of the probability of finding a photon at a particular space–time point. The major modification necessary is to interpret the wave intensity as a photon probability density. In the following chapters, we describe the photons and their many nonlinear optical effects in applied photonics.

Properties of the Photon

Max Plank was the first to suggest that light consists of particles called photons. Some prefer to regard photons as localized packets of energy. A photon has zero rest mass and carries electromagnetic energy and momentum. It travels at the speed of light *in vacuo*, but its speed is retarded in matter. Photons have a dual particle/wavelike property. They travel like a wave and interfere and diffract like matter. The emission and absorption of photons occur in quanta of energy. In other words, energy and momentum of a photon are quantized.

Photon Energy

The energy of a photon can be expressed as $E = h\nu$, where h is Plank's constant and ν is the frequency of the mode of the harmonic oscillator to which the photon belongs. As $h = 6.63 \times 10^{-34}$ J-s and is a constant, energy can be added or substracted in steps only; that is, energy is quantized. The wavelike nature of a photon is revealed in the frequency ν, which can be related to the wavelength by the equation $\nu\lambda = c$, the velocity of light *in vacuo* as in classical theory. The unit of energy is joules. Its practical unit is electron-volts (eV). But a convenient unit is reciprocal wavelength (cm^{-1}). Thus, wavelength $\lambda = 1$ μm has energy 1.24 eV; its reciprocal wavelength is 10,000 cm^{-1}. Other values can be calculated and plotted.

The relationships between photon frequency ν (Hz), wavelength λ (μm), energy E (eV), and reciprocal wavelength $1/\lambda$ (cm^{-1}) are shown in Figure 2.1.

Description of the Electromagnetic Field
Consistent with the Photon

The basis of the quantum-mechanical description of the field is that the total energy of the radiant field can be described like the energy of a harmonic oscillator. In the same way that the quantum-mechanical oscillator has discrete, equally spaced energies for its energy levels, the photon field has equally spaced energy levels indicating the number of photons of basic energy unit $h\nu$.

From the statistical probability distributions of the state of the photon, we can calculate the variances for the expected values and then show that the coherent field has the minimum uncertainty of any possible state. Thus we are allowed to treat the coherent state almost as a classical stable wave, using the simplest description of the classical E and B fields.

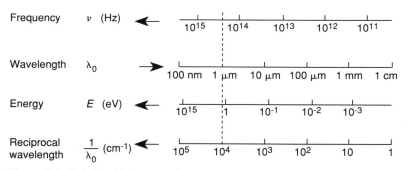

Figure 2.1 Relationship between frequency, wavelength, energy, and reciprocal wavelength. Frequency ν and wavelength λ are related by $\nu/\lambda = c$, the speed of light in the medium. Energy $E = \hbar\nu = \hbar\omega$, where h is Planck's constant, $h = 6.63 \times 10^{-34}$ J-s, and $\hbar = h/2\pi$. E is also $= (2\pi c\hbar)/\lambda$ in reciprocal wavelength units. $E(eV) = 1.24/\lambda$ (μm).

Let us start with a classical plane wave traveling in the r direction, written as

$$\psi = A \exp[i(\mathbf{k} * \mathbf{r} - \omega t)]\mathbf{1r} \tag{2.1}$$

where $\mathbf{k}$, the propagation wavevector of the wave, satisfies the equation

$$\mathbf{k} = \left(\frac{1}{h}\right)\mathbf{p} \tag{2.2}$$

where h is Planck's constant and $\mathbf{p}$ is its momentum. To modify Eq. (2.1) to represent a particle wave traveling in the z direction with an angular frequency ω ($\omega = 2\pi\nu$), the particle must have a well-defined momentum p that satisfies the de Broglie relation $\lambda = h/p$ [5].

The momentum of a photon is a vector quantity. It is expressed as

$$\mathbf{p} = \hbar\mathbf{k} \tag{2.3}$$

where $\hbar = h/2\pi$, and $\mathbf{k}$ is the wavevector. The magnitude of the momentum is $p = h/\lambda$.

Photons possess angular momentum depending on the nature of radiation. In electric dipole radiation, the orbital angular momenta differ from each other by $\pm\hbar$ unit and are also quantized. The $+$ and $-$ signs before h indicate the left and right circularly polarized photons, respectively. A linearly polarized photon is equivalent to the superposition of a right and a left circularly polarized photon, each with a probability 1/2.

Energy and momentum are then related by

$$\mathbf{p} = \left(\frac{E}{c}\right)\mathbf{1}r \tag{2.4}$$

The polarization of a photon is that of its mode. We can prescribe its polarization as plane polarization by modifying A in Eq. (2.1) as $A = Ax\,\mathbf{1}x + Ay\,\mathbf{1}y$, where $Ax = (1/\sqrt{2})(Ax - Ay)$, and $Ay = (1/\sqrt{2})(Ax + Ay)$, respectively. For a circularly polarized photon, a similar expression can be derived.

The Uncertainty Principle

It is impossible to describe with absolute certainty the position and momentum of a photon. Instead, we must speak of the expected values of these quantities. This is because position and momentum do not exist separately. Heisenberg's uncertainty principle states that in any measurement of the position and momentum of a particle, the uncertainties in the two measured quantities will be related by $(\Delta x)(\Delta p) > \hbar$. Similarly, the uncertainty in energy measurement will be related to the uncertainty in time at which the measurement was made by $(\Delta E)(\Delta t) > \hbar$. Thus, we cannot properly speak of the position of a photon, but instead the probability of finding a photon at a certain position. This is an important result of quantum mechanics. A probability density function can be obtained for a photon in a certain environment, and this function can be used to find the expected values of important quantities such as position, momentum, and energy.

The statistical distribution of the number of photons depends on the nature of the light source. Light can be produced by thermal heating. In this case, the probability distribution of finding n photons is governed by the Boltzmann distribution and is given by

$$P_n = A \exp\left\{\frac{-En}{(kT)}\right\} \tag{2.5}$$

where A is a constant, and k is Boltzmann's constant ($k = 1.38 \times 10^{-23}$ J/°K); however, for coherent radiative light, the Poisson distribution applies.

$$P_n = \mathbf{n}^n \exp\frac{(-n)}{n!} \qquad n = 0, 1, 2, \ldots \tag{2.6}$$

where $\mathbf{n}$ is the mean value of n, a random number.

The preceding is a brief review of the theory of quantum mechanics and can be found in many standard textbooks [6].

Interaction of Photons with Atoms

The wave nature of photons is expressed in Eq. (2.1). A bundle of photons in motion constitutes a beam of light. The electric field of the light wave can interact with the valence electron(s) of the atom in matter. The force exerted on the atom causes the atom's electrons to vibrate or accelerate. Conversely, vibrating electric charges emit light. The interaction process is governed by the rules of quantum mechanics which allow only specific levels of energy to participate in the exchange.

Two processes may be observed in this interaction, absorption and emission. The latter can again be divided into spontaneous and stimulated emission.

Absorption. The photon may impart energy to the matter or atom if its energy matches the difference between two energy levels. The atom thus gains energy and the photon is said to be absorbed. The atom is thereby raised to a higher energy level. Conversely, the interaction between photon and matter may result in the transition of the atom to a lower energy level, resulting in emission of a photon of energy equal to the difference in energy between the levels. In all these transitions, the law for conservation of energy holds true.

Spontaneous emission. If the atom is initially in the upper energy level, it may drop spontaneously to a lower level and release its energy as a photon. As the transition is independent of the number of photons that may already be in the mode, it is called spontaneous emission. It is a random phenomenon. The probability of spontaneous emission is proportional to the transition cross section which is a function of the frequency.

Stimulated emission. If an atom is in the upper energy level and the mode already contains a photon, the atom may be stimulated to emit another photon that duplicates itself in frequency, the direction of propagation and polarization. This emission is coherent. The probability of emission is also proportional to the transition cross section, but this may be multipled by n if n photons are present.

We refer to these processes often in the following chapters.

Interaction of Light with Light

This section is written for the purpose of clarifying the sometimes very contradictory statement that light beams do not interfere with each other.

According to the classical electromagnetic theory of light, the behavior of a light beam is prescribed solely by the wavelength of the light and its velocity in the medium. The transmission, refraction, and reflection of a beam of light in a transparent medium are not affected by the presence of a second beam. We were also taught that the principle of superposition holds true in optics and that the frequency of the light cannot be altered by passage through a medium. These statements are all true in normal situations with ordinary light intensity where the absorption coefficient α and the index of refraction n are regarded as material constants that do not change with light intensity.

Interaction between light beams may, however, take place through changes in the dielectric properties of the medium. The refractive index and the absorption coefficient of dielectric media could change under intense incident light, as evidenced by many experiments.

The electric field of sunlight at the surface of the earth has an amplitude of about 10 V/cm. This intensity can change the refractive index of glass by only one part in 10^{15}. The magnetic field can change the refractive index by the Faraday effect. But the magnetic field of sunlight is about 1/30th of a gauss, less than one-tenth the strength of the earth's magnetic field. The change in the refractive index of glass effected by this field is only about one part in 10^{12}. No instrument in the world can detect these changes yet. Thus, for these effects to be readily observable, a great deal of energy must be concentrated in a narrow band of wavelengths. The statement that light beams do not interfere holds true under these conditions; that is, the classical theory is still true under weak field intensities. With the intensities now available in lasers, however, the refractive index of matter can be changed. The Kerr effect, the Raman scattering effect, and other nonlinear optical effects have been observed under high-light-intensity conditions. Thus, since the invention of laser, higher light intensities have become available, and nonlinear effects in optical materials have become very important in many optical experiments. Among these are harmonic generation, parametric amplification, and others, many applications of which are described in later chapters.

Nonlinear Optics

The focused light from certain lasers has an electric field in excess of 10^7 V. Pulsed lasers used in fusion experiments may have much higher intensities. Such strong fields are comparable to the cohesive local electric fields in crystals. Interaction of photons with atoms in matter is therefore expected.

The nonlinear optical effect usually stems from the nonlinearity of an optical medium, such as changes in refractive index and absoption coefficient under intense incident light. In a nonlinear medium, a change in the speed of light on intensive light excitation is expected. Light can also alter its frequency as it passes through a nonlinear medium. In addition, the principle of linear superposition no longer applies to nonlinear optics. Finally, light can interact with light within these media [7].

In classical electromagnetic theory, the electric flux density **D** and the magnetic flux density **B** satisfy Maxwell's wave equations. In linear dielectric material,

$$\mathbf{D} = \epsilon_0 \mathbf{E} + \mathbf{P} \qquad (2.7)$$

where ϵ_0 is the permittivity in free space, and **P** is the polarization vector such that

$$\mathbf{P} = \epsilon_0 \chi \mathbf{E} \qquad (2.8)$$

where χ is the dielectric susceptibility of the material.

In nonlinear dielectric material, the dominant consideration is the nonlinear susceptibility. The nature of this quantity is that it constitutes the response function of the material on an impressed external electromagnetic field. In general, for a sufficiently large electric field, the nonlinear susceptibility becomes a tensor and therefore the polarization vector can best be represented by an infinite power series expansion of the form

$$\mathbf{P}_i = \epsilon_0 \{ \Sigma \, \chi_{ij} \mathbf{E}_j + \Sigma \, \chi_{ijk} \mathbf{E}_j \mathbf{E}_k + \Sigma \, \chi_{ijkl} \mathbf{E}_j \mathbf{E}_k \mathbf{E}_l + \ldots \} \qquad (2.9)$$

where χ_{ij} is a dimensionless second-rank tensor called the linear susceptibility coefficient that characterizes the electrodynamic properties of electron/ion dioples, and χ_{ijk} and χ_{ijkl} are the second- and third-rank tensors, respectively. $\mathbf{E}_j$, $\mathbf{E}_k$, and $\mathbf{E}_l$ are the electric fields in the directions j, k, and l, respectively. Note that Eq. (2.9) is very complex if all components of the tensor exist. Fortunately, for most crystals or nonlinear media, symmetry exists. The electric field can be oriented such that the susceptibility of only one particular direction will play an important role in characterizing the second- and third-order nonlinearities. This simplifies the expression of the polarization vector and it can be rewritten as

$$\mathbf{P} = \epsilon_0 \{ \chi^{(1)}(x)\mathbf{E}(x) + \chi^{(2)}(x)\mathbf{E}(x)\mathbf{E}(x) + \chi^{(3)}\mathbf{E}(x)\mathbf{E}(x)\mathbf{E}(x) + \ldots \} \qquad (2.10)$$

where $\chi^i_j(x)$ are the ith-order susceptibility of the material. The first term on the right-hand side of Eq. (2.9) is the polarization vector of linear dielectrics and has been replaced by $\chi^{(1)}$. The second term χ_{ijk} is replaced by $\chi^{(2)}$, and the third term χ_{ijkl} has also been replaced by $\chi^{(3)}$. The electric field components are replaced by $\mathbf{E}_1$, $\mathbf{E}_2$, and $\mathbf{E}_3$, respectively.

In the case in which the susceptibility differs in the x, y, and z dimensions, the χ_{ij} format should be retained. Any crystal anisotropy can be taken care of using this format. For isotropic media, only χ exists. The most important outcome of the diople moment described in Eq. (2.8), which is a simplified version of the first part of Eq. (2.10), is the creation of a second electromagnetic wave having the same frequency as the original light wave. This second wave affects the propagation of the original wave in many ways. Most linear optical properties, such as absorption, birefringence, dispersion, and index of refraction, can be deduced from this expression.

The second and third terms represent second- and third-order nonlinearity, respectively. From the second term of the power series in Eq. (2.10), we may study the properties of optical rectification, Pockel's effect, second-harmonic generation (SHG), sum- and difference-frequency mixing, and optical parametric amplification (OPA) or oscillation (OPO).

The third term in Eq. (2.10) leads us to study the Kerr effect, third-harmonic generation (THG), four-wave mixing, third-order sum- and difference-frequency mixing, coherent anti-Stokes Raman scattering (CARS), stimulated Raman scattering, stimulated Brillouin scattering, phase conjugation, self-focusing, self-modulation (soliton), and two-photon absorption, ionization, and emission.

When the P–E relationship is plotted, a deviation from a straight line is an indication of nonlinearity. In the following discussion of nonlinear dielectrics, the effects of anisotropy and inhomogeneity of the material are neglected for simplicity.

The study of nonlinear dielectric material is also complicated by its structural difference. If now we consider space inversion from x to $-x$, we have symmetry constraints that dictate the presence or absence of specific $\chi^i(x)$ terms. For example, the condition for the presence of second-order nonlinear susceptibility is seen by examining

$$\mathbf{P}_2(x) = \epsilon_0\chi^{(2)}(x)\mathbf{E}(x)\mathbf{E}(x) \tag{2.11}$$

If we write $\mathbf{P}_2(-x) = \epsilon_0\chi^{(2)}(-x)\mathbf{E}(-x)\mathbf{E}(-x)$, then it is clear that $-\mathbf{P}_2(x) = \epsilon_0\chi^{(2)}(-x)(\chi)\mathbf{E}(x)\mathbf{E}(x)$. Therefore,

$$\mathbf{P}_2(\mathbf{x}) = \epsilon_0\chi^{(2)}(-x)\}\mathbf{E}(x)\mathbf{E}(x)$$

or

$$-\chi^{(2)}(-x) = \chi^{(2)}(x)$$

It is thus the material with no center of inversion that will yield $\chi^{(2)}(x)$ as the leading nonlinear contribution. In a similar manner, we can show that a material system with a center of inversion will not have $\chi^{(2)}(x)$ or any other even-ordered susceptibilities; the leading nonlinear susceptibility becomes $\chi^{(3)}(x)$.

Second-Order Nonlinearity

Second-Harmonic Generation

Assume that a dielectric medium contains only second-order nonlinearity, or $\mathbf{P} = \chi^2(x)\mathbf{E}(x)\mathbf{E}(x)$. The response of this medium to an alternating electric field of angular frequency ω has an even symmetry. This is shown in Fig. 2.2. Thus, a second-order nonlinear optical medium will create a polarization field with a dc component and a second-harmonic component, 2ω. A second-harmonic frequency component has been generated. Further studies established the fact that the amplitude of the

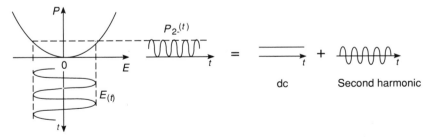

Figure 2.2 Second-harmonic generation scheme. A sinusoidal electric field of angular frequency ω in a second-order nonlinear optical medium creates a polarization with a component at 2ω and a steady component.

radiated second-harmonic optical field is proportional to the incident optical intensity squared, to the square of the coefficient $\chi(x)$ [in Eq. (2.10)], and to the inverse fourth power of the wavelength. The efficiency of the second-harmonic generation is proportional to W/A, where W is the incident power and A is the cross-sectional area. Many bulk nonlinear crystal such as KDP, as well as optical waveguides, have been explored for the purpose of achieving second-harmonic generation.

The direct-current component observed in Fig. 2.2 can be used to generate dc voltage across a nonlinear crystal when exposed to intense light radiation. This is, however, a very expensive way to produce a direct current. But it can serve as a bias for the polarization field and will affect other properties of the medium.

This nonlinearity can also be used for rectification.

Electrooptic Effect

Pockel's effect can also be derived from second-order polarization. If the incident light contains a large dc component E_0, the resulting polarization will contain dc component P_0, a component containing the original frequency P_ω and a second-harmonic frequency component $P_{2\omega}$. If the input light intensity of frequency ω is such that $E_0 \gg E_\omega$, then by neglecting $P_{2\omega}$, making $P_\omega = \epsilon_0 \chi E_\omega$, we have Pockel's effect. The refractive index changes in proportion to the applied electric field and is known as the linear electrooptic effect. The resulting change in the refractive index can be expressed as $\Delta n = (\chi/n\epsilon_0)E_0$, which can also be written as $r = -2\chi^{(2)}/\epsilon_0 n^4$, which is known as Pockel's coefficient.

Frequency Mixing

Let the input light initially contain two optical waves of optical frequencies ω_1 and ω_2. In passing through the medium containing second-order nonlinearity, the resultant output then contains five frequencies, 0, $2\omega_1$, $2\omega_2$, $\omega_+ = \omega_1 + \omega_2$, and $\omega_- = \omega_1 - \omega_2$, plus all other higher-order components. This process is called frequency mixing. By mixing two optical waves of different frequencies, a third wave of the sum or difference frequencies can be generated if other frequency components can be filtered out. If the difference frequency is chosen, known as down conversion, then a new wave of lower frequency will be generated. This process is also known as a three-photon interaction. The generation of these frequency components must also satisfy another condition, the phase matching requirement. That is,

$$\mathbf{k}_3 = \mathbf{k}_1 + \mathbf{k}_2 \tag{2.12}$$

and as

$$\omega_3 = \omega_1 + \omega_2 \tag{2.13}$$

Equations (2.12) and (2.13) are called the phase and frequency matching components, respectively.

A   B

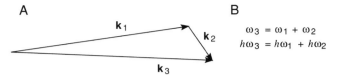

$$\omega_3 = \omega_1 + \omega_2$$
$$\hbar\omega_3 = \hbar\omega_1 + \hbar\omega_2$$

Figure 2.3 A three-wave mixing phase matching condition showing $\omega_3 = \omega_1 + \omega_2$, and $\mathbf{k}_3 = \mathbf{k}_1 + \mathbf{k}_2$. (A) Wavefronts of the waves represented by their respective wavevectors, the phase-matching condition. (B) The frequency matching and energy relations are stated.

Consider the case of a field $E(t)$ comprising two harmonic components at frequencies ω_1 and ω_2 for second-order nonlinearity, we can write the expressions for waves 1 and 2 according to Eq. (2.1) as

$$\mathbf{E}(\omega_1) = A_1 \exp(-i\mathbf{k}_1 * \mathbf{r})$$
$$\mathbf{E}(\omega_2) = A_2 \exp(-i\mathbf{k}_2 * \mathbf{r})$$

Then, in accordance with Eq. (2.11), $E(\omega_3)$ can be written as

$$\mathbf{E}(\omega_3) = \chi^{(2)} A_1 A_2 \exp(-i\mathbf{k}_3 * \mathbf{r}) \qquad (2.14)$$

Thus, Eq. (2.12) results. Replacing $\mathbf{k}_1$, $\mathbf{k}_2$, and $\mathbf{k}_3$ each by their equivalent identities, respectively,

$$\left(\frac{n\omega_3}{c_0}\right) = \left(\frac{n\omega_1}{c_0}\right) + \left(\frac{n\omega_2}{c_0}\right)$$

one can immediately see that the frequency condition in Eq. (2.13) is followed. Thus, when the phase condition is satisfied, the frequency condition is automatically satisfied. The plane wavefronts and their phase relationship can be plotted as in Figs. 2.3A and B, respectively. A closed triangle of the wavevectors is shown.

Parametric Amplification

On the basis of Eqs. (2.12) to (2.14), schemes using a laser to pump another light source for parametric amplification have been developed. If the signal light has a frequency ω_1, and a pump light of frequency ω_3 is used to supply energy for amplification, then, for a parametric amplifier to work, an idler wave of freqeucy ω_2 must be provided by the system to satisfy both phase and frequency matching conditions.

Third-Order Nonlinearity

In dielectrics of symmetric structure, the second-order nonlinear polarization term is missing, leaving the third-order nonlinearity as the dominant term. Then, $\mathbf{P} = s\mathbf{E}^3$, where $s = \epsilon_0 \chi^{(3)}$ in Eq. (2.10) and is called a Kerr coefficient.

Third-Harmonic Generation

Material with third-order nonlinearity responds to optical fields by generating third- and higher-harmonic frequencies. In response to an input light wave $E(\omega)\exp(i\omega t)$, the nonlinear polarization vector contains frequency components at ω and 3ω. The 3ω component is the third-harmonic signal generated by the nonlinearity.

Kerr Effect

Kerr expressed the change in the refractive index of the dielectric when it is exposed to high-intensity light as

$$n(I) = n_0 + n_2 I \qquad (2.15)$$

Equation (2.15) shows that n is now a function of I, the intensity of the input light, and n_2 includes the nonlinearity effect induced by the third-order polarization effect. As light intensity I is proportional to the square of the electric field, E ($I = E^2(\omega)/2$), the Kerr effect is often called a quadratic electrooptical effect. And n_2 is the Kerr coefficient, $n_2 = (\eta_0/n^2\epsilon_0)\chi^{(3)}$, where $\eta_0 = \sqrt{\epsilon_0/\mu_0}$.

Self-Focusing Effect

The phase shift incurred by an optical beam traveling in a nonlinear medium is given by $\theta = 2\pi n(I)L/\lambda$, where L is the distance of travel, and λ the optical wavelength *in vacuo*. In the presence of third-order nonlinearity, from Eq. (2.15), the phase shift is proportional to its own intensity, or it undergoes self-phase modulation.

The phase velocity of the wave in the optical Kerr effect depends on its own intensity, or it is self-induced. Thus, this effect can be used to self-focus the beam. Figure 2.4 shows such a scheme. A light beam whose intensity has a maximum at the center is directed on a plane sheet material of thickness d and contains a third-order nonlinear medium. The index of refraction changes with the beam intensity, thus impinging a phase shift that varies with the intensity of the beam. The sheet acts as a

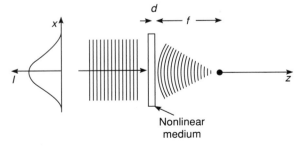

Figure 2.4 A self-focusing beam. A third-order nonlinear medium acts as a lens, the focusing power of which depends on the intensity of the incident beam. d = nonlinear crystal plate. After B. E. A. Saleh and M. C. Teich, *Fundamentals of Photonics* © 1991, John Wiley & Sons.

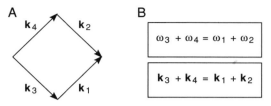

A

B

$$\omega_3 + \omega_4 = \omega_1 + \omega_2$$

$$k_3 + k_4 = k_1 + k_2$$

Figure 2.5 A four-wave mixing phase-matching condition. (A) Vector diagram showing the phase-matching condition. (B) The energies of the waves are related.

graded-index medium, forcing the wavefront to bend accordingly. Under certain conditions, self-focus is obtained as shown.

Phase Conjugation

With four-wave mixing experiments, a very interesting proposition is found that if for a degenerative case, the frequency matching condition and the phase conjugation relationship are both satisfied; that is,

$$\omega_1 = \omega_2 = \omega_3 = \omega_4 \tag{2.16}$$
$$k_3 = -k_4 \tag{2.17}$$

many useful devices can be built for fiberoptic applications. In Eqs. (2.16) and (2.17), wave ω_2 is generated in response to the input light of ω_1, which is proportional to the conjugate version of wave ω_1, thus a phase conjugator. Waves ω_1 and ω_2 are counterpropagating. Or wave ω_2 retraces the wave ω_1 toward the source. Waves ω_3 and ω_4 are a pair of counterpropagating pump waves. The phase matching condition indicates that the wavevectors form a closed parallelepiped structure as shown in Fig. 2.5.

We devote Chapter 10 to an illustration of phase conjugation. The ideas mentioned in this chapter provide device possibilities. Many have already been built. Others await development.

Frequency Mixing

The response of superposition of two or more waves in dielectric materials containing second- or third-order nonlinearity is also interesting. Thus, the second-order nonlinear medium can be used to mix two optical waves of different frequencies and generate a third wave at either the sum or different frequency. In fact, the fiber laser is one recent development in fiberoptics based on frequency mixing. We devote Chapter 4 to a description of this technique.

Nonlinear Crystals

The symmetry of crystals affects the optical properties of dielectric media in many interesting ways. Crystals that display inversion symmetry do not display second-order harmonics, as the third-rank susceptibility χ_{ijk} is zero. The fourth-rank susceptibility tensor does, however, occur in such media. Using mathematical group theory, scientists are able to classify the crystals into various classes. Study of the crystal symmetry helps to select the desired property of nonlinear media as discussed under Second-Order Nonlinearity and Third-Order Nonlinearity, respectively. For further information on nonlinear crystals, see Butcher and Cotter [8] and Boyd [9].

Advances in crystal growth technology continue to develop and new materials have been reported. Table 2.1 is a partial list of the nonlinear crystals under investigation at present. Only the names and their optical susceptibilities are listed. In Table 2.1, only dominant susceptibilities in definite directions are listed. In selecting crystals for certain application, beside choosing the largest susceptibility that the crystal can offer, other properties such as the size of the crystal, its coupling compatibility with other elements, its availability, and its cost should also be considered.

Research into electrooptic polymers has progressed rapidly. Thin-film devices have been fabricated with nonlinear organic molecules attached as pendant side

TABLE 2.1
Nonlinear Optical Susceptibilities[a]

Material	Susceptibility d_{im} ($\times 10^{-13}$ m/V)	Material	Susceptibility d_{im} ($\times 10^{-13}$ m/V)
Quartz	$d_{11} = 4.02$	KD_2PO_4 (KD*P)	$d_{36} = 5.28$
	$d_{14} = 0.08$		$d_{14} = 5.28$
$Ba_2NaNb_5O_{15}$	$d_{31} = -146.62$	CdS	$d_{33} = 360.25$
	$d_{32} = -146.62$		$d_{31} = 377.01$
	$d_{33} = 201.07$		$d_{36} = 418.90$
$LiNbO_3$	$d_{22} = 31$	Ag_3AsS_3 (proustite)	$d_{22} = 284.85$
	$d_{31} = 58.65$		$d_{31} = 150.80$
	$d_{33} = 410.52$	$CdGeAs_2$	$d_{36} = 4566.01$
$BaTiO_3$	$d_{15} = -171.75$	$AgGaSe_2$ (purargurite)	$d_{36} = 337.31$
	$d_{31} = -180.13$	$AgSbS_3$ (pyrargyrite)	$d_{31} = 125.67$
	$d_{33} = -67.02$		$d_{32} = 134.05$
$NH_4H_2PO_4$ (ADP)	$d_{14} = 5.03$	β-BaB_2O_4 (β-barium borate)	$d_{11} = 19.27$
	$d_{36} = 5.03$	$KTiOPO_4$ (KTP)	$d_{31} = 65$
KH_2PO_4 (KDP)	$d_{14} = 5.03$		$d_{32} = 50$
	$d_{36} = 4.61$		$d_{33} = 137$
$LiIO_3$	$d_{35} = -54.46$		$d_{24} = 76$
	$d_{36} = -41.89$		$d_{15} = 61$
CdSe	$d_{15} = 309.99$	LiB_3O_5 (LBO)	$d_{32} = 8.4$
	$d_{31} = 284.85$		$d_{15} = 7.1$
	$d_{33} = 544.57$		

[a] Data taken from Higgins [10].

chains onto a variety of different backbones. They possess Pockel's constants exceeding that of $LiNbO_3$, having low dielectric constants and low loss tangents. These materials have been used as quasi-phase-matching, second-harmonic-generating devices [11, 12].

Interaction of Light and Sound

Light waves are electromagnetic in nature. Acoustic waves are elastic or particle in nature. The energy of a harmonic oscillator relative to electromagnetic radiation is quantized, and the quantized particles are called photons. Similarly, the corresponding particles in the elastic wave are called phonons. Thus the interaction of light and sound may also be called photon–phonon interaction.

The existence of an interaction between optical and acoustic waves was predicted in the 1920s by Brillouin [13]. Experimental verification followed in the 1930s by Debye and Sears and others [14]. Practical applications started in the 1960s, when radar signal processing using acoustooptic interactions was introduced [15]. Since then, ultrasounic imaging processing in the medical field has taken over and progressed very rapidly [16, 17].

The Bragg Cell

Optical and acoustic waves can interact with each other on a common ground, usually through a crystal transparent to the optical wave and with an acoustic property that varies with the strain applied to the medium by the acoustic wave through a transducer. Typical electrooptic crystals such as tellurium oxide (TeO_2), lithium niobate ($LiNbO_3$), and gallium phosphide (GaP) can be used as the common ground constructed as a Bragg cell. An acoustic wave is launched into the crystal through one or more piezoelectric transducers attached to its ends. The Bragg cell can either be a bulk, a thin-film, or a surface-type crystal. Figure 2.6 shows the structure of a Bragg cell. In its simplest form, it consists of a plane crystal directed in the x direction. A transducer is attached to the bottom (or top or both) of this body. An optical wave is incident from the left side at an angle to the horizontal as shown in Fig. 2.6A. For a thin-film cell, only the surface of the cell is active. The interaction is confined to its surface only. The width of the cell may or may not be specified.

When an acoustic wave is fed through the transducer to the crystal as shown, it sets the particles of the crystal in motion as it travels along the x axis and creates an elastic wave pattern along its way. The periodic occurrence of compressions and rarifications of the medium produces a density variation in the form of a moving grating in the crystal. Now when an optical wave is incident in the direction to the crystal as shown, photon–phonon interaction takes place at the moving acoustic wavefronts. The light beam sees a practically stationary grating, is reflected in a new direction, and changes its frequency as well. This is because the optical frequency is so much higher than the acoustic frequency so that during the flight time of the opti-

A

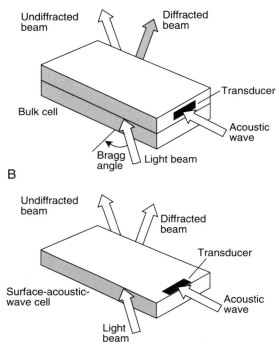

B

Figure 2.6 Two types of Bragg cells: (A) bulk type; (B) thin-film type. Orientations of the optical and acoustical waves are assigned. After E. Goutzoulis, Digital electronics meets its match, *Spectrum* **25**, 24. © 1988. *IEEE.*

cal beam, the acoustic wave remains as if it were stationary. Brillouin was the first to discover this interference effect. In this section, we introduce the general description of the interaction effect conceptually and leave the applications to the following chapters.

Wave Concept of Photon–Phonon Interaction

Let us describe the two waves in the interaction process as follows.

An optical wave with a free space wavelength λ_0 is incident on the crystal at an angle θ with the z axis. The crystal is oriented in the x direction. The frequency of the optical wave is ν and its angular frequency is $\omega = 2\pi\nu$. The wavevector of this wave in the crystal is $\mathbf{k} = n\omega/c_0$, where n is the index of refraction of the medium, and c_0 is the vacuum speed of light. Similarly, the acoustic wave launched into the crystal via a transducer has wavelength Λ. The sound speed is v_s, so that the acoustic frequency is f and its angular frequency is $\Omega = 2\pi f$. The wavevector of this wave is $q = 2\pi/\Lambda$.

Let S_0 be the acoustic wave amplitude; then the intensity of the elastic wave can be specified as

$$I_s = \tfrac{1}{2}\rho v_s^3 S_0^2 \tag{2.18}$$

where ρ is the mass density of the medium. For a transparent medium before the sound wave is applied, the change in refractive index is

$$n(x, t) = -\tfrac{1}{2}pn^3 s(x, t) \tag{2.19}$$

where p is the photoelastic constant. The negative sign indicates that for a positive strain s applied to the crystal, the refractive index decreases. The magnitude of the refractive index change is therefore

$$\Delta n_0 = \tfrac{1}{2}pn^3 S_0 \tag{2.20}$$

Expressed in terms of I_s,

$$\Delta n_0 = \left(\frac{W}{2}\right)^{1/2} I_s^{1/2} \tag{2.21}$$

where $W = (p^2 n^6/2\rho v_s^3)$ is a parameter representing the effectiveness of the acousticoptic effect.

As for the optic wave, it sees a change in n as

$$n(x) = n - \Delta n_0 \cos(qx - \Phi) \tag{2.22}$$

compared with that due to the acoustic wave,

$$n(x, t) = n - \Delta n_0 \cos(\Omega t - qx) \tag{2.23}$$

Because they are in the same medium, it is required that $\Phi = \Omega t$.

We have established that the acoustic wave creates a set of parallel planes from which the optical beam will be reflected. We could proceed to calculate the intensities of the two adjacent optical waves and find the condition for constructive interference of the reflected waves. Instead, we prefer to use Bragg's X-ray analog and borrow its result. Figure 2.7 shows the X-ray diffraction experiment. Two planes are separated by an acoustic wavelength Λ apart. Consider a beam of optical wave incident on the plane and making an angle θ with the acoustic grating planes. The reflected waves from two adjacent parallel planes interfere constructively only when their optical paths differ by a integral multiple of the wavelength. This occurs when $\theta = \theta_B$, that is, at the Bragg angle. From the geometry of the diagram in Fig. 2.7, the path difference is $2\Lambda \sin \theta$. The Bragg condition becomes

$$2 \sin \theta_B = \frac{q}{k} = \frac{\lambda}{\Lambda} \tag{2.24}$$

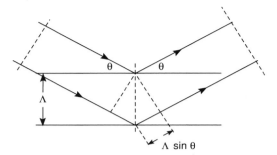

Figure 2.7 X-ray diffraction pattern showing the path difference in the reflected optical waves from a pair of parallel planes separated by Λ.

The Bragg angle θ_B in Eq. (2.24) refers to the angle observed inside the medium of sound propagation. Outside the medium, Snell's law changes the value of θ_B to θ_B', and

$$\sin \theta_B' = \frac{n\lambda}{\Lambda} = \frac{\lambda}{\Lambda} \tag{2.25}$$

The frequency of the diffracted wave suffers a Doppler shift. For wavefronts of the sound wave that move away from the incident wave making an angle θ with the z axis in Fig. 2.8A, the incident wave suffers -1 order diffraction. The frequency is downshifted to

$$\omega_- = \omega_0 - \Omega_a \tag{2.26}$$

Similarly, the diffracted wave can have $+1$ order as shown in the bottom curve of Fig. 2.8B. The frequency is upshifted to

$$\omega_+ = \omega_0 + \Omega_a \tag{2.27}$$

Narrow Bragg Cell

In the Bragg cells shown in Fig. 2.6, we used a width of the cell (or width of the transducer) of $L \gg \Lambda$, the wavelength of the acoustic wave. What if L is decreased? Debye and Sears suggested a diffraction pattern for this column acoustic grating [14]. As L, the width of the transducer, is decreased, the column of sound in the medium looks less and less like a plane wave. The radiation pattern of this column broadens so as to make available additional plane waves of sound for a wide range of interactions. As seen in Fig. 2.9A, narrowing L brings both up- and downshifts in frequency simultaneously. For a normal incidence of optical wave to the sound column, the diffraction pattern may include many high-order diffractions if L is made small enough, each different by the order $\pm 1, \pm 2, \ldots, \pm m$. Each ray is separated from

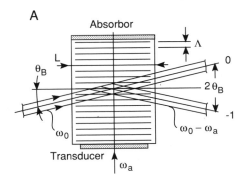

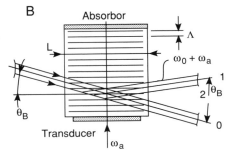

Figure 2.8 Reflections from layers of an inhomogeneous medium. Diffraction in the Bragg region showing the downshift in frequency; the − order in (A), and the upshift, the + order in (B), resulting from the interactions. After B. Korpel, *Proc. IEEE* **69**, 49. © 1981, *IEEE.*

the next order by an angle λ/Λ. A second effect of narrowing L is that the Bragg condition is modified such that

$$\sin\left(\frac{\theta}{2}\right) = \frac{\lambda}{2\Lambda} \quad \text{or} \quad \theta \approx \frac{\lambda}{\Lambda} \tag{2.28}$$

as the Bragg condition is satisfied if the reflected wavevector $\mathbf{k}$ makes an angle $\pm\theta$.

Figure 2.9B shows the vector diagram of the wavenumbers $\mathbf{k}$'s of the radiation pattern of the transducer. The wavevectors of the light and acoustic waves satisfy the relation

$$^-\mathbf{k} = {}^-\mathbf{k}_0 \pm {}^-\mathbf{k}_a \tag{2.29}$$

Narrow Bragg cells will be used as switches and imaging processors in later chapters.

A thin acoustooptic Bragg cell can also be constructed using thin film of electrooptic materials deposited on dielectric plates. This is a surface-acoustic-wave (SAW) cell. In SAW cells, the interaction takes place at the thin surface. Usually two

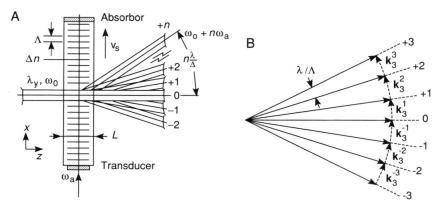

Figure 2.9 Diffraction in a narrow Bragg cell. (A) Many generated orders, both + and −. (B) Wavevector diagrams showing the multiple scattering waves. After B. Korpel, *Proc. IEEE,* **69,** 49. © 1981. *IEEE.*

transducers, one at each end (top and bottom), are attached to the cell to support two oppositely directed acoustic waves on the same SAW cell simultaneously. Many applications can use this construction advantageously.

(We used a thin Bragg cell for SAW to distinguish it from the narrow bulk Bragg cell.)

Bragg Cell Parameters

The important performance parameters of a Bragg cell are diffraction efficiency bandwidth, time aperture, and resolution.

Diffraction efficiency ranges from 1 to 200% per watt of acoustic power, depending on the material. The 200% comes from the fact that in some samples, 100% of the light can be diffracted with only 0.5 W of acoustic power.

The time aperture of a cell is a measure of the time available for sound and light to interact. Different applications may demand different time apertures. For example, spectrum analyzes may need a small time aperture, on the order of 0 to 300 ns, to restore pulsed signals precisely. For correlators, on the other hand, a larger time aperture is required to handle long relative delays. Time apertures as long as 1 to 80 μs are commercially available.

The resolution or time–bandwidth product represents the amount of information in a cell. Again, the requirements are different for different applications. For spectrum analyzers, a time–bandwidth product (TB) of 1000 is enough. For correlators, a larger value of the product is required to analyze various signal parameters precisely.

Summary

We have reviewed the properties of photons. We have also highlighted some non-linear effects that will be referred to in future chapters. These are the second-order nonlinear effects which include harmonic generation, rectification, frequency mixing, and parametric amplification; and the third-order nonlinear effects, including third-harmonic generation, Kerr effect, self-induced phase modulation, self-focusing, and phase conjugation. We are going to see how these effects can be put into practice to build useful devices for telecommunications and other applications.

An introduction to the acoustooptic effect was also included in this chapter. Many applications of this effect are described in Chapters 11, 12, and 14.

Before we close this chapter, we should mention one important aspect that contributes to the development of nonlinear optics but is beyond the scope of this book. The development of nonlinear crystals and of fine-tuning techniques for manufacturing these crystals has dazzled us in recent years. Readers interested in this subject are advised to watch closely for new developments in current publications. We may have touched only a very small portion of the developing nonlinear optics, but already we can foresee the vast possibilities for device applications. The future is wide open for further advancement.

References

1. M. Planck, Ueber das Gesert zer Enerieverteilung im Normalspectrum. *Ann. Phys.* **4**, 553 (1901).
2. A. Einstein, Die Plansche Theorie der Strahlung und die Theorie der spezifischen Warme. *Ann. Phys.* **22**, 180 (1907).
3. A. H. Compton, Wave-length measurements of scattered x-rays. *Phys. Rev.* **21**, 715 (1923); The spectrum of scattered X-rays. *Phys. Rev.* **22**, 409 (1923).
4. T. Young, discovered the principle of optical interference around year 1800.
5. L. de Broglie, A tentative theory of light quanta, *Philos. Mag.* **47**, 446 (1926).
6. R. H. Diche and J. P. Wittke, *Introduction to Quantum Mechanics,* Addison–Wesley, Reading, MA, 1960.
7. A. Yariv, *Quantum Electronics,* 3rd ed., Wiley, New York, 1989.
8. P. Butcher and D. Cotter, *The Elements of Nonlinear Optics,* Cambridge Univ. Press, London/New York, 1990.
9. R. Boyd, *Nonlinear Optics,* Academic Press, San Diego, 1992.
10. T. V. Higgins, Nonlinear crystals: Where the colors of the rainbow begin. *Laser Focus World,* pp. 125–133 (Jan. 1992).
11. G. R. Mohlmann, Development and application of optically nonlinear polymers. In *Electronic Properties of Conjugated Polymers. III. Basic Models and Applications,* Proceedings of an International Winter School (H. Kuzmany, M. Mehrings, and S. Roth, Eds.), pp. 232–236, Springer-Verlag, Berlin, 1989.
12. D. Haas, Recent developments in the application of polymers to electro-optic and second-harmonic devices. In *Photonic Networks, Components and Applications* (S. L. Chin, Ed.), Vol. 2, pp. 98–109. Series in Optics and Photonics, World Scientific, Montebello, 1990.
13. L. Brillouin, Diffusion de la lumiere et des rayons X par un corps transparent homogene, *Ann. Phys. (Paris)* **17**, 88–122 (1992).

14. P. Debye and F. W. Sears, On the scattering of light by supersonic waves, *Proc. Natl. Acad. Sci. USA* **18,** 409–414 (1932).
15. J. H. Collins, F. G. H. Lean, and H. J. Shaw, Pulse compression by Bragg diffraction of light with microwave sound, *Appl. Phys. Lett.* **11,** 240–242 (1967).
16. *Proceedings of the 12th International Symposium on Acoustical Imaging, London, England,* July 1982. Plenum Press, New York.
17. M. Noble, Acousto-optic devices: Diverse applications. *Laser Optronics,* pp. 27–29 (Mar. 1992).

CHAPTER 3

Recent Advances in Semiconductor Laser Technology

Introduction

Semiconductor lasers are considered the most suitable light source to be operated in conjunction with optical fibers. They are comparable in size to and as simple to operate as optical fibers. The basics of semiconductors have been treated in elementary textbooks and are not repeated here. We emphasize recent trends and developments in semiconductor lasers that are particularly designed for applications to optical fibers in telecommunications and other related fields. In these applications, we need a laser as a light source that possesses the following desired characteristics: desired wavelength range, 1.3 to 1.55 μm; high optical power, from milliwatts to several watts; room temperature operation; high quantum efficiency; low threshold current, in the milliampere range; single-frequency, single-mode, and stable operation; short pulse, say, in the femtosecond range; if tunable, an adequate tuning range, say, a few nanometers; and cost effectiveness.

Several recent developments have become prominent: single-quantum-well (SQW) and multiple-quantum-well (MQW) lasers; distributed feedback lasers; distributed reflector (DR) lasers; and surface-emitting (SE) lasers. We review the findings leading to the development of these lasers.

Review of Semiconductor Lasers

What has already been done in laser oscillators? Let us first refresh ourselves on lasers in general.

Historically, a semiconductor laser was made of a $p-n$ homojunction planar diode by diffusing impurities of the p-type, say, into a bulk n-type semiconductor. When the junction or the diode is forwardly biased, carriers of electricity, electrons and holes, are said to have been injected into the respective side of the junction. If the bias is high enough so as to cause carrier inversion, that is, the number of electrons occupying the higher energy states in the conduction bands becomes higher than the number below, stimulated emission can result. Stimulated emission is the result of a recombination process in which most of the electrons in the upper bands depopulate simultaneously to recombine with holes in the valence bands and emit photons in the process. Unlike the spontaneous recombination process, which occurs randomly, stimulated recombination emits coherent radiation. If a resonant cavity is provided by cleaving the end facets of the diode, sustaining oscillation is obtained. The wavelength of the photon emission is inversely proportional to the bandgap energy of the semiconductor material used.

In this section, we point out what has been learned from previous laser experiments, and what can be done to improve the properties of lasers for certain applications.

First, we have observed that it is very difficult to initiate laser emission with a homojunction diode laser. Such a device requires an extremely large threshold current density to lase, even at liquid nitrogen temperature. And even then, only pulse operation is possible. Earlier lasers needed a threshold current density of about 10^5 A/cm^2 [1].

Second, we have also observed that a simple laser of this kind possesses many modes of oscillation and that the oscillation is unstable [2].

As early as 1967, researchers realized that the diffused homojunction of the laser could account for most of these troubles. Suggestions were made to replace the homojunction with a heterojunction by alloying two semiconductors of different energy bandgaps or by another processing technique that makes abrupt junctions [3, 4].

We divide the discussion into three parts: (1) laser materials and processing; (2) laser structure tailoring; (3) laser performance improvements.

Semiconductor Materials and Processing for Long-Wavelength Lasers

Semiconductor Compounds and Lattice Matching

The choice of semiconductor materials for a laser depends on the wavelength required for the operating system. The trend in modern telecommunication systems is toward the use of longer wavelengths, where the fiber loss is lower. For silica optical fibers, this occurs at about 1.5 μm. That makes the use of GaAs crystals for lasers less suitable. More complex compounds must be chosen. These compounds include ternary and quaternary semiconductors [5, 6]. An epitaxial technique for making junctions is also recommended [7]. For wavelengths above 1 μm, quaternary compounds such as InGaAsP/InP and AlGaAsSb are recommended [8–10]. In particular,

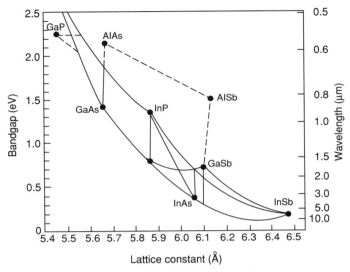

Figure 3.1 Bandgap (eV) versus lattice constant (Å) of group III–IV semiconductor materials. Near-vertical solid lines show lattice matching possibilities. After T. P. Lee [59], with permission.

a compound often used in the range 0.92–1.65 μm is In$_{1-x}$Ga$_x$As$_y$P$_{1-y}$/InP. Here, x and y are the percentage compositions of each element in the two groups, respectively. Variations of x and y are required by the need to adjust the bandgap energy (or the desired wavelength) and for better lattice matching. We present here two graphs, energy gap versus lattice constants for III–V semiconductors and alloys (Fig. 3.1) and energy gap versus composition of InGaAsP quaternary alloy (Fig. 3.2). For the previously mentioned alloy on an InP substrate whose lattice constant is 5.87 Å, a perfect lattice match with GaAs can be found in Fig. 3.1 by drawing a line from the point marked InP at 5.87-Å until it meets the solid line marked GaAs. If, however, a certain percentage of mismatch can be tolerated, an appropriately slanted line may be drawn from that point. The desired wavelength range can be read from the right-

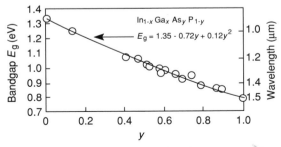

Figure 3.2 Variation of the bandgap energy in a quaternary semiconductor alloy, In$_{1-x}$Ga$_x$As$_y$P$_{1-y}$, as a function of y. After T. P. Lee [59], with permission.

hand scale. In this case, it is from 0.92 to 1.65 μm. It has been shown [11] that for the best lattice match to InP, $y = 2.2x$ must be chosen. The bandgap energy can be calculated by the equations $E_g = 1.35 - 0.72y + 0.12y^2$ eV, and the wavelength in micrometers $= 1.24/E_g$. This curve is shown in Fig. 3.2.

Lattice matching is very important. Lasers made with a matched lattice semiconductor compounds have a longer life. The number 1.24 is the result of unit conversion for hc when λ is expressed in micrometers and E_g in electron-volts.

Epitaxial Processing

Epitaxial growth of semiconductor materials can result in abrupt junctions that are highly recommended for building double heterojunction (DH) laser diodes. Very thin junctions can be grown with this technique. Liquid phase epitaxy (LPE) was introduced first [12]. It produces excellent junctions and is still in use for some special junctions in industry. But because of the slow growing process and size limitation in growth, LPE is not suitable for mass production. Other epitaxial processes, such as vapor-phase epitaxy (VPE) [13] and molecular beam epitaxy (MBE) [14], are preferable in industry. In recent years, thinner, large-area, and uniform abrupt interface epitaxial layers have been achieved by organometallic vapor-phase epitaxy (OMVPE) compared with the epitaxial technique [15]. OMVPE is also known as MOCVD. As the demand to grow thinner layers for MQW lasers increases rapidly, OMVPE becomes increasingly important in semiconductor material processing.

Compound Semiconductor Materials

Many semiconductor compounds have been used to make lasers. In the following partial list are compounds, and the wavelength ranges from near ultraviolet to the far infrared, that have been successfully used to make lasers. Many compounds require special care, however; for instance, some lasers may require special cooling and some may require optical or electron beam pumping.

Compound/ composition	Wavelength range (μm)	Compound/ composition	Wavelength range (μm)
$Cd_xZn_{1-x}S$	0.32–0.51	$In_{1-x}Ga_xAs_yP_{1-y}$	0.97–1.08
CdS_xSe_{1-x}	0.5–0.7	$Cd_xPB_{1-x}S$	1.03–4.0
$(Al_xGa_{1-x})_yIn_{1-y}P$	0.6–0.8	$Cd_xHg_{1-x}Te$	1.2–17
$In_xGa_{1-x}As$	0.56–3.0	$InAs_xSb_{1-x}$	3.7–5.4
$GaAs_{1-x}P_x$	0.6–0.9	$Pb_xS_{1-x}Se$	4.2–8.2
$Al_xGa_{1-x}As$	0.7–0.9	$Pb_xSn_{1-x}Te$	≥6.8
$(Al_xGa_{1-x})_yIn_{1-y}As$	0.9–1.3	$Pb_xSn_{1-x}Se$	≥8.5
$InAs_xP_{1-x}$	0.93–3.0	$Bi_{1-x}Sb_x$	≥50
$GaAs_xSb_{1-x}$	0.95–1.06		

The exact fractions of x and y depend on the manufacturer's specifications.

Tailoring the Laser Structure

In tailoring the laser structure, we expect to fulfill two purposes: to control the mode structure and to regulate the current flow across the junction.

Research results suggested that the reason for multimodes in laser oscillation must be related to cavity dimensions. How much can we reduce cavity dimensions to achieve single-mode operation? Structural change may also affect current density. The goal is to increase the injected carrier density without increasing the threshold current unduly. Reducing the chance of spontaneous emission could also improve lasing efficiency. These objectives can be reached by carrier confinement. To achieve better carrier confinement in the laser structure, the resonant cavity must be designed properly.

For a rectangular parallelepiped laser cavity, with the dimensions x, y, and z oriented in thickness, width, and length directions, respectively, we find that historically in laser diodes, these dimensions are all large compared with the wavelength of the photon emitted. Even in the x direction, the thickness is usually larger than the diffusion length of the carriers (typically a few micrometers). The dimensions in the width direction y and in the length direction z may both be hundreds or thousands of wavelengths long, respectively. The laser may choose any appropriate cavity size to lase that satisfies the relationship of any multiple of a half-wavelength. Usually, all modes are excited simultaneously. This laser thus possesses many modes. Moreover, individual modes are not stable. The output may jump from one mode to another and back again dependent on the operating conditions. They are often competing among themselves, resulting in a multimode laser with unsteady operation.

We distinguish these laser modes as the transverse mode (in the x direction), the lateral mode (in the y direction), and the longitudinal mode (in the z direction). We discuss the techniques for limiting the number of modes in each direction so as to achieve single-mode laser operation.

Transverse Mode Control

Transverse mode control can be achieved by limiting the thickness of the active layer, the layer within which the process of recombination takes place. Abrupt $p-n$ junctions, grown with a DH structure, can keep the active layer thickness to less than 1 μm. It was found that if the thickness d is kept below 0.5 μm, only the lowest mode, thus a single transverse mode, is possible [16]. Heterojunction structure can also confine the injected carriers in the respective regions and prevent most carrier leakage. Efforts at carrier confinement greatly reduce the threshold current density.

Lateral Mode Control

If the lateral dimension (y direction) of the active layer of a laser is unlimited, many lateral modes can exist. Often, many preferred modes can coexist. They hop around and form "kinks" in the output characteristics, as evidenced by the filamentary pattern of the output [17]. The laser output spectrum becomes complex and the threshold current density high. Control of the lateral modes can be achieved by the technique of gain control [18] or index control [19], or both. Gain-guided lasers are simpler to make. They can be used for high-power applications because the light from the laser is dispersed over a large enough area of the facet to avoid surface damage. But their weaker optical confinement limits beam quality and makes it difficult to obtain stable output in a single longitudinal mode. The index-guiding property of the structure is derived from the fact that in a DH structure, the potential barriers produced by the semiconductor junctions of higher-bandgap materials have lower indexes of refraction, thus providing the active layer with a waveguide property for guiding the propagation. The current flow through the laser is restricted to a narrow strip. Index-guided lasers offer better beam quality, are more efficient, and usually have a lower threshold current in GaAlAs diodes. They are used for most diode laser applications. Figure 3.3 shows some functionally similar types of index-guided laser structures on the market. These include the buried heterostructure which, has a narrow active region surrounded by material of lower refractive index (Fig. 3.3A); the channeled-substrate planar structure, which has an active layer grown on a substrate with a channel etched into it (Fig. 3.3B); the ridge waveguide structure, which has a ridge above the active stripe (Fig. 3.3C); and the dual-channel planar buried heterostructure, which has an isolated active stripe in a mesa by etching two parallel channels around it, then grow-

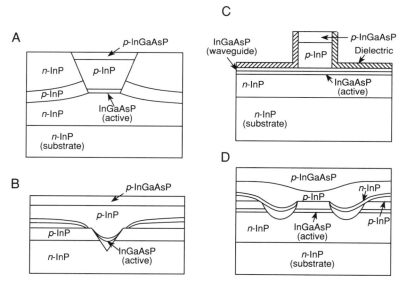

Figure 3.3   Index-guided laser structures. (A) Buried heterostructure (MSB). (B) Channeled substrate planar (CSP). (C) Ridge waveguide structure. (D) Dual-channel buried structure (DC-PBH). After T. P. Lee [59], with permission.

ing material around it to bury the heterostructure. Brief descriptions of various lateral mode control schemes can be found in the literature [20].

By keeping the width dimension below 0.5 μm, a lateral mode can be reduced to a single mode. Also, by limiting the width dimension, further lowering of the threshold current density can be achieved.

Longitudinal Mode Control

The usual length of a laser cavity formed by cleaving the end faces of a diode is about 250 μm, long enough to allow many longitudinal modes to be excited simultaneously. At 1.3 μm, the spectrum of the laser can spread over several tens of nanometers, with 1-nm separation between each spectra. This is a highly multimode scheme. For ordinary applications where direct detection is employed and where the transmission rate is low, say, below 100 Mb/s, this does not matter. For high-speed, long-haul signal transmission, where the use of a single-frequency laser is essential, methods of improvement must be adopted. Multimode operation leads to pulse spread during propagation along a dispersive optical fiber. In systems employing heterodyne detection, multifrequencies can never be tolerated.

The key to longitudinal mode control is to provide a gain differential between the desired and side modes in the cavity. The simplest way to increase mode discrimination is to shorten the cavity length L. Lee *et al.* have shown that by reducing L from 250 to 25 μm, the side mode separation can be increased from 1 to 10 nm [21]; however, a cavity shorter than 50 μm is very difficult to fabricate. Shorter cavities also cause a reduction in power capacity. Other methods will be introduced in sections involving single-frequency lasers.

Attempts to Improve Laser Performance

After many years of research, involving tens of thousands of scientists and engineers, we can finally compile a list of the properties a laser should have for a particular application. Then, we will suggest ways to achieve these objectives. This list could never be complete; it may be extended at any time in the future.

Threshold Current Density

Although diode current begins to flow in a $p-n$ diode as soon as it is forward biased, the diode will not lase until a critical value, known as the threshold current density, is reached. At this level, the injected carrier density is said to have reached the inversion stage so that stimulated emission dominates. The equation for this current density can be derived formally. But intuitively, it can be written as $J_{th} = qN_{th}v$, where q is the electronic charge, N_{th} is the threshold carrier density, and v is the carrier velocity when moving across the active layer, which is d/τ, where τ is the carrier lifetime and d the active layer thickness. The important factor is now N_{th}, which has been derived [22] as

$$N_{th} = N_0 + \frac{\left[\alpha_c + \left(\frac{1}{L}\right)\ln\left(\frac{1}{R}\right)\right]}{(\Gamma A)} \tag{3.1}$$

where N_0 is the carrier density required to reach optical transparency (carrier inversion), α_c is the cavity loss per unit length, L is the length of the cavity, R is the facet reflectivity, Γ is an optical confinement factor, and A is a function involving the differential gain. The derivation of N_{th}, which is omitted here, involves the dynamics of carrier motion in the active layer, the material of the junction, and the structure of the cavity. Both parameters N_0 and A are material constants. On the basis of this discussion, the threshold current density equation can be written as

$$J_{th} = \frac{qdN_{th}}{\tau} \tag{3.2}$$

Equation (3.2) can then be used to discuss how to reduce the threshold current density.

A simple example based on the preceding equation, when applied to a homojunction $p-n$ diode, will show the magnitude of the calculation. For a GaAs diffused junction diode, $d = L_n = \sqrt{[D_n\tau_n]} = 0.93$ μm, N_{th} can be approximated by the density of energy states within $\Delta E = 0.5$ eV, which is about $8.79 \times 10^{17}/\text{cm}^3$. (Here, L_n is the diffusion length, D_n is the diffussion coefficient, and τ_n is the carrier lifetime.) Using the recombination lifetime of the injected electron (into the p side) as 0.56 ns, we find $J_{th} = 23.5$ kA/cm², a huge current density. We need to reduce this number by several orders of magnitude to a few tens of amperes per square centimeter.

To reduce the threshold current density, one can (1) reduce N_{th}, (2) reduce the active layer thickness d, and (3) increase the recombination lifetime of the injected carrier.

1. To reduce N_{th}, we first have to choose a semiconductor material that gives a smaller N_0 and a larger A, the differential gain value. $A = dg/dN$ and is related to the optical gain g by $g = A[N - N_0]$. In a later section, we introduce the use of the quantum size effect of a thin active layer to alter the band structure of a semiconductor so as to increase greatly the value of A and to reduce N_0. The epitaxial technique to make a DH laser with better lattice matching also helps to reduce N_{th} because it reduces the interfacial trap density and maximizes stimulated emission. Trap density at the interface encourages the unwanted recombination of carriers. A large confinement factor Γ and a low-loss reflective facet both can contribute to a smaller N_{th}.

2. An epitaxial technique that makes a thin layer or small d, together with a small cross-sectional area of the current flow, gives a smaller volume and thus a lower threshold current.

3. A DH structure improves the carrier confinement. As the active layer has a lower bandgap energy, and is sandwiched between two layers of higher-bandgap materials, an optical waveguide structure is formed, as the active layer has a higher index of refraction to guide the optical signal.

At present, layers thinner than 100 Å have been reported [23].

Carrier Lifetime

Carrier lifetime can be increased if unwanted recombinations are eliminated. DH structure, thin active layer, and good lattice matching can combine to increase the lifetime of useful carriers. Spontaneous recombinations are unwanted and should be limited as much as possible. Quantum size effect can provide an energy structure that encourages stimulated emission. This effect will be used in quantum-well (QW) lasers.

Optical Power Output

When the laser is biased to a current I, larger than the threshold current I_{th}, optical power begins to appear at the output. The optical power can be expressed as [22]

$$P_0 = \eta_d \left[\frac{h\nu}{2q} \right] [I - I_{th}] + P_{sp} \tag{3.3}$$

where η_d is the external quantum efficiency, ν the optical frequency, h Planck's constant, and P_{sp} the power contributed by spontaneous recombination of the carriers. In a well-designed laser, P_{sp} is kept very low and is usually neglegible. Thus it can be seen that as the operating current I increases, the optical output power increases, at first nearly proportionally and then more slowly, until it finally saturates at higher current values. The saturation effect could be attributed to a decrease in quantum efficiency due as a result of thermal heating and other causes.

Remember also that the optical power of a laser is the total sum of a larger number of modes, known as multiple longitudinal modes. The distribution of power between the different modes is not uniform. It decreases as the mode order increases, as shown in Fig. 3.4, where optical power is plotted against current for various mode numbers in a typical laser operation.

Note also that the optical power of the higher-order modes saturates more readily than the zeroth-order mode, indicated by lower amplitudes, a condition that favors higher current operation. This trend is also clearly shown in Fig. 3.5, where the total optical power is plotted against the injection current. At several injection currents above the threshold current, the spectra of the modes are shown in the insets, where the peak power of the individual mode is plotted against the wavelength for a finite current I. Peak power of the longitudinal modes occurs at a wavelength where the optical gain is closest to the loss.

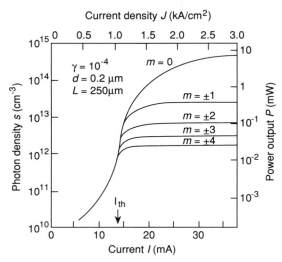

Figure 3.4 Photon density versus injection current of a laser for various mode numbers m. After T. P. Lee [59], with permission.

This is the normal condition of multimode laser operation where the laser cavity length L is long, several hundred times the optical wavelength. Notice that at higher currents, more power is concentrated to the zeroth-order mode. If the envelope of these spectra is drawn, a resonance-like curve is obtained. The spectral width of the laser oscillation is defined as the half-width of the resulting curve. Sideband wavelengths are separated from each other by approximately 1 nm and decreases in amplitude as they move away from the center wavelength. With an increase in I, as shown on the main curve, the optical gain of the center mode increases, causing the curve to be more peaked and decreasing the spectral width. Spectral width becomes more narrow at higher currents [24]. More and more optical power is now concentrated to the zeroth-order mode. Thus this system works in favor of higher injection currents. In a system employing heterodyne detection, or in high-speed signal transmission, where only a single mode is desired, it becomes important to operate a laser at a high current; however, a higher operating current will produce more heat. If it is not properly dissipated, the extra heat increases the device temperature, which decreases its external quantum efficiency and thus the power output. An elevated device temperature also causes I_{th} to increase, thus cutting the power output further. An optimum value of the operating current must be compromised.

In modern telecommunication systems, a laser is required to operate as a single-freqency, single-mode oscillator. This means a single peak in the mode diagram. Even the width of this single peak needs to have a narrow linewidth.

A recent advancement in material technology has helped to ease the problem of power dissipation in semiconductor lasers [25]. Diamond is a highly thermally con-

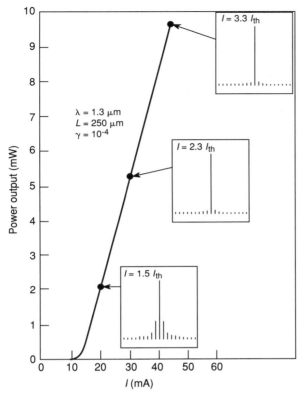

Figure 3.5 Laser power output (mW) versus injection current (mA) of a typical operation. The spectral distribution of the output at three different operating levels is sketched to show the improvement in spectral purity at higher operating current or power. After T. P. Lee [59], with permission.

ductive material. When it is used as substrate for a device, it can increase the heat dissipation capacity manyfold, thus increasing the power-handling capability of the laser. But diamond is not a good substrate for making lasers. A technique has been developed to deposit the laser structure first on GaAs substrate and then lift the thin sheet off and redeposit it on diamond [26]. Others use buffer layers to affect the lattice matching [27].

Modulation and Frequency Response of Diode Laser

The fact that a laser diode can be so conveniently pumped by current injection makes it easy to modulate the optical power as required for signal transmission. The question is: Can laser modulation follow the signal variation in a synchronous fashion?

The problem of how fast a laser can be modulated depends on the interplay between the stimulated emission and the injected carriers. Actual derivation of the modulation speed involves the interplay between the rate equations of carrier inversion and photon emission in the active region of the laser. The theory behind the interplay can be found in other books [28]. Suffice here to state that the response of the transfer characteristics $\Delta P_m/\Delta J_m$ will be resonant at a frequency ω, such that

$$\omega^2 = \frac{AP_0}{\tau_P} \tag{3.4}$$

where ΔP_m is the peak modulation power, P_0 is the power output, ΔJ_m is the peak modulation current density, A is a factor containing many quantities, such as the confinement factor, the gain factor, and others, and τ_P is the photon lifetime [29]. The modulation bandwidth can be derived from Eq. (3.4). It can be said that the resonant frequency depends on the square root of the laser power, the gain coefficient A, and inversely τ_P.

Frequency Chirp

When a single-frequency laser is current modulated directly, its light output may suffer frequency chirping. Frequency chirping is the result of the variations in the refractive index of the medium where carrier density is changed as a result of current modulation. The frequency of the power output of the laser becomes time dependent, thus defeating the purpose of using a single-frequency laser [30]. The cleaved-coupled-cavity laser, once highly praised as the laser of the future, suffers from this drawback.

Although frequency chirping can be minimized by properly choosing the laser material, the laser structure, and even the pulse shape, the most effective way is to avoid it. This can be done by using external modulation.

Linewidth of Semiconductor Laser

The resonant cavity of a semiconductor laser with mirrors as end reflectors can support many modes. A special arrangement to discourage other modes in favor of only one mode is needed to obtain a single-mode (longitudinal) laser. The shape of this single mode is still a broad line with a typical 3-db width of 100 MHz. Broad linewidth in a laser poses a problem in coherent transmission and receiving systems, particularly for modulated lasers, that requires a two-orders-of-magnitude lower, or less than 1-MHz, spectral width.

The broadening of semiconductor laser linewidth is due primarily to the random spontaneous emission into the lasing mode which introduces fluctuations in phase shift. The linewidth of a single-mode semiconductor laser can be expressed as

$$\Delta \nu = \left(\frac{K}{4\pi N}\right)(1 + \alpha^2) \tag{3.5}$$

where K is the average spontaneous emission rate per unit volume, N is the photon density in a laser cavity, and α is the linewidth enhancement factor.

For a Fabry–Perot laser, N is proportional to $I - I_{th}$, and K varies as N_{th}, the carrier density at threshold. Thus for a narrow linewidth, the laser needs to operate at a high current, $I \gg I_{th}$, and a smaller α factor.

Advanced Lasers for Long-Haul Telecommunications

Rapid development of single-mode optical fibers for high-bit-rate telecommunication systems ensures an ever-increasing demand on the quality of light sources. These demands including single frequency, narrow linewidth, room temperature operation, stable frequency, high modulation speed, low noise, high power, and low threshold current. Wavelength tunability is also a popular demand. Special lasers of great interest are the multiple-quantum-well (MQW) laser; the distributed feedback (DFB) laser; the surface emitting laser; the dynamic-single-mode (DSM) laser; and the distributed reflector (DR) laser.

Multiple-Quantum-Well Lasers

In lasers with DH structure, the active layer can be made very thin ($1000-3000$ Å), thin enough to confine electrons and the optical field; however, the properties of laser remain the same as in the bulk material and further improvement is necessary. Why is the MQW laser so special that it deserves further discussion? The answer is that quantum size effect begins to play a role in QW lasers if the active layer thickness d is reduced under 500 Å [31]. Basically, when d (hereafter, we shall call it L_z) is made comparable to the de Broglie wavelength ($\lambda = h/p$), propagation of electrons is restricted in the transverse direction, and the allowed energy density of the semiconductor undergoes a drastic change, from the normal parabolic three-dimensional distribution to being quantized in a stepwise two-dimensional distribution as shown in Fig. 3.6 [32].

Density-of-Stated Distribution

The carrier energy eigenvalues can be approximated by two components, a component (z) normal to the layer and the normal unconfined Block function components (x, y) in the plane of the layer. This is expressed as

$$E(n, k_x, k_y) = E_n + \left(\frac{h^2}{8\pi^2 m^*_{n,p} (k_x^2 + k_y^2)} \right) \tag{3.6}$$

where E_n is the nth confined particle whose energy eigenvalue is the z component of the Hamiltonian in a box, $E_n = (h^2/2m^*_e)(n\pi/L_z)^2$ for $n = 1, 2, 3$, and $m^*_{n,p}$ is the effective mass of the electron or hole, h is Planck's constant, and k_x and k_y are the

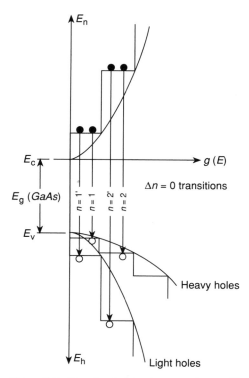

Figure 3.6 Density of states for a quantum-well heterostructure. The continuous solid parabolic lines indicate the energy states of bulk GaAs. After N. Holonyak, Jr., *et al.,* **QE-16,** pp. 170–185. Q. W. Heterostructure lasers. © 1980 *IEEE.*

crystal momenta in the x and y directions, respectively. Similar expressions apply to the holes, either light or heavy holes, Thus, E_n can be E_1, E_2, or E_3 for electrons; E_{hh1}, E_{hh2}, or E_{hh3} for heavy holes; and E_{lh1}, E_{lh2}, or E_{lh3} for light holes. As each energy band has its respective constant energy, it is called a subband with a constant density of states. E_1 is the lowest subband, followed by E_2, E_3, etc. This is shown in Fig. 3.6. The steplike density of states is shown as heavy lines. The parabolic curves plotted as thin lines along the steps are energies for the respective electrons and holes in their conduction and valence bands for the bulk device [32]. Thus, for a considerable range of energy, the density of states remains constant and is labeled E_n, $n = 1$, 2, and 3, respectively.

Recombination Transitions

Recombination transitions take place between a bound state in the conduction band E_n and the valence band E_{hhn} or E_{lhn} and release photons of energy given by the total energy difference. The gain of the laser is proportional to the injection carrier density. Injected carrier densities of bulk and quantum-well structures can be compared by calculating the number densities of the respective energy states, as $n_c = \int f(E)p(E)dE$, where $f(E)$ is the Fermi–Dirac distribution function and $p(E)$ is the density-of-states function. It is found [32] that for a step energy density function, the distribution of the injection carrier density is concentrated at the band edges, whereas for the uniformly distributed case, as in parabolic energy states (in bulk semiconductor), the distribution is spread over the complete energy band. Thus, recombination in a bulk sample occurs throughout the parabolically varying densities of states, which accounts for the much lower intensity compared with that which occurs at a fixed energy for a quantum-well structure. Therefore the maximum gain of a QW laser is much larger than that of a conventional bulk DH laser. This accounts for another adavantage of the quantum-well structure over the bulk sample. A considerable number of electrons can be gathered into this nearly constant energy state at E_1 (say), ready to recombine with a group of holes at a fixed energy E_{hh1} to contribute to the lasing action. This leads to a larger optical gain than in a simple DH laser. The QW device also reduces the probability of spontaneous emission. The threshold current density is thereby also reduced. I_{th} values as low as 65 Å/cm² have been reported [33]. In addition, line width is narrower.

Another advantage of QW over DH devices is the phonon-assisted recombination effect. In a bulk sample with parabolic density energy states, injected carriers at high energy scatter downward to lower energy states along the parabolic energy curve through thermalization. The process is often constrained. This is not the case in a QW structure where the steplike density of states is constant. In fact, the basic electron–phonon interaction is enhanced and, possibly, stimulated phonon emission can even occur [34]. The carrier thermalization effect for the quantum-well device can transfer electrons to well below E_1 such that laser operation at energy $h\nu < E_g$ becomes possible.

Low threshold current is another advantage of a MQW laser over the conventional DH laser. But to minimize its threshold current density, we must optimize the number of quantum wells in a MQW [35].

Multiple-quantum-well lasers have been developed for the 1.3- and 1.5-μm wavelength regions progressively in recent years [36–40]. In an InGaAs/InGaAsP MQW system at 1.5 μm, output powers of 190 mW [41] and narrow linewidths of 250 kHz at 4 mW [42] have been obtained.

Linewidth is a problem that needs to be addressed in single-frequency lasers. Linewidth broadening is the result of change in lasing frequency with optical gain caused by changes in the index of refraction when spontaneous emission is present. In MQW lasers, because of their higher optical gain, the linewidth is narrower, about 250 kHz.

Quantum-Well Structures

The basic quantum-well structure consists of an ultrathin (<200 Å) active layer of direct-bandgap semiconductor material sandwiched between suitable waveguide

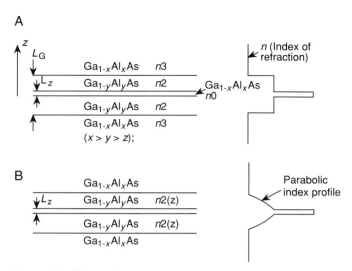

Figure 3.7   Schematic diagram of a quantum-well configuration with (A) a separate carrier and optical confinement heterostructure (SCH) and (B) a graded-index and separate carrier and optical confinement heterostructure (GRIN-SCH). After M. Asada *et al.* **QE-22,** pp. 1915–1921. © 1986, *IEEE.*

structures. This is a single-quantum-well (SQW) structure. Diagrams of various quantum-well configurations are shown in Figs. 3.7A and B. The index of refraction profile sketched on the side of the diagram indicates the optical confinement structure. In Fig. 3.7B, a GRIN (a parabolic index profile) is shown. A graded-index, separate confinement heterostructure single-quantum-well, wide-strip laser has been reported that operates at 780 nm [43]. It is an AlGaAs ternary alloyed structure. This wavelength (780 nm) is important for application in optical imformation processing systems such as optical disk file equipment. This laser can deliver an output power over 60 mW at 50° C. Stable operation for more than 1000 hours was achieved.

To minimize the threshold current density of a QW laser, we need to use multiple quantum wells. Thus double-quantum-well (DQW), triple-quantum-well (TQW), and multiple-quantum-well structures (MQW) were tried. Depending on the model gain required, there is usually a maximum number of wells that is the best for each design [44].

A diagram of an InGaAs/InGaAsP MQW laser structure is shown in Fig. 3.8A. The General layout of various epitaxial layers is outlined. Figure 3.8B shows the details of the waveguide layers. Here, four quantum wells of GaInAsP barrier layers are sandwiched between InP layers to form the waveguide and also serve as the confinement structure.

A ridge-type stripe quantum-well laser is illustrated in Fig. 3.9 [45]. It is a MQW laser. The active layer of this laser is of n-type and composed of ten GaInAs wells

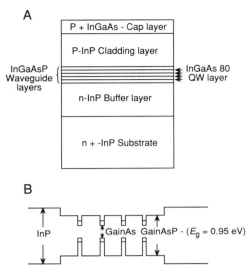

Figure 3.8 (A) Diagram of an InGaAs/InGaAsO multiple-quantum-well laser structure showing the different layers. (B) Four-cell quantum-well structure. After T. P. Lee [59], with permission.

(thickness $L_z = 80$ Å each) and nine AlInAs barriers (thickness $L_b = 20$ Å each). Other details are shown in the diagram. It is a single-longitudinal mode laser operating at 1.55 μm. The linewidth seems to become narrow, about half that of the InGaAsP/InP DH lasers.

Strained-Layer Multiple-Quantum-Well Lasers

Under A Review of Semiconductor Lasers, the importance of lattice matching in selecting semiconductor materials for making diode lasers was mentioned. Epitaxial layer compositions must be selected not only to have a specific bandgap and refractive index, but also to match the lattice constant of the substrate. The interatomic spacing of all layers must be within about 0.1% of that in the substrate. For GaAlAs/ GaAs epilayers, as shown in Fig. 3.1, the match can be accomplished easily; however,

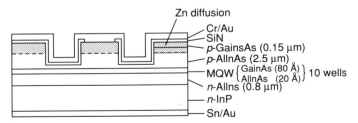

Figure 3.9 Ridge-type stripe quantum-well laser. After Y. Matsushima et al. [46].

for InGaAsP epilayers, stable lattice matching to the substrates is often difficult. This makes it hard to design a semiconductor laser at certain wavelengths between 0.9 and 1.1 μm.

It is very interesting to note that for semiconductor layers below a critical thickness, say, a few nanometers as often used in fabricating quantum-well lasers, the mismatching becomes less critical. In fact, the strain developed as a result of lattice mismatching may even improve laser operation and enhance its lifetime. Applying tensile or compressive strain via built-in lattice mismatching in InGaAs/InP MQW devices improves the laser's properties. The valence band degeneracy is removed, allowing for reduced Auger recombination and intravalence band absorption. Auger recombination is a process involving the transfer of energy from the electron in the conduction band to another free electron or hole, thus reducing the photon emission. Compressive strain lowers the threshold current (≤ 1 mA) of the device. Tensile strain provides higher power (≥ 200 mW CW) for a SQW device. Operating at 4800 nm, this laser has been used as pump source for an Er-doped fiber amplifier in a telecommunication system.

High-power strained-layer InGaAs/GaAs GRIN single-quantum-well lasers at an emission wavelength of 980 nm have been fabricated [46]. A light power as high as 270 mW and a maximum front power conversion efficiency of 51.5% have been obtained. The strained layer is installed such that the heavy hole subband of the quantum layer can be reduced by compressive strain, which results a reduction in the threshold current and improvement in other areas of laser performance.

Surface-Emitting Lasers

Surface-emitting QW lasers are attractive for possible applications in optical computing and image processing. They operate at a low threshold current and high differential quantum efficiency. In a typical design [47], the top-surface-emitting QW laser consists of a vertical $P-i-n$ junction, grown on a Si-doped n^+-GaAs substrate by the MBE method. The top mirror is Be-doped and the bottom mirror is the substrate. Electrical current is injected through the bottom and top mirrors as shown in Fig. 3.10. The top and bottom mirrors also serve as a Fabry–Perot resonator for the laser. The active layer of AlAs/AlGaAs consists of four 100-Å-thick quantum wells separated by three 70-Å-thick AlGaAs barriers. Injection current flow and laser output are shown in the diagram. The CW threshold currents are 3.5 to 8.0 mA for 10- to 30-μm-diameter lasers designed to operate at 845 nm. Initial CW slope efficiencies of 1.2 mW/mA up to 0.6 mW CW output power were reported. The measured spectral linewidth is 0.023 Å.

The University of California, Santa Barbara, reported an exciting CW submilliampere surface emitter at the XIIth International Semiconductor Laser Conference, held in Davos, Switzerland, in 1990 (see Conference Digest). Figure 3.11 illustrates a planarized vertical-cavity surface-emitting (VCSE) laser that emits light through the substrate. The area of the device is 6 × 6 μm. The threshold current is only 0.7 mA.

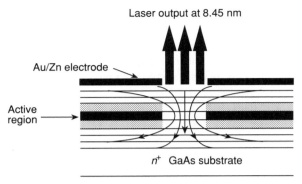

Laser output at 8.45 nm

Au/Zn electrode

Active region

n^+ GaAs substrate

Figure 3.10 Top-surface-emitting quantum-well laser structure. After T. P. Lee *et al., Electron. Lett.* **26,** 710–711. © 1990, *IEE.*

Feedback is obtained from multilayer semiconductor-stack reflectors, shown as mirrors in the figure. Light propagates perpendicularly to the active layer. The device has a strained-layer single quantum well, which emits at 980 nm.

Vertical-cavity-surface-emitting lasers may provide many advantages: the device can be fabricated by a fully monolithic process and can be densely packed into a two-dimensional array, the operation will be single mode as the mode spacing is large (about 20 nm), and a circular beam is achievable.

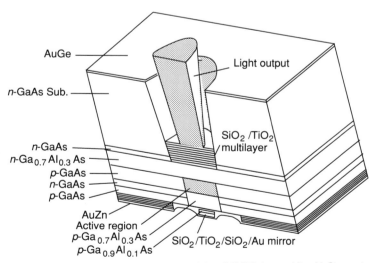

AuGe

Light output

n-GaAs Sub.

n-GaAs
n-Ga$_{0.7}$Al$_{0.3}$As
p-GaAs
n-GaAs
p-GaAs

SiO$_2$ /TiO$_2$ multilayer

AuZn
Active region
p-Ga$_{0.7}$Al$_{0.3}$As
p-Ga$_{0.9}$Al$_{0.1}$As

SiO$_2$/TiO$_2$/SiO$_2$/Au mirror

Figure 3.11 Vertical-cavity surface-emitting (VCSE) laser. After Y. Suematsu, *et al.* Proc. Vol. 80. © 1992, *IEEE.*

1.3-μm-Wavelength Lasers

Recently the focus of laser development has been shifted toward these emitting 1.3 μm. This is especially true in Europe, as most of the existing fiber cables in Europe operate on this wavelength. Promising results have been reported by Mohrle *et al.* [45]. Fundamental characteristics of InGaAsP MQW Separate confinement heterostructure (SCH) lasers emitting in the 1.3-μm wavelength range have been investigated. The laser structures were grown by the low-pressure MOVPE method. The design parameters included extremely thin (<2 nm) well layers, cavity lengths L greater than 200 μm, and four or more wells to maintain the low threshold current, $J = 780$ Å/cm^2.

Quantum Wire and Quantum Box

In MQW structures, the electrons are confined in two dimensions as described in under Advanced Lasers of Choice for Long-Haul Telecommunications. We mentioned the many advantages for using this laser structure. Would it have further advantages, if we confine the electrons even more, by slicing the planes into wires or even boxes as shown in Figs. 3.12B and C, respectively? (Figure 3.12A is a redrawing of a QW structure in three dimensions for comparison.) The respective energy state diagrams are shown directly below each case [48]. The dramatic change in energy state localization can be seen as the state density becomes stronger with higher dimension, this suggesting that quantum wire and quantum box devices would increase the efficiency of operation, lower the threshold current, and reduce

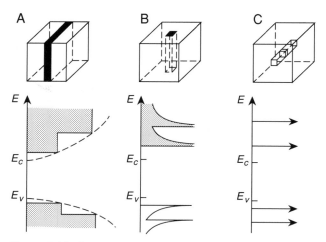

Figure 3.12 Structures and density of states for (A) quantum well, (B) quantum wire, and (C) quantum box. After M. Asada *et al.*, **QE-22**, pp. 1915–1921. © 1986. *IEEE.*

the temperature dependence. It has been shown that the optical gain of GaAlAs/InP quantum-well structures at the same injection carrier density ($N = 3 \times 10^{18}$/cm^3) increases to approximately 2 times (QF), 4 times (QW), and 16 times (QB) that of a bulk medium. The original quantum-well structure is called quantum face (QF) here.

Fabrication of QW and QB lasers is difficult as the demand for semiconductor material of fine size on the order of 10 nm cannot be met easily at present.

Wavelength Tuning in Quantum-Well Lasers

Reviewing the properties of a MQW laser as described, we shall say that all desirable characteristics we set out to improve have been reached except for wavelength tunability.

An external cavity tuning technique can be used to tune the wavelength of a QW laser. External to the laser chip, a diffraction grating is added, which serves both as a mirror and as a narrow band filter. Rotation of the grating changes the lasing frequency. Fine-tuning can be achieved by axial displacement of the grating. A tuning range of 55 to 105 nm, centered about a 0.8-μm SQW laser has been reported [49]. Other schemes, such as acoustooptically tunable semiconductor lasers [50] and electrooptically tunable semiconductor lasers have also been the subjects of successful experiments.

External cavity tuning has a disadvantage in that extreme mechanical stability of the system is required.

Distributed Feedback and Distributed Bragg Reflection Lasers

Advanced communication systems demand single-mode laser sources with stable and spectral purity beyond the capability of ordinary semiconductor lasers with Fabry–Perot resonators. Ordinary semiconductor lasers have a wide spectral gain; the net gain difference between various longitudinal modes is very small, leading to multimode lasing as described earlier under Longitudinal Mode Control.

An elegant and simple way to build single-frequency lasers is to integrate a wavelength-selective structure, such as a distributed Bragg grating, into a laser cavity. If the period of the grating Λ equals the Bragg wavelength λ_B, only the mode near the Bragg wavelength will build up to lase; other modes will be suppressed from oscillating [51]. If the grating structure is built into its gain region (pump region), the laser is called a distributed feedback (DFB) laser, shown in Fig. 3.13A. If the pumped region covers only a part of the distributed grating, and the unpumped grating is coupled to a low-loss waveguide, it is called a distributed Bragg reflector (DBR) laser, shown in Fig. 3.13B. In both cases, the usual cleaved mirror on one or both ends of the resonator is replaced by a grated reflector. Wavelength selectivity is ob-

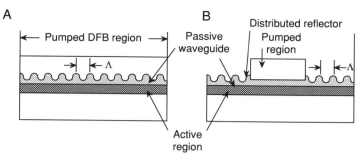

Figure 3.13   Diagrams of (A) distributed feedback laser and (B) distributed Bragg reflection laser. After J. E. Bowers and M. A. Pollack [20], with permission.

tained by designing the grating's spatial period Λ that is closest to the Bragg wavelength, $\lambda_B = 2n_e\Lambda/l$, where n_e is the effective refractive index of the mode and l is the integer order of the grating. Figures 3.13A and B show the design differences.

Quarter-Wavelength Shifted Distributed Feedback Laser Diode

The wavelengths of the longitudinal modes for a DFB laser with an ideal grating can be expressed as

$$\lambda = \lambda_B \pm \left[\frac{(m + 1)\lambda_B^2}{2nL}\right]$$

where m is the mode index, L is the effective grating length, and λ_B is the Bragg wavelength. Ideally, the mode wavelengths are spaced symmetrically around the Bragg wavelength. The lowest modes, corresponding to $m = 0$, have two frequencies; therefore, the laser is not a single-mode oscillator. Kogelnik and Shank [52], using couple mode theory, identified the two-mode property of a DFB laser. A phase shift of a quarter-wavelength, preferably placed in the middle section of the grating, will break the two-mode degeneracy and bring the laser into single-mode operation. The performance of $\lambda/4$-shifted DFB lasers is superior to that of conventional DFB lasers in terms of dynamic-single-mode (DSM) stability and low partition noise at high modulation speeds (at multi-gigabit/second). The linewidth is narrow, about 3 MHz under CW operation [53].

Distributed Reflector Lasers

Conventional $\pi/2$ phase ($\lambda/4$)-shifted and phase-adjusted DFB lasers can be replaced by one-sided distributed reflector (DR) lasers. A DR laser consists of both active and passive distributed reflectors, with only one facet serving as the output. A schematic of the structure of the DR laser, a 1.55-μm GaInAsP/InP SCH-BIG-DR laser, is

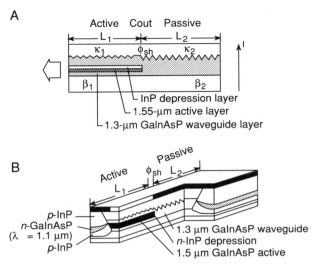

Figure 3.14 Distributed reflector (DR) laser. (A) Sketch of a GaInAsP/InP SCH-BIG-DR laser. (B) Actual layout. After Y. Suematsa *et al.* [60], with permission.

shown in Fig. 3.14 [54]. (BIG stands for bundle-integrated-guide.) Note that the grating is asymmetric and the active region extends only halfway into the structure. Only one end has an antireflection coating. The schematic refractive index distributions along the transverse direction of active and passive regions are shown in Fig. 3.14A, and the structure of the laser is shown in Fig. 3.14B. With this structure, which is only 10 to 20% longer than that of the DFB laser, it can deliver 16 times more power, double the quantum efficiency, with about the same threshold current. The linewidth–power product of the DR laser was 10 MHz · mW, compared with 50 MHz · mW for the DFB laser in the same active region [54].

Wavelength-Tunable Distributed Feedback Lasers

Wavelength tunability is a desirable property in a laser in coherent transmission and receiving systems. In multichannel frequency division multiplex (FDM) systems, the tuning range of the tunable laser used as a local oscillator limits the maximum number of channels. The popular structures for single-longitudinal-mode operation in DFB and DR lasers have to be modified to meet the required continuous tunability.

An external cavity structure can be used to tune the laser; however, it is difficult to realize continuous wavelength tuning with an external cavity structure.

Because the refractive index of a semiconductor can be easily changed by applying a curent or an electric field, wavelength tuning by electricity becomes most attractive for practical applications in DFB and DBR lasers.

Sectional Wavelength-Tunable Distributed Bragg Reflection Lasers

An ideal structure for wavelength tuning in DBR lasers is shown in Fig. 3.15. This structure has separate gain, phase control and DBR sections. Each section has an individual electrode so that the current into each section can be controlled separately [55]. The lengths of gain (first section), phase control (second section), and DBR (third section) sections are 300, 200, and 300 μm. Let the lengths of the sections be L_a, L_P, and L_d, and the injection currents I_a, I_P, and I_d, respectively. For a DBR laser, by keeping I_a nearly constant, and the reflectivity of DBR sufficiently high, increasing either I_a or I_P (keeping I_d unchanged) can contribute to wavelength tuning. A maximum tuning range of 6.2 nm has been reported [55].

Spectral broadening occurs as the wavelength is tuned toward the lower end, possibly as a result of an increase in absorption loss in the external cavity. This can be avoided if the gain section is replaced by using the MQW active layer [56].

In DFB lasers, if the equivalent reflective index of the whole cavity can be changed without a gain shift, an ideal continuous wavelength tuning can be achieved. The tuning range is limited only by the maximum index change attainable. A DFB laser structure that is doing just that has been reported by Amann *et al.* [57]. It is a twin-guide DFB laser consisting of two sections, the active section and the guide section, both controlled individually by current injection. The maximum continuous tuning range is about 7.1 nm. The linewidth broadening effect also exists, possibly as a result of an increase in leakage current and absorption loss in the guide section.

Wavelength tunability can be improved by monolithically integrating the DFB structure with two or three electrodes [57, 58]. By applying a large current to one electrode and a smaller current to the other, one can create separate pumping level regions along the grating. At a lower injection current, slightly below the threshold

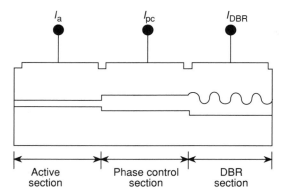

Figure 3.15   Three-section distributed Bragg reflection laser showing the different controls. After Y. Kotaki *et al.* [53], with permission.

current, the grating serves as a Bragg reflector, resulting in a large change in refractive index and substantial wavelength tuning. The high-current-pumped section contributes to photon generation and output. The tuning is continuous and has a range of 1.9 nm. DFB cavity length can be long, 1.2 nm in this laser. A linewidth as low as 500 kHz was obtained.

Summary

In this chapter, we reviewed what has been done to improve the characteristics of lasers in general; defined what we wanted for a laser in long-haul telecommunications; and discussed the present trend in the choice of lasers in telecommunications. Quantum-well lasers and DFB and DBR lasers and various combinations of those schemes are the most celebrated for this application. Practically, the use of lasers has expanded into applications in LANs, MANs, and ISDN and loop systems. Two outstanding papers reviewing recent advances in semiconductor lasers are highly recommended [59, 60]. You can expect to see more fruitful results and broad applications of semiconductor lasers in the future.

References

1. N. Holonyak, Jr., and S. F. Becavqua, Coherent (visible) light emission from $Ga(As_{1-x}P_x)$ junctions. *Appl. Phys. Lett.* **1**, 82–83 (1962).
2. M. I. Dumke, G. Burbs, F. H. Dill, Jr., and G. Lasher, Stimulated emission of radiation from GaAs $p-n$ junctions. *Appl. Phys. Lett.* **1**, 62–64 (1962).
3. H. Kroemer, A proposed class of heterojunction injection lasers. *Proc. IEEE* **51**, 1782–1783 (1963).
4. H. Kressel and J. K. Butler, *Semiconductor Lasers and heterojunction LEDs,* Academic Press, New York, 1977.
5. H. C. Casey and M. B. Panish, *Heterostructure Laser Devices,* Part A and B, Academic Press, New York, 1978.
6. J. J. Hsieh, J. A. Possi, and J. P. Donnelly, Room temperature CW operation of GaInAsP/InP double-heterostructure diode lasers emitting at 1.1-μm. *Appl. Phys. Lett.* **28**, 709–711 (1976).
7. J. M. Woodall, H. Rupprecht, and G. D. Pettit, Efficient electroluminescence from epitaxially grown $Ga_{1-x}Al_x$ $p-n$ junctions. *IEEE Trans. Electron Devices* **ED-14**, 630 (1967).
8. C. J. Neuse, G. H. Olsen, M. Eittenberg, J. J. Gannon, and T. J. Zameroski, Room temperature $In_xGa_{1-x}As/In_yGa_{1-y}P$ 1.06- μm lasers. *Appl. Phys. Lett.* **29**, 807–809 (1976).
9. R. E. Nahory, M. A. Pollack, E. D. Beebe, J. C. DeWinter, and R. W. Dixon, Continuous operation of 1.0-μm wavelength $GaAs_{1-y}Sb_y/Al_xGa_{1-x}Sb$ double-heterostructure injection lasers at room temperature. *Appl. Phys. Lett.* **28**, 19–21 (1976).
10. W. B. Joyce, R. W. Dixon, and R. L. Hariman, Statistical characterizaion of the lifetimes of continuously operated AlGaAs double heterostructure lasers. *Appl. Phys. Lett.* **28**, 684–686 (1976).
11. R. E. Nahory, M. A. Pollack, W. D. Johnston, and R. L. Barnes, Bandgap versus composition and demonstration of Vegard's law for $In_{1-x}Ga_xAs_{1-y}P_y$ lattice matched to InP. *Appl. Phys. Lett.* **33**, 659–661 (1978).

12. R. A. Logan and F. K. Reinhart, Integrated GaAs–Al$_x$Ga$_{1-x}$As double-heterostructure laser with independently controlled optical output divergence. *IEEE J. Quantum Elecon.* **QE-11**, 461 (1975).
13. G. H. Olsen, Vapor-phase epitaxy of GaInAsP. In *GaInAsP Alloy Semiconductors* (T. P. Pearall, Ed.), p. 11, Wiley, New York, 1982.
14. A. Y. Cho and I. Hayashi, Growth of extremely uniform layers by rotating substrate holder with molecular beam epitaxy for application to electrooptic and microwave devices. *J. Appl. Phys.* **42**, 4422 (1971).
15. G. B. Stringfellow, Organometallic vapor-phase epitaxial growth of III–V semiconductors. In *Semiconductors and Semimetals* (W. T. Tsang, Ed.), Vol. 22A, p. 209, Academic Press, New York, 1985.
16. G. P. Agrawal and N. K. Dutta, *Long-Wavelength Semiconductor Lasers,* Van Nostrand–Reinhold, New York, 1986.
17. G. H. B. Thompson, *Physics of Semiconductor Laser Devices,* Wiley, Chichester, 1980.
18. P. A. Kirby, A. R. Goodwin, A. R. Thompson, G. H. B., and P. R. Selway, Observations of self-focusing in stripe geometry semiconductor lasers and the development of a comprehensive model of their operation. *IEEE J. Quantum Electron.* **QE-13**, 705–719 (1977).
19. K. Aiki, M. Kuroda, T. Umeda, J. Ito, R. Chinnone, and M. Maeda, Transverse mode stabilized Al$_x$Ga$_{1-x}$As injection lasers with channeled-structure-planar structure. *IEEE J. Quantum Electron.* **QE-14**, 89–94 (1978).
20. J. E. Bowers and M. A. Pollack, Semiconductor lasers for telecommunications. In *Optical Fiber Telecommunications II* (S. E. Miller and I. P. Kaminow, Eds.), Academic Press, New York, 1988.
21. T. P. Lee, C. A. Burrus, R. A. Linke, and R. J. Nelson, Short-cavity single-frequency InGaAsP buried heterostructure lasers. *Electron. Lett.* **19**, 82–84 (1983).
22. G. P. Agrawal and N. K. Dutta, *Long-Wavelength Semiconductor Lasers,* Van Nortrand–Reinhold, New York, 1986.
23. F. Sholz, P. Weidmann, K. W. Benz, G. Trankle, E. Lach, A. Forchel, G. Laube, and J. Weidlein, GaInAs–InP multiquantum well structures grown by metalorganic gas phase epitaxy with adducts. *Appl. Phys. Lett.* **48**, 911 (1986).
24. R. J. Nelson, R. B. Wilson, P. D. Wright, P. A. Barns, and N. K. Dutta, CW electrooptical properties of InGaAsP buried heterostructure laser diodes. *IEEE J. Quantum Electron.* **QE-17**, 202–207 (1981).
25. J. Narayon, Fanning the hope for flat diamond. Reported by Iran Amato. *Science,* **252**, p. 375 (1991).
26. E. Yablonovitch, D. M. Hwang, T. J. Gmitter, L. T. Florez, and J. P. Harbison, Van der Waals bonding of GaAs epitaxial liftoff films onto arbitrary substrates. *Appl. Phys. Lett.* **56**, 2419–2421 (1990).
27. H. Shimizu, K. Ito, M. Woda, T. Sugino, and I. Teramoto, "Improvement in operation lives of GaAlAs visible laser by introducing GaAsAs buffer layers." *IEEE J. Quantum Electron.* **QE-17**, pp. 763–767 (1981).
28. J. T. Verdeyen, *Laser Electronics,* 2nd ed., Prentice-Hall, Englewood Cliffs, NJ, 1989.
29. K. Y. Lau, C. Harder, and A. Yariv, Ultimate frequency response of GaAs injection lasers. *Opt. Comm.* **36**, pp. 472–474 (1981).
30. T. L. Kock and R. A. Linke, Effect of nonlinear gain reduction on semiconductor laser wavelength chirping. *Appl. Phys. Lett.* **48**, 613–614 (1986).
31. W. T. Tsang, Quantum confinement heterostructure semiconductor lasers. In *Semiconductors and Semimetals* (R. K. Willardson and A. C. Beer, Eds.), Vol. 24, Ch. 7, Academic Press, New York, 1987.
32. C. Weibuch, Fundamental properties of III–V semiconductor two dimensional quantized structure. The basis for optical and electronic device applications. In *Semiconductors and Semimetals* (R. K. Willardson and A. C. Beer, Eds.), vol. 24, Ch. 1, Academic Press, New York, 1987.
33. C. A. Wang, OMVPE grown of low threshold high efficiency GaAs/AlGaAs and strained layer In-GaAs/AlGaAs diode lasers. In *Integrated Photonic Research Conference, TUGI, Hilton Head, South Carolina, 1990,* p. 97.
34. N. Holonyak, Jr., *et al.,* Phonon-assisted recombination and stimulated emission in quantum-well Al$_x$Ga$_{1-x}$As/GaAs heterostructures. *J. Appl. Phys.* **51**, 1328–1337. (1980).
35. Y. Arakawa and A. Yariv, Quantum-Well lasers, gain, spectra, and dynamics. *IEEE J. Quantum Electron.* **QE-22**, 1887 (1986).

36. W. T. Tsang, Ga$_{0.47}$In$_{0.53}$As/InP double-heterostructure multiquantum well lasers grown by chemical beam epitaxy. *J. IEEE Quantum Electron.* **QE-23**, 936–942 (1987).

37. U. Koren, B. I. Miller, Y. K. Su, T. L. Koch, and J. L. Bower, Low internal loss separate confinement heterostructure InGaAs/InGaAsP quantum well laser. *Appl. Phys. Lett.* **51**, 1744–1746 (1987).

38. K. Kasukawa, Y. Imajo, and T. Makino, 1.3-μm GaInAsP/InP buried heterostrucure graded index separate confinement multiple quantum well (BH-GRIN-SC-MQW) lasers entirely grown by metallorganic chemical vapor deposition. *Electron. Lett.* **25**, 104–105 (1989).

39. Y. Arakawa and T. Takahashi, Effect of nonlinear gain on modulation dynamics in quantum well lasers. *Electron. Lett.* **25**, 169–170 (1989).

40. A. Kasukawa, I. J. Murgatroyd, Y. Imajo, T. Namegaya, H. Okamoto, and S. Kashima, 1.5-μm GaInAsP graded index separate confinement heterostructure multiple quantum well laser diode grown by metallorganic chemical vapor deposition (MOCVD). *Electron. Lett.* **25**, 659–661 (1989).

41. D. M. Cooper, C. P. Selter, M. Aylett, D. J. Elton, M. Harlow, H. Wickes, and D. L. Murrel, High-power 1.5 μm all-MOVPE buried heterostructure graded index separate confinement multiple quantum well lasers. *Electron. Lett.* **25**, 1635–1637 (1989).

42. H. Yamazaki, T. Sasaki, N. Kida, M. Kitamura, and I. Mito, 250 kHz linewidth operation in long cavity 1.5 μm multiple quantum well DFB-LD with reduced enhancement factor. Presented at the Optical Fiber Communication Conference, PD-33, San Francisco, California, January 22–26, 1990.

43. T. Takeshita *et al.*, High-power operation in 0.98-μm strained-layer InGaAs/GaAs single-quantum-well ridge waveguide lasers. *IEEE Photonic Technol. Lett.* **2**, 849–851 (1990).

44. S. Yamashita, S. Nakatsuka, K. Uchida, T. Kawano, and T. Kajimura, High-power 780 nm AlGaAs quantum-well lasers and their reliable operation. *IEEE J. Quantum Electron.* **27**, 1544–1549 (1991).

45. M. Mohrle, M. Rosenzweig, H. Duser, and D. Grutzmacher, Fundamental characteristics of InGaAs/InGaAsP MQW-SCH-lasers emitting in 1.3 μm wavelength range. *IEE Proc.* **139**, 2932 (1992).

46. Y. Matsushima *et al.*, Narrow spectral linewidth of MBE-grown GaInAs/AlInAs MQW lasers in the 1.55 μm range. *IEEE J. Quantum Electron.* **QE-25**, 1376–1380 (1989).

47. Y. Arakawa and A. Yariv, Quantum well lasers, gain, spectra, dynamics. *J. Quantum Electron.* **QE-22**, 1887–1899 (1986).

48. D. Mehuys, M. Miiitelstein, A. Yariv, R. Sarfaty, and J. E. Unger, Optimized Rabry–Perot (AlGa)As quantum well lasers, tunable over 105 nm. *Electron. Lett.* **25**, 143–145 (1989).

49. G. Coquin, K. W. Cheung, and M. M. Choy, Single- and multiple-wavelength operation of acousto-optical tuned semiconductor lasers at 1.3 microns. In *11th IEEE International Semiconductor Laser Conference Digest*, pp. 130–131, IEEE, Boston, 1988.

50. F. Heismann, R. C. Alferness, L. L. Buhl, G. Eistenstein, S. K. Korotky, J. J. Veselka, L. W. Shutz, and C. A. Burrus, Narrow-linewidth, electrooptically tunable InGaAsP-Ti:LiNb03 extended cavity laser. *Appl. Phys. Lett.* **51**, 164–165 (1987).

51. H. Kogelnik and C. V. Shank, Stimulated emission in a periodic structure. *Appl. Phys. Lett.* **18**, 152–154 (1971).

52. H. Kogelnik and C. V. Shank, Coupled-wave theory of distributed feedback lasers. *J. Appl. Phys.* **43**, 2327–2335 (1972).

53. Y. Kotaki, S. Ogita, M. Mstauda, Y. Kuwahara, and H. Ishkawa, Tunable, narrow-linewidth with high-power λ/4-shifted DFB laser. *Electron. Lett.* **25**, 990–991 (1989).

54. J. I. Shim, K. Komori, A. Arai, I. Arima, Y. Suematsu, and R. Somchai, Lasing characteristics of 1.5 μm GaInAsP/InP SCH-BIG-DR lasers. *Trans. IEEE J. Quantum Electron.* **QE-27**, pp. 1736–1745 (1991).

55. Y. Kotaki and H. Ishikawa, Wavelength tunable DFB and DBR lasers for coherent optical fibre communications. *IEE Proc.* **138**, 171–177 (1991).

56. M. Fukuda, K. Sato, Y. Konodo, and M. Nakao, Continuously tunable thin active layer and multisection DFB laser with narrow linewidth and high power. *IEEE J. Lightwave Technol.* **7**, 1504–1509 (1989).

57. M. C. Amann, S. Illek, C. Schanen, W. Thulke, and H. Lang, Continuously tunable single-frequency laser diode utilising transverse tuning scheme. *Electon. Lett.* **25**, 837–839 (1989).

58. E. Yamamoto, M. Hamada, K. Suda, S. Nogiwa, and T. Oki, Wavelength tuning characteristics of DFB lasers having twin-guide structures modulated by injection current or electric field. *IEE Proc.* **139,** 24–28 (1992).

59. T. P. Lee, Recent advances in long-wavelength semiconductor lasers for optical fiber communication. *Proc. IEEE* **79,** 253–276 (1991).

60. Y. Suematsu, K. Iga, and S. Aral, Advanced semiconductor lasers. *Proc. IEEE* **80,** 383–397 (1992).

CHAPTER 4

Fiber Lasers

Introduction

The fiber laser represents an innovative use of optical fibers to produce light. It is a natural by-product of the fiber amplifier used to enhance signal transmission in optical fiber systems. In electronics courses, we were taught that any device that amplifies can be made to oscillate if feedback is provided. Such is the case with fiber amplifiers.

At present, we are not sure what to call it: fiber laser or optical parametric oscillator (OPO). To qualify it as a laser, the operation of the device must involve the process of stimulated emission. But the operating principle of the fiber laser does not necessarily involve such an emission process. At one time, the term *optical parametric oscillator* seemed to fit this emission better. But until it is announced that amplified spontaneous emission (ASE) produces oscillations that make no use of the parametric amplification, we will remain confused. We use fiber laser for short.

Do we need another source of light in fiber optics? The answer is yes. As we have aleardy learned from the development of laser light sources in previous chapters, laser structures of the type suitable for long-haul transmission at present are very complex. First, there is the choice of semiconductor materials. Only those with the correct bandgap energy to generate the wavelength needed can be used. Then there are the processes involved in building the laser by epitaxial heterojunction layers and in shaping the structure so that carrier confinement and waveguide properties are satisfied. Again, proper mode selection is exercised. The result is that a suitable laser may become the most expensive single piece of equipment in the system. Yet we still have to worry about the reliability and lifetime of the laser. Any progress that improves and simplifies the light source at low cost is welcomed. The fiber laser seems to fit these requirements nicely.

After the discovery in 1989 of an Er-doped optical fiber that amplifies the signal at 1550 nm in optical transmission systems, scientists began to wonder about the possibility of fiber oscillators and worked on them diligently. When a reasonable length of fiber is chosen in combination with a resonator, the fiber amplifier should be able to oscillate at the required wavelength designed for the fiber. Fiber amplifiers are discussed in Chapter 8.

What do we expect from using a fiber laser? The following are the prospective advantages: (1) Only a finite short section of fiber is required. Some designs use fibers less than 1 m; others use fibers as long as several tens of meters. The single-mode fiber has been designed for a particular wavelength, so that the process of material selection can be eliminated. (2) There are no special structural design considerations, such as carrier confinement, waveguiding property, and mode selection mechanism; many of these characteristics have been built into the fiber already. (3) Fiber loss per kilometer is not a concern as only a short length of fiber is used. (4) No special coupling device is needed to introduce the light to the fiber. (5) The noise is expected to be low, as little extra noise will be introduced along the process.

What do we need to make a fiber laser? We need (1) a resonant system that can feed back and regenerate the amplified signal to a level sustaining oscillation; (2) a pump power to start the laser, both frequencywise and in power level (a semiconductor diode laser with a power capacity large enough to supply the system and be cost effective is preferred); (3) an external modulation system to impress signal modulation to the system; and (4) a coupling section to bring in the pump laser to excite the fiber. In some cases, a polarization controller may have to be used.

Considering what we need and what we can get, it becomes obvious that the fiber laser is worth the effort of research. Refinement will always be needed to build a better light source.

Operating Principle of Fiber Lasers

There are at least two ways to obtain fiber laser: by feeding a portion of the amplified signal to a resonator to sustain the oscillation or by pumping the fiber parametrically in a resonator to generate the signal frequency.

Amplified Spontaneous Emission

Amplified spontaneous emission (ASE) is the unwanted product of a fiber amplifier that appears as noise when the fiber amplifier is operating; we try to limit its presence. But it can also be put into constructive use by purposely cultivating it in a resonator to make it oscillate.

The lasing effect in rare-earth-doped glass rods was demonstrated as early as 1961 [1]. Monomode fiber lasers have been developed [2]. Obviously, the fiber amplifier and fiber laser must have something in common. As the excited carriers at upper energy levels relax to lower levels and emit a photon(s), energy is gained from the pump supply. It is important to know to what extent the gain in the amplifier can be increased without producing spurious lasing. On the other hand, one wishes to determine the onset mechanism of lasing or the threshold of lasing in the case of fiber oscillator. We need theoretical guidance to control parameters for fiber laser design. Theoretical discussion on this subject is complicated by the fact that many assumptions used in the study of ordinary lasers may not be valid for fiber lasers. For example, as described by Le Flohic *et al.* [3], the uniform decay rate, usually considered in frequency dependent cavity losses, does not apply to Er-doped fibers in the resonant cavity, and the homogeneous line assumption applies strictly to monomode operation and does not apply to the Er-doped fiber. At present, the author is not aware of any complete theory or model that describes the operation of fiber lasers satisfactorily.

Experiments have proceeded rapidly, as can be seen in the literature [4, 5]. Duling *et al.* used a single-mode fiber of length l with end reflectors that was end pumped with a diode laser [5]. Feedback from reflectors at the fiber ends starts to build up and lasing begins. Both positive- and negative-going ASE powers were used. Fiber lengths up to 50 m were used. The pump source is a diode laser at 807 nm. The fiber source started to lase when pump power of about 40 mW was applied. Output power rises almost linearly to about 80 mW at 280-mW pump power. It is found that high output power (several tens of milliwatts) can be obtained with good wavelength stability and long life. The output wavelength is at 1060 nm. These sources are of potential use as fiberoptic gyroscopes. The authors suggested a model on which power output can be calculated, but they failed to predict the lasing threshold. The lasing threshold was determined experimentally. The measured threshold value showed a strong dependence of the pump power involved. Note that the authors called the device a *fiber source* instead.

Parametric Fiber Lasers

The basic idea behind the fiber laser can also be derived from a parametrically pumped fiber amplifier if a resonator is provided. Figure 4.1 shows such a scheme. A length of optical fiber is fitted into a pair of reflectors R_1 and R_2 (or mirrors) at the ends that serves as a resonator, as shown. The optical fiber, which is designed to transmit an optical signal frequency at f_1, acts as a weak nonlinear element within the resonator. The fiber is now pumped at frequency f_3 through the input reflector R_1. Because of the nonlinear effect of the fiber, frequency mixing between the pump frequency f_3 and the molecular vibrational frequency of fiber f_2 (the idler) takes place to generate a new frequency f_1, which exits from the far end of the resonator through the reflector R_2 after moving back-and-forth across the resonator a number of times. As described in Chapter 2, the relations between frequency matching, such that $f_1 = f_3 - f_2$, and the phase-matching condition (to be discussed in the next section) are satisfied, and a parametric downconversion light source is obtained.

Here f_2 is the idler frequency corresponding to the molecular vibrational freqency of the fiber constituents. For an Er-doped silica fiber, if the signal wavelength is

A

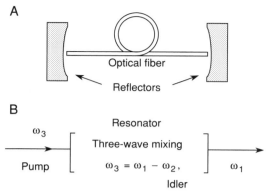

B

Figure 4.1   Scheme of a parametric oscillator. (A) A short section of optical fiber is enclosed in a resonator consisting of two mirrors at the ends. (B) The laser amplifier is pumped at one end and the output is taken from the other mirror end.

1.55 μm, and it is pumped at 1.48 μm, the idler frequency is $f_2 = 0.0915 \times 10^{16}$ Hz, corresponding to a wavelength of 32.77 μm. For other Er-doped fibers co-doped with Ge, a diode laser such as Ti:sapphire may be used as a pump. Usually, f_1 is determined by the system need and the choice of fiber type. Idler frequency f_2 is fixed by the material constants of the fiber. One is then to find a pump frequency that satisfies the parametric relationship for pumping. As there exists more than one single vibrational frequency in the medium, many possible pump frequencies may be used. Normally, the most efficient one is chosen for pumping.

Frequency Mixing in Nonlinear Medium

The operation of parametric pumping in fiber amplifiers involves nonlinear optics, which can best be explained by the process of frequency mixing involving nonlinear optical phenomena introduced in Chapter 2. Again, as we will see, a stimulated emission process may not be involved.

As explained under Nonlinear Optics in Chapter 2, the nonlinearity of a dielectric medium can be expressed in terms of the polarization vector **p** in Eq. (2.9). For a slightly nonlinear dielectric medium, assume that only the first two terms in **p** are used to express nonlinearity, that is, $\mathbf{p}_{n1} = \epsilon_0 \chi \mathbf{E}^2$. Thus the polarization vector contains the square of the signal electric field $E(t)$, traveling along the z direction. In a two-frequency mixing process, the sum of these signal waves can be expressed as

$$e(t) = A_1 \exp(-ikz)\exp(-i\omega_1 t - i\phi)$$
$$+ A_2 \exp(-ikz)\exp(-i\omega_2 t - i\phi) \tag{4.1}$$

Take the real part of Eq. (4.1) as

$$E(t) = E(\omega_1)\cos \omega_1 t + E(\omega_2)\cos \omega_2 t \tag{4.2}$$

The second-order nonlinear polarization, which contains the square of E, becomes

$$P = p_0 + p(2\omega_1) + p(2\omega_2) + p(\omega_+) + p(\omega_-) \quad (4.3)$$

where

$$p_0 = [E(\omega_1)^2 + E(\omega_2)^2]$$
$$p(2\omega_1) = E(\omega_1)^2$$
$$p(2\omega_2) = E(\omega_2)^2$$
$$p(\omega_+) = 2E(\omega_1)E(\omega_2)$$
$$p(\omega_-) = 2E(\omega_1)E^*(\omega_2)$$
$$\omega_3 = \omega_+ = \omega_1 + \omega_2 \quad (4.4a)$$

or

$$\omega_3 = \omega_- = \omega_1 - \omega_2 \quad (4.4b)$$

In second-order linearity, the interaction of two optical waves generates a third wave at the difference frequency which is called the downconversion. If the third wave generated has the sum frequency, it is called the upconversion. In frequency mixing, all five frequencies—$0, 2\omega_1, 2\omega_2, \omega_1 + \omega_2$, and $\omega_1 - \omega_2$—may exist. But only those that satisfy the additional phase condition can be generated. If we choose ω_1 as the desired or signal frequency, then the pump and idler frequencies are ω_3 and ω_2, respectively. The frequency-matching condition, Eq. 4.4a, has been met automatically. To see whether the condition of phase matching has been met, let us rewrite only the amplitudes of Eq. (4.1). The amplitudes of the electric field are respectively

$$E(\omega_1) = A_1 \exp(-ik_1 z)$$
$$E(\omega_2) = A_2 \exp(-ik_2 z)$$

Write

$$E(\omega_3) = E(\omega_1)E(\omega_2) = A_1 A_2 \exp(-ik_3 z) \quad (4.5)$$

One can deduce from this that

$$k_3 = k_1 + k_2 \quad (4.6)$$

Thus, the phase-matching condition is also satisfied. Plotting the wave vectors in Eq. (4.6) resulted in a closed triangle, as had been done in Fig. 2.3. Conservation of photon momentum holds true by multiplying Eq. (4.6) by h to yield

$$hk_3 = hk_1 + hk_2 \quad \text{or} \quad p_3 = p_1 + p_2 \quad (4.7)$$

where h is Planck's constant. The phase-matching condition is met.

Nonlinear Materials for Optical Parametric Oscillators

In this section, we divert our attention to the parametric generation of frequencies by other materials. In early OPO work, the nonlinear material of choice was lithium

TABLE 4.1
Properties of Nonlinear Materials for Optical Parametric Oscillators

Material	Transmission range (μm)	Phase-matching range (μm)					Nonlinear figure of merit C^2 (GW)$^{-1}$	Optical damage threshold (GW/ cm^2)
		0.266-μm pump	0.355-μm pump	0.532-μm pump	1.064-μm pump	2.05-μm pump		
BBO (BaB_2O_4)	0.190–2.6	0.3–2.5	0.41–2.5	0.67–2.5	—	—	40	~1.5
LBO (LiB_3O_5)	0.16–2.6	0.3–0.41 0.75–2.5	0.41–2.5	0.67–2.5	—	—	5.4	~2.0
KNB ($KNbO_3$)	0.35–4.2	—	—	0.61–4.2	1.43–4.2	—	44	~1.2
KTP ($KTiOPO_4$)	0.35–4.0	—	—	0.61–4.0	1.45–4.0	—	45	~1.5
LNB ($LiNbO_3$)	0.35–4.3	—	—	0.61–4.3	1.42–4.3	—	15	~0.20
$AgGaS_2$	0.8–9.0	—	—	—	1.2–9.0	2.6–9.0	75	~0.040
$AgGaSe_2$	1.0–15	—	—	—	—	2.4–15	100	~0.040
$ZnGeP_2$	2.0–8.0	—	—	—	—	2.7–8	270	~0.050

Reprinted with permission from Bosenberg *et al.* [30].

niobate ($LiNbO_3$). Since 1983, more crystalline materials have become available. These materials collectively provide good coverage of the entire wavelength range from the infrared to the visible blue as shown in Table 4.1. Listed in this table are the properties of these nonlinear crystals, including the range over which the material is transparent, OPO output wavelength range paired with several pump wavelengths, figure of merit, and damaging threshold for each material.

The parametric tuning range in a given crystal is determined by the pump wavelength, the transparency of the crystal, and the range over which phase matching is possible. The figure of merit indicates the magnitude of the nonlinear coefficient. Most of the materials listed have significantly higher nonlinear coefficients than $LiNbO_3$. Also, most crystals may be pumped by more than one pump wavelength.

Resonant Cavity for a Fiber Oscillator

Bragg Grating Reflector Cavity

Return now to the fiber laser. We wish to discuss the resonant cavity used for a fiber laser. Although a pair of reflecting mirrors external to the fiber can be used as a resonator cavity, it is more convenient to build the cavity right on the fiber proper, to save space and avoid extra coupling devices. An example of such a setup is the Bragg

A

B

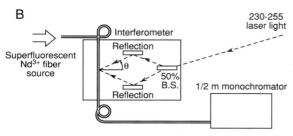

Figure 4.2 Typical Bragg phase grated fiber. (A) The embedded BPG ends of the fiber which form the resonant cavity of the fiber laser. (B) A method used to write BPG on fiber holographically. After G. A. Ball, W. W. Morey, and W. H. Glenn [7].

phase grating. Meltz *et al.* showed a way to build an intracore Bragg phase grating on the core of a single-mode optical fiber as follows [6]. First, a section of the fiber end (0.5 m of Er-doped fiber) is stripped off the buffer. Then this section of the fiber is exposed transversely to a coherent source containing a two-beam interference pattern. The ultraviolet wavelengths of the two-beam interference pattern are chosen to lie within the oxygen vacancy defect band of germanium, which would form a permanent photorefractive variation in the core of the fiber, duplicating the sinusoidally varying radiation intensity. The wavelength of the signal reflected by the Bragg phase grating (BPG) is determined by the grating periodicity. The reflector's bandwidth and amplitude are determined by the magnitude of the index change and the length of the grating.

In one example, reported by Ball *et al.,* two reflectors were fabricated on opposite ends of a length of Er-doped fiber to form a 0.5-m cavity [7]. The gratings were written by the method described earlier, approximately 10 cm from each cleaved fiber end. The Er-doped germanoaluminosilicate fiber has a 5-μm core diameter and a differential refractive index of 0.021. The reflectivity is about 80 to 100%.

Setup of a typical Bragg phase grated fiber laser is illustrated in Fig. 4.2. Figure 4.2A illustrates the embedded sections of the BPG laser at the ends of the fiber. Figure 4.2B demonstrates the method used to write BPG holographically.

Ring Fiber Laser

In another example, a ring fiber laser cavity is built. This is perhaps the most simple structure of a fiber laser. It consists of a finite length of fiber designed for the desired wavelength and coupled through waveguide-type directional couples (DCs) to a di-

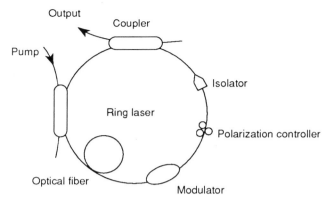

Figure 4.3   Ring fiber laser scheme, consisting of a short length of erbium-doped fiber with directional couplers, to use the diode pump and to extract output power.

ode pump laser and modulator, isolator, and polarization controllor to form a ring. The output is taken from another DC as shown in Fig. 4.3. When the pump laser is activated, lasing at the desired wavelength can be expected. The length of the fiber used for the laser is determined by the doping concentration in the core of the fiber. A higher doping concentration allows a shorter fiber length to absorb the pump energy. For an erbium-doped fiber, usually a low doping concentration is used. The fiber length is designed so that the system gain is just enough to offset the cavity loss. Fibers longer than necessary will leave a portion of the fiber not being pumped and thus absorbing the laser energy.

Rare-earth-doped fibers have very small core size, about several micrometers in diameter. The tight confinement of light leads to high small-signal gain, up to 10,000, so fiber lasers can tolerate high cavity losses. The cleaved face of the fiber provides adequate feedback for lasing.

Semiconductor diode lasers or solid-state lasers can be used for pumping fiber lasers.

Typical Fiber Lasers

Many rare earth elements have been used to dope glass fibers to produce fiber lasers. Examples of these experiments, besides the Er-doped fiber, include the Nd^{3+}-doped germanosilicate fiber, the Pr^{3+}-doped fluoride fiber, and the Nd^{3+}-ZBLAN fiber. For different rare-earth-doped fibers, different pumping schemes are suggested.

Er^{3+}-Doped Fiber Laser

Fiber lasers did not follow far behind the development of Er^{3+}-doped fiber amplifiers. We describe two different schemes in the following.

The energy-level diagram of erbium-doped silica glass can be either a two-level or a three-level system depending on the pump excitation. When pumped with 1480 nm, the pump energy excites the impurity ions to the $^4I_{13/2}$ state. The transition from $^4I_{13/2}$ to $^4I_{15/2}$ produces a laser line in the range 1520–1560 nm. The lasing bandwidth centered at 1540 nm is about 40 nm. When it is pumped with 980 nm, it behaves as a three-level system. The pump energy excites the ions first to $^4I_{11/2}$, which decays to $^4I_{13/2}$ and lases in the same range as the first case.

The first experiment was described by Iwatsuki [8], who used a single-mode Er-doped superfluorescent fiber, about 90 m long, coiled between a resonator consisting of a pair of dichroic mirrors. The Er-doped fiber had a core diameter of 6 μm, a cutoff wavelength of 0.97 μm, and a refractive-index difference of 0.32. The Er dopant concentration was 30 ppm. The dichroic mirror had a reflectivity of ≈100% at 1.48 μm and more than 95% at 1.5 μm. Both ends of the Er-doped fiber were polished at 8° angles to avoid laser oscillations. The system was pumped by a 1.48-μm laser diode, and the output power and the spectral profile were measured and analyzed. The output power-versus-pump power curve showed that a minimum pump power (threshold power) was required before output power rose rapidly. This threshold power is about 46 mW of pump power. Power output reached 3.8 mW at a pump power of 80 mW. A stable spectral profile of a 2-nm optical bandwidth at 1.53-μm peak wavelength is obtained. The emitted light is found to be unpolarized. A typical power outpower-versus-pump power relationship is illustrated in Fig. 4.4.

Zyskind et al., described another Er-doped fiber laser [9]. A single mode, high concentration (2500 ppm) Er-doped fiber of only 3–4 cm in length was used in a fiber laser. BGP gratings on each ends were used as reflectors. The resonant cavity length is only 1–2 cm. Diode pumping at either 980 or 1480 nm was used to generate 1540 nm laser output. Output of 122 μW for a pump power of 34 mw at 1480 nm, and 181 μW with 61 mW at 980 nm was reported.

The fiber laser was end-pumped with a Ti:sapphire laser at 980 nm. The pump radiation was coupled to the fiber through a microscope objective. The light output was measured at the other end of the reflector.

The laser was observed to lase at 1548 nm. The slope efficiency was about 27%.

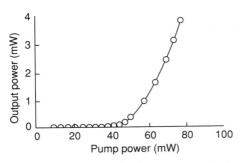

Figure 4.4 Er-doped fiber laser: power output versus pump power. Note that power output begins to rise after a minimum pump power is reached. After C. A. Miller et al. [11].

A peak power of 5 mW was observed at a pump power of 50 mW. A linewidth of less than 47 kHz was measured.

In the examples described in the preceding two sections, erbium-doped silicate fibers either with or without a codopant such as aluminum or germanium were used. The purpose of using a codopant is to improve lattice matching between the erbium element and the host fiber material as will be described in Chapter 8. The addition of codopant also changes the natural molecular vibrational frequency of the fiber, and so the absorption and therefore the pump frequency for effective pumping are also altered. Improved pumping efficiency can be achieved.

Erbium-doped fibers are designed for operation at a wavelength of 1550 nm where the fiber loss is at its minimum. This type of fiber is to be installed in the new generation of optical fibers in the future.

Pr^{3+}-Doped Fluoride Fiber Laser

An alternative fiberoptic transmission system that has already been installed and operational since the 1980s is at the 1.31-μm band. The scientific community, particularly in Europe, seemed intent on developing the 1310-nm window to save the existing system. For this purpose, optical amplification with high gain is required in 1300-nm telecommunication systems. Praseodymium-doped fluoride fiber seems to have the potential to parallel the erbium-doped silica fiber in the 1550-nm window.

A Pr^{3+}-doped single-mode fluoride fiber laser has been developed by Ohishi et al. [10]. The fiber has a core diameter of 4.5 μm and is about 10 m long. The dopant concentration is 1000 ppm. The differential refractive index is 0.6%. The fiber loss measured 0.87 dB/m at 1017 nm. The fiber is designed for a signal wavelength at 1310 nm. The cavity of the laser consists of a waveguide coupler on the input end and a mirror on the other end. The fiber is pumped with an argon-pumped Ti:sapphire laser at 1017 nm. The output is taken from the mirror end. Lasing starts at about 360 mW of pump power. The lasing frequency is at 1310 nm and remains unchanged at higher pumping power. The slope efficiency is about 9.86%. A maximum power of 26 mW is obtained at a pump power of 370 mW. No saturation effect was observed at this pumping power. The design was not optimized to obtain low threshold and high slope efficiency. It could be improved by careful design. But the large output power is impressive.

Neodymium–ZBLAN Fiber Laser

Another fiber laser has been reported to operate at 1345 nm [11]. A rare-earth-doped zirconium fluoride glass (ZBLAN) fiber is used with the objective of developing sources compatible with standard single-mode fiber transmission systems in the 1300-nm window. The acronym ZBLAN stands for zirconium, barium, lanthanum, aluminum, and sodium. The host material is doped with neodymium. The multimode fiber used for the laser structure has a core diameter of 40 μm and a length of 35 cm.

The core was doped with neodymium ions of 1000 ppm concentration. The core index was raised by 0.01% above the cladding by the addition of PbF_2. The fiber was designed to operate at 1345 nm. When pumped with a GaAlAs diode laser at 795 nm, lasing starts at about 60 mW of pumping power. The slope efficiency is 57%. A maximum power output of 30 mW was reported at 120 mW of pumping power. No saturation was in sight. The output power can be coupled to a single-mode fiber at the output mirror.

With improved mirror design and improved couping efficiency, it is possible to excite the fiber at 1300 nm where zero dispersion occurs.

Nd^{3+}-Doped Fiber Laser

Attempts have been made to design fiber lasers for operation at wavelengths other than 1310 and 1550 nm.

An efficient integrated Nd^{3+} fiber laser that operates at 1088 nm has been reported [12]. The aim was a rugged, diode-pumped, high-efficiency, low-threshold, tunable fiber laser. A single-mode germanosilicate fiber with a core diameter of 3.5 μm is used. The fiber is doped with neodymium of 500 ppm and has a NA of 0.2. The fiber loss is about 0.05 dB/m at 1088 nm. The fiber laser is end-pumped with a Ti : sapphire laser at 810 nm through the 80% grating using a microscope objective. The cavity is formed by a pair of Bragg phase gratings holographically written on each end of a 90-m fiber. Lasing starts at about 5 mW of pumping power and increases with a slope efficiency of about 12.3%. A maximum power of 2.6 mW is obtained at a pump power of about 34 mW.

The laser is tunable by varying the operational temperature. The temperature tuning is about 0.012 nm per degree centigrade in laser wavelength. Wavelength tuning in excess of 3 nm could be achieved by heating the Bragg reflectors to 250°C.

Pump Source Requirements
for End-Pumped Lasers

An ideal pump source for optical parametric oscillators should deliver good beam quality, high peak power and pulse energy, and high pulse repetition frequency to yield high average output power. Also, narrow bandwidth is critical for pumping narrowband OPOs.

Two schemes of pump are usually used to supply energy to the laser, the end pump scheme and the parallel pump scheme. In the end pump scheme, one pump laser is used in series with the laser to be pumped; thus the pump energy is limited. The parallel pump scheme could use an array of pump lasers, all supplying energy to one laser at the same time and thus providing much more energy to the laser. Usually, array lasers have to be mode-locked to supply energy all at one time.

Mode Locking in Fiber Lasers

A laser can oscillate on many longitudinal modes, with frequencies that are equally separated by the intermodal spacing frequency $v_F = c/2d$, where c is the speed of light in the medium and d is the length of the resonator cavity. If the bandwidth of the spectral distribution is B, then the number of possible laser oscillation modes is $m = B/v_F$. With a bandwidth B of 1500 MHz and cavity length of 150 cm, at least 15 or more modes could oscillate simultaneously. Optical power is divided among the different oscillating modes. The distribution of the amplitudes of the modes is centered at a frequency v_o where the loss is the lowest. The amplitudes decrease further away to both sides of this frequency; however, the number of modes that actually carry optical power is far less than the number of possible modes. Still, the desired mode shares only a part of the available optical power. Figure 4.5 represents the frequency distribution of a multimode laser operation. Figure 4.5A shows the envelope of the amplitude distribution with a frequency spread of Δv around the center frequency v. Figure 4.5B illustrates the mode distribution in mode lines as $+m$ and $-m$ numbers around the center 0. These lines are separated by $c/2d$, where d is the length of the resonator; $m = 0, \pm 1, \pm 2, \pm 3$.

A multimode laser can operate on a single mode by introducing an element inside the resonator to provide sufficient loss to discourage oscillation of the undesired modes. An alternative way is to lock all the modes into a single mode and extract the total optical power.

Pulse operation is often desirable for signal transmission. The most direct method of obtaining a pulsed light beam from a laser is to use a CW laser in conjunction with an external switch or modulator that transmits the light only during selected short time intervals. This scheme is very inefficient, and the peak power is limited to the steady power of the CW source.

Although pulsed light can easily be achieved by installing a modulator or by using a Q-switch (by inserting a saturable absorber) inside the resonator cavity to turn off the laser output periodically, the most effective method is to use the mode-locking

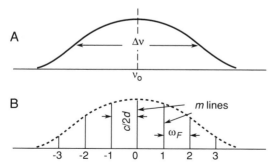

Figure 4.5 Longitudinal modes in laser operation. (A) Overall bandwidth of the laser output Δv. (B) Fine structure of the components showing the different modes.

method. Mode locking can be achieved by locking the phases of each longitudinal mode of a multimode laser together, forcing the laser to form a periodic pulse train.

Consider each laser mode to be represented by a uniform plane wave traveling in the z direction with velocity $c = c_o/n$. We wish to see, if the modes were locked, how the waveform and wave amplitude would change. Methods for locking the modes are discussed later.

The general equation describing the classical electromagnetic field for the total complex wavefunction of the field as a sum of these individual waves is

$$U(z, t) = \sum_q A_q \exp\{i\omega_q(t - z/c)\} \tag{4.8}$$

where

$$\nu_q = \nu_o + q\nu_F, \quad q = 0, \pm 1, \pm 2, \ldots \tag{4.9}$$

and ν_q = the frequency of mode q. Substitute Eq. (4.9) into Eq. (4.8) and assign ν_o as the center frequency of the laser lineshape at $q = 0$. Equation (4.8) can be rewritten as

$$U(z, t) = A\{t - z/c)\exp\{i2\pi\nu_o(t - z/c)\} \tag{4.10}$$

where

$$A(t) = \sum_q A_q \exp(i2\pi qt/T_F) \tag{4.11}$$

$$T_F = 1/\nu_F = 2\,d/c \tag{4.12}$$

It can be shown that the complex envelope $A(t)$ in Eq. (4.11) is a periodic function of period T_F, and $A(t - z/c)$ is a periodic function of z of period $cT_F = 2d$. The optical intensity, which is proportional to the square of the amplitude, can then be expressed as

$$I(t, z) = A(t - z/c)^2 = m^2A^2\{\text{sinc}^2(t - z/c)/T_F\}/\{\text{sinc}^2[t - z/c]/T_F\} \tag{4.13}$$

where m is the number of modes and

$$\text{sinc } x = \sin(\pi x)/\pi x.$$

Figure 4.6A represents the frequency distribution of a complex wavefunction U_ν that expresses I, the intensity of a complex M of waves of equal phase, and Fig. 4.6B represents the time distribution of the intensity when mode-locked. The mode number m and the period T_F of the pulse are as shown. Note that the waveform is periodic and in pulse form. The peak pulse intensity is now m times stronger and the pulse width $1/m$ times shorter, respectively.

If the waveshape is analyzed, the following results can be drawn:

Temporal period	$T_F = 2d/c$, where d is the cavity length
Pulse width	$\tau = T_F/m = 1/m\nu_F$
Spatial period	$2d$
Pulse length	$d = c\tau = 2d/m$
Mean Intensity	$I = mA^2$
Peak intensity	$I_p = m^2A^2$

As an example, consider a Nd^{3+} : glass laser operating at 1060 nm. The linewidth $\nu = 3 \times 10^{12}$ Hz is expected. If the resonator is 10 cm long and the mode separation is $\nu_F = c/2d = 1$ GHz, then $m = \nu/\nu_F = 3000$ modes. With mode locking, the peak intensity would be 3000 times stronger. The pulse width $\tau = 1/3000 \times 3 \times 10^{12} = 10^{-14}$ s, and the pulse would occupy only $d_p = 0.3$ mm of space.

Mode-Locking Methods

Mode locking can be achieved by either passive or active methods. An absorber, whose absorption coefficient decreases as the light intensity passes through it increases, can act as a passive mode-locking device when placed within the laser cavity. Only light of high intensity can pass through the absorber. This happens when different modes are all in phase. Modes whose phases are different from each other are discriminated against and are absorbed. Active mode-locking devices involve the use of electrooptic and acoustooptic switches inside the cavity to modulate the phases into locking position. When the switch is open (or activated), it passes light only for the duration of the pulse. Only those modes that have equal phases can lase. And once the oscillations start, they continue to be locked. Modes of equal phase are summed up and a giant pulse can be formed. In the absence of phase locking, the individual modes having different phases are dependent on the random conditions at the onset of their oscillation and are totally or partially blocked, thus adding to the loss. As shown in Fig. 4.6B, the intensity of the pulse can be m times larger and the pulse length $1/m$ times smaller.

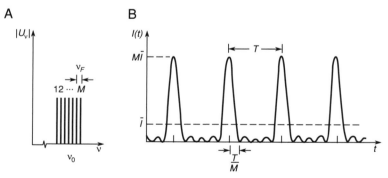

Figure 4.6 Mode locking to obtain giant pulses of shorter duration. (A) Frequency distribution of the intenstiy of a complex of M waves of equal intensity and phase. (B) Time dependence of the complex waves in mode-locking condition. Note that giant pulses are formed. Each pulse has a width that is M times smaller than T_F and a peak intensity that is M times greater than the mean intensity. After Saleh and Teich, *Fundamentals of Photonics,* with permission. © 1991, John Wiley & Sons.

Active Mode Locking

Er-doped fibers represent an attractive means of generating broadly tunable, high-average-power oscillations at a wavelength suitable for long-haul communication systems. By actively mode-locking an Er-doped fiber laser, it is possible to generate high-repetition-rate wavelength tunables for amplification of solitons. Wigley *et al.* described such a system [13]. Figure 4.7 is a diagram of their experiment. The erbium-doped fiber is 3 m long, with an outer diameter of 115 μm, a differential refractive index of 0.08, and a cutoff wavelength of 1.17 μm. The fiber is placed in a cavity consisting of a mirror *M* at one end and a diode laser at the other end. The mirror transmits 90% at a pump wavelength of 514 nm from a CW Ar laser and reflects 50% at 1550 nm. The diode laser serves as an inline modulator that is controlled by applying both a dc bias and a rf source to activate the laser. When no current is injected, the diode laser exhibits absorption. It becomes transparent when carriers are injected and its forward current increased. The radio frequency source at 500 MHz is used to drive the diode for pulse operation. The output of the fiber laser is taken via the mirror *M*. Other necessary components are indicated on the diagram as shown. The system is operating at a direct current at 20 mA and with 600 mW of rf power to obtain mode-locked operation of the erbium-doped fiber laser. The rf drive frequency can be adjusted for minimum pulse width from the laser system. A typical output pulse is about 37 ps at an average power of over 20 mW. This relatively inexpensive method of modulation has the potential to produce substantially shorter pulses and higher peak powers than previously demonstrated.

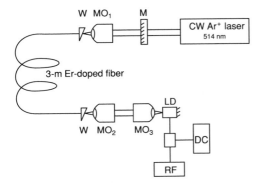

Figure 4.7 Active mode-locking scheme. Fiber: 3-m-long Er-doped silica fiber. Resonator: between the mirror M and the diode laser. Modulator: with DC and RF controls. W is a quartz wedge for index matching. MOs are microscope objectives. The laser is pumped by an argon ion laser at 414 nm. Output at 1550 nm is taken from the mirror with a beam splitter (not shown). After P. G. J. Wigley *et al.* [13].

Passive Mode Locking

Haberl *et al.* demonstrated a passively mode-locked fiber laser [14]. They used a 40-cm fiber doped with 1700 ppm Nd^{3+} that had a NA of 0.17 and a core diameter of 5 μm. This device was placed into a Fabry–Perot cavity with an intracavity polarizer, a dispersive delay line, and a weakly linearly birefringent fiber attached to one of the two cavity mirrors as a passive mode-locking device. A CW operating krypton laser supplied the 350 mW of pump power. The fiber delivered a power of 20 mW at 1064 nm with a 60-fs pulse at a repetition rate 48 mHz. The device can be made to deliver low-noise operation up to 5 dB below the amplitude noise of the pump laser. Active frequency stabilization reduces the time jitter to less than 130 fs in a frequency band from 10 Hz to 100kHz, and for periods less than 1 ms a timing jitter less than 5 fs is obtained.

High-Power Fiber Lasers

Diode-pumped lasers seem to have opened the door to a variety of new lasers including high-power and new-frequency lasers. With the addition of multipumped schemes [15] and new nonlinear materials, many high-power and different-frequency lasers are possible. With the frequency doubling technique, lasers in the ultra-violet range become available. As power and pulse energy from diode-pumped solid-state lasers increase, applications broaden.

Diode Laser Pumping

The term *diode laser pumping* has appeared many times in the preceding text, showing that diode lasers are being used to supply output light power, as in parametric amplifiers, or to initiate lasing in another laser, as in power lasers. We wish to amplify the idea of using diode lasers for these purposes and to explore its importance in future developments in photonics.

In laser applications, we usually select a wavelength that operates best for the desired task; for example, for long-haul communication systems using optical fibers, we choose 1.55 μm because at this wavelength, the fiber loss is lowest. In Chapters 5 and 6, we describe the technique of using diode laser pumping to achieve high optical power, or to obtain wavelengths otherwise inaccessible or impractical to obtain directly, such as to reach the ultra-violet range. We also see the advantages of using fiber amplifiers over electronic ones in these systems as described in Chapter 8. Fiber amplifiers gain their power from the pump, which is another laser. In each case diode laser pumping plays a major role in the success of these applications. One can see that by using diode laser pumping in combination with other lasers or amplifiers, many desirable characteristics can be gained. These include attainment of a wavelength of interest, gain of a higher power output, improvement of the conversion efficiency of the system, and achievement of lower cost of installation and operation of the system. The diode laser is compact and potentially inexpensive. Diode laser pumping thus opens the door to future photonic applications.

GaAlAs laser diodes which emit 800 nm are used to pump neodymium lasers. The InGaAsP diode have been used to pump lasers in the range 1300–1550 nm. The appealing features of diode pumping are simple implementation, high quantum efficiency, and low dissipation of thermal heat to the system. The price you have to pay for it, at present, is the high cost of the semiconductor diodes. But in time, and particularly as the demand for diode lasers increases, mass production may help to lower the cost considerably. The current price for semiconductor diode lasers includes the design of single-mode, stabilized, single-frequency devices, which may not be needed for other operations. Solid-state diode lasers are designed to fill these gaps.

Diode laser pumps are usually applied to a parametric oscillator by means of end pumping; that is, the pump power is connected in series with the laser cavity through the end mirror. The geometric configuration of this scheme makes it difficult to use more than one pump as a power source, thus limiting the power-carrying capability of the end-pumped systems. At present, for example, for end-pumped lasers at 1064 or 1047 nm, the maximum end-pumped power is limited to about 2 W.

A side-pumping technique can be applied to laser systems if high power is desired. In this case, an array of lasers are grouped for pumping, which are applied to the laser cavity in parallel. Usually, the technique of injection locking of the pumping laser array is used to ensure high power supply. But the quantum efficiency of side-pumped lasers is generally lower than that of end-pumped lasers.

Solid-State Lasers

The discussion of solid-state lasers in this section is limited to their application to pumping fiber amplifiers and/or fiber oscillators. This topic is discussed further in Chapter 5.

The first optical laser demonstrated by Maiman in 1960 was a solid-state ruby laser (Cr^{3+}-doped into sapphire, Al_2O_3) [16]. This invention revolutionized optical science and turned it into the most rapidly growing area of modern technology. Although semiconductor lasers were at the forefront in the development of communication systems as optical fibers and integrated microelectornics were introduced, the development of solid-state lasers was never far behind. Diode solid-state lasers have been developed to cover a broad range of wavelengths and power capabilities. Here we mention only a few samples, in relation to fiber lasers in particular.

Materials and their processing are the central problem in fabricating solid-state lasers. The search for new materials for solid-state lasers is similar to that for semiconductor lasers: looking for a combination of elements that have similar band-energy levels between the host and doping materials. The function of the host material may be quite different from that in semiconductor lasers. The combination of host crystals and impurity doping ions is chosen such that a divalent or trivalent impurity ion can substitute for a host ion of the same state for a structural match and that energy absorbed by the impurity ions when pumped up to higher levels can be transferred to the host ions readily. We can find many rare earth elements that satisfy these requirements. Of course, the primary purpose of the combination is to achieve a desirable wavelength and/or to obtain a power level of choice. Lattice matching is also an important criterion. The host crystal should also be able to dissipate the heat produced in the operation.

Many times, lattice mismatching problems can be alleviated by using suitable codopants.

The host crystal or glass is usually an insulator. The impurity doping ions must have an energy level structure for easy pumping. They must have the long-lived metastable states close to some energy levels of the host material. Many such pairs are available. A few combinations of elements are listed here for reference. (See Chapters 5 and 6 for details.)

Ruby Cr^{3+}: Al_2O_3 (0.69 μm)
Ti^{3+}: Al_2O_3 (0.68–1.13 μm) [17]
Cr^{3+}: Al_2BeO_4 (0.72–0.79 μm) [18]
Nd^{3+}: YAG   ($Nd^{3+}_xY_{3-x}Al_5O_{12}$) [19], (1.06–1.32 μm)
Nd^{3+}: YLF   (1.047–1.06 μm) [20]
Er^{3+}: Glass   (1.06, 1.54 μm) [21]
Cr^{3+}: Mg_2SiO_4 (1.35–1.61 μm) [22]
Co: MgF_2 (1.75–2.5 μm) [23]
Tm: YAG   (2.01 μm) [24]
Ho: YAG   (2.01 μm) [25]
Er^{3+}: $LiYF_4$ (2.8 μm) [26]
Er: YAG   (2.94 μm) [27]

Solid-state diode lasers can be used to pump other semiconductor lasers or other lasers.

Solid-state optical lasers can provide optical power output of different magnitudes suitable for virtually all different applications, from scientific, to industrial, to medical, to home electronics. Through material research, new materials and new processing techniques will be added continuously [28]. The future is unlimited.

Harmonic Generation

Harmonic generation of higher frequencies can be achieved by pumping nonlinear crystals with lasers of established lower frequencies. Frequency doubling and tripling have been reported in the literatures [29].

In nonlinear crystals, as introduced in Chapter 2, the polarization density of the crystal becomes a nonlinear function of the electric field. Second-order nonlinearity (quadratic) produces frequency doubling; third-order nonlinearity generates frequency tripling. For example, the KDP crystal, when pumped with a ruby laser at 694 nm, gives a second-harmonic frequency at 347 nm in the ultraviolet range [30]. Ge- and P-doped glass fibers, when pumped by a Nd^{3+} laser at 1060 nm (infrared), produce a second-harmonic frequency at 530 nm in the visible green range [31].

If the parametric amplification principle is applied to nonlinear crystals, the generation of even more frequencies becomes possible. This is known as frequency conversion, the mixing of two monochromatic waves to generate a third wave whose frequency is the sum or difference of the frequencies of the original wave [32–34]. Table 4.1 lists the properties of nonlinear materials that can be used for different transmission ranges of wavelengths.

Tunable Lasers

Tunability is a desirable property of solid-state lasers. Optical parametric converters provide tunable output by nonlinear conversion of a fixed frequency pump. Two factors have contributed to the rapid advancement of optical parametric oscillators. First, continual advances in the growth and fabrication of nonlinear materials encourage research aimed at extending the frequency range of the laser operation. Second, the performance of solid-state pump lasers in driving optical parametric oscillators has been improving continuously since the first Nd : YAG laser was introduced many years ago. This topic is further discussed in Chapters 5 and 6.

Summary

In this chapter, the principle of the oscillator is restated and applied to fiber amplifiers. The fiber oscillator is structurally simple and seems to fit into fiber telecommunications naturally. Much research has been devoted to the development of these devices. From the recent flood of publications in this field, it seems that we have opened the doors in many directions. First, research is renewed into doping host fiber materials with different rare-earth dopants, each of which could be used for a particular wavelength range for different applications. We give a few examples of each kind in this chapter. Then for each type of doped fiber, the research extends into the choice of pump wavelength, which naturally affects the choice of doping material and the doping percentage. The parametric principle has been used to explore the possibility of developing fiber lasers for use at unexplored wavelength ranges, such as lasers in the ultraviolet range and other long infrared wavelengths. For fiber lasers, as only a short length of fiber is needed, the loss factor of the fiber is not a concern; one is free to search for materials and methods untouched at present time. Experiments on double-pumping schemes further push the search beyond present limits. There could be no limit to further development.

References

1. E. Snitzer, Optical maser action of Nd in a barium crown glass. *Phys. Rev. Lett.* **7**, 444–446 (1961).
2. J. Stone and C. A. Burrus, Neodymium-doped silica lasers in end-pumped fiber geometry. *Appl. Phys. Lett.* **23**, 388–389 (1973).
3. M. Le Flohic, P. L. Francois, M. J. Y. Allain, F. Sanchez, and G. M. Stephan, Dynamics of the transient building-up of emission in Nd-doped fiber lasers. *IEEE J. Quantum Electron.* **QE-27**, 1910–1921 (1991).
4. K. Liu, M. Digonnet, H. J. Shaw, B. J. Ainslie, and S. P. Craig, 10 mW superfluorescent singlemode fiber source at 1060 nm. *Electron. Lett.* **23**, 1320–1321 (1987).
5. I. N. Duling III, R. P. Moeller, W. K. Burns, C. A. Villarreal, L. Goldberg, E. Snitzer, and H. Po, Output characteristics of diode pumped fiber ASE sources. *IEEE J. Quantum Electron.* **QE-27**, 995–1003 (1991).
6. G. Meltz, W. W. Morey, and W. H. Glenn, Formation of Bragg gratings in optical fibers by a transverse holographic method. *Opt. Lett.* **14**, 823–825 (1989).
7. G. A. Ball, W. W. Morey, and W. H. Glenn, Standing-wave monomode erbium fiber laser. *IEEE J. Photon. Technol. Lett.* **7**, 613–615 (1991).
8. K. Iwatsuki, Er-doped superfluorescent fiber laser pumped by 1.48-μm laser diode. *IEEE Photon. Technol. Lett.* **2**, 237–238 (1990).
9. J. L. Zyskind, V. Mizrahi, D. J. DiGiovanni, and J. W. Sulhoff, Short single frequency erbium-doped fibre laser, *Electron. Lett.,* pp. 1385–1387, **28** (15) (1992).
10. Y. Ohishi, T. Kanamori, and S. Takashashi, Pr-doped fluoride single-mode fiber laser. *IEEE Photonic Technol. Lett.* **3**, 688–690 (1991).
11. C. A. Millar, S. C. Fleming, M. C. Brierley, and M. H. Hunt, Single transverse mode operation at 1345 nm wavelength of a diode-laser pumped neodymium–ZBLAN multimode fiber laser. *IEEE Photon. Technol. Lett.* **2**, 415–417 (1990).
12. G. A. Ball and W. W. Morey, Efficient integrated Nd fiber laser. *IEEE Photon. Technol. Lett,* **3**, 1077–1078 (1991).
13. P. G. J. Wigley, A. V. Babushkin, J. I. Vukusic, and J. R. Taylor, Active mode locking of an erbium-doped fiber laser using intracavity laser diode device. *IEEE J. Photon. Technol. Lett.* **2**, 543–545 (1990).
14. F. Haberl, M. H. Ober, M. Hofer, M. E. Fermann, E. Winter, and A. J. Schmidt, Low-noise operation modes of a passively mode-locked fiber laser. *IEEE J. Photon. Technol. Lett.* **3**, 1071–1073 (1991).
15. P. Becker, A. G. Processer, T. Jedju, J. D. Kafka, and T. Baer, High-intensity and high-repetition-rate Q-switched diode-pumped Nd:YLF-pumped femtosecond amplifier. *Opt. Lett.* **16**, 1847–1849 (1991).
16. T. H. Maiman, Stimulated optical rediation in ruby masers. *Nature (London)* **187**, 493 (1960).
17. P. F. Moulton, Spectroscopic and laser characteristics of Ti:Al$_2$O$_3$. *J. Opt. Soc. Am.* **B3**, 125–133 (1986).
18. J. C. Walling, H. P. Jenssen, R. C. Morris, E. W. O'Dell, and O. G. Petersen, Tunable laser performance in BeAl$_2$)$_4$:Cr^{3+}. *Opt. Lett.* **4**, 182–183 (1979).
19. T. Y. Fan and R. L. Byer, Diode-laser pumped 946 nm Nd:YAG laser at 300 K. *Tech. Dig. CLEO '87,* Apr. 26–May 1, 1987.
20. G. J. Kintz, R. Allen, and L. Esterowitz, CW and pulsed 2.8 μm laser emission from diode-pumped Er3$^+$:LiYF$_4$ at room temperature. *Appl. Phys. Lett.* **50**, 1553–1555 (1987).
21. T. Yamashita, Nd- and Er-doped phosphite glass for fiber laser. In *Fiber Laser Sources and Amplifiers* (M. J. F. Digonnet, Ed.). *Proc. SPIE* **1171**, 291–297 (1990).
22. M. G. Livshits, B. I. Minkov, and A. P. Shkadarevich, Laser and flashing lamp pumped Mg$_2$SiO$_4$ lasers. In *Advanced Solid State Lasers,* Tech. Dig. Ser., pp. 1–4, Optical Society of America, Washington, DC, (1986).
23. D. Welford and P. F. Moulton, Room temperature operation of a Co:MgF$_2$ laser. *Opt. Lett.* **13**, 975–977 (1988).

24. R. C. Stoneman and L. Esterowitz, Efficient, broadly tunable, laser pumped Tm:YAG and Tm:YSGG CW lasers. *Opt. Lett.* **15**, 486–488 (1990).

25. R. Allen, L. Esterowitz, L. Goldberg, J. R. Weller, and M. Storm, Diode pumped 2 μm holmium laser. *Electron. Lett.* **22**, 947 (1986).

26. A. R. Harmer, A. Linz, and D. R. Gable, Fluorescence of Na^{3+} in lithium yittrium fluoride, *J. Phys. Chem.* **36**, 1483–1491 (1969).

27. P. Urquhart, Review of rare-earth-doped fiber lasers and amplifiers. *Proc. IEEE* **135**, 385 (1988).

28. P. P. Sorokin and M. J. Stevenson, Solid state optical masers using samarium in calcium fluoride, *IBM. J of Research and Development* **5**, 56–58 (1961).

29. W. Bosenberg, D. Guyer, D. D. Lowenthal, and S. E. Moody, Parametric optical generation: From research to reality. *Laser Focus World* **28**, 165–170 (1992).

30. W. R. Bosenberg, W. S. Polouch, and C. L. Tang, High efficiency and narrow linewidth operation of two-crystal Beta-BaB_2O_4 optical parametric oscillators. *Appl. Phys. Lett.* **55**, 1952 (1989).

31. M. E. Ebrahimzadeh, A. J. Henderson, and M. H. Dunn, An excimer-pumped β-BaB_2O_4 optical parametric oscillator tunable from 354 nm to 2.370 μm. *IEEE J. Quantum Electron.* **QE-26**, 1241 (1990).

32. L. R. Marshall, J. Kasinski, A. D. Hays, and R. Burnham, Efficient optical parametric oscillator at 1.0 micrometer. *Opt. Lett.* **16**, 681–683 (1991).

33. Y. E. Fan, R. C. Eukardt, J. Nolting, and R. Wallenstein, Visible BaB_2O_4 optical parametric oscillator pumped at 355 nm by a single-axial-mode pulsed source. *Appl. Phys. Lett.* **53**, 2014 (1988).

34. D. H. Jundt, G. A. Magel, M. M. Fejer, and R. L. Byer, Periodically poled $LiNbO_3$ for high frequency second harmonic generation. *Appl. Phys. Lett.* **59**, 2657 (1991).

CHAPTER 5

Solid-State Lasers

Introduction

The history of the laser begins with the invention of the solid-state optical laser in 1960 [1] made with Cr-doped sapphire. This device, which consisted of a ruby rod pumped by a flash lamp, was very bulky. Semiconductor lasers were introduced in 1962 with a small $p-n$ junction diode made of GaAs alloy and direct-current injection [2]. The development of semiconductor lasers was so rapid as their applications expanded into many fields that they quickly overshadowed solid-state lasers. Actually, the development of solid-state lasers is growing steadily and seems to have accelerated in recent years into power lasers covering a wide range of frequencies. Their applications in the medical and industrial fields are growing rapidly. In this chapter, a short review of solid-state lasers is given.

Fundamental Concepts of Solid-State Lasers

The fundamental process of the laser is the interaction of atoms at different energy levels in matter and their feedback action within the resonant cavity. These processes include population inversion by absorption, spontaneous and stimulated emission, and the enhancement of oscillation by resonance. For solid-state lasers operating on stimulated transitions, the interaction is confined to the electronic levels of ions of the activators, which are the impurity doping ions contained in solid crystalline or glassy media or the host material. Most activators belong to divalent or trivalent transition metal groups of the Periodic Table. Any laser host crystals can be used as a host medium.

Population Inversion

If we assume that the discrete energy levels in atoms are E_1, E_2, E_3, and so on, such that $E_1 < E_2 < E_3$, and E_1 is called the ground state, then at thermal equilibrium temperature T, the relative population densities N_n of two different levels n and m are given by Boltzmann's relation

$$\frac{N_n}{N_m} = \exp\left\{\frac{(E_n - E_m)}{kT}\right\} \tag{5.1}$$

where Boltzmann's constant $k = 1.38 \times 10^{-23}$ J/°K. For visible wavelengths, the number of atoms above the ground state is ordinarily very small; however, the population in upper energy states can be increased momentarily if some atoms are subjected to injection or pumping so that absorbed energies from the pumping sources cause electrons to be raised from lower to upper energy levels. When the population density of the upper state increases beyond that of a lower-state population, inversion is achieved. In optically pumped media, this is brought about by an absorption process, where energy equaling the difference energy between two states is absorbed, or $E_j - E_i = h\nu_{ij}$.

Spontaneous and/or Stimulated Emission

When the pump energy is removed, electrons at higher levels begin to relax back to the lower level.

If an electron in energy state E_j above the ground state relaxes to a lower energy energy level E_i, it emits a photon with energy $E_j - E_i$, with arbitrary phase and direction. This is spontaneous emission.

On the other hand, a resonant photon may stimulate the excited atomic system to emit another photon of the same energy, direction of propagation, polarization, and phase, and this radiation is therefore coherent. This is the process of stimulated emission.

Assume the instantaneous populations of E_1 and E_2 to be N_1 and N_2, respectively. According to Eq. (5.1), under thermal equilibrium conditions, $N_1 \gg N_2$. No photon emission exists. Most electrons are in the lower energy level. If the atoms exist in a radiation field of photons with energy $h\nu_{12}$, such that the energy density of the field is $p(\nu_{12})$, where $p(\nu_{12})$ is the probability density (per second), the probability of an emission taking place in a time interval between t and $t + \delta t$, then stimulated emission can occur. The rate of change of atoms by stimulated emission can be written as $B_{21}N_2 p(\nu_{12})$, where B_{21} is the rate of stimulated emission (a proportional factor). The rate of absorption is $B_{12}N_1 p(\nu_{12})$, where B_{12} is the rate of absorption (a proportional factor). Furthermore, the rate of atomic change by spontaneous emission is $A_{21}N_2$, where A_{21} is the rate of spontaneous emission. At steady state, these rates must balance:

$$B_{12}N_1 p(\nu_{12}) = A_{21}N_2 + B_{21}N_2 p(\nu_{12}) \tag{5.2}$$

The ratio of the rates of spontaneous to stimulated emission is equal to $8\pi h(v_{ji}/c)^3$, where $c = c_0/n$, c_0 is the speed of light *in vacuo,* and n is the index of refraction. Similarly, the ratio of stimulated emission to absorption is $(B_{21}/B_{12})(N_2/N_1)$.

Other physical characteristics of general interest in laser design are spectral linewidth, gain cross section, and lifetime of the upper laser level.

Linewidth

Linewidth comes into the picture because the energy levels in solids are really not sharp lines due to thermal lattice vibrations and/or lattice defects. This effect (interactions with phonons of the host medium) causes line broadening, which for some crystals may have a Lorentzian lineshape. The cross section for stimulated emission is defined as

$$\sigma_{21} = \frac{(A_{21}\lambda_{21}^2)}{4\pi^2 n^2 \delta\lambda} \tag{5.3}$$

where $\delta\lambda$ is the width between half-intensity points in the broadened line.

The Gain Equation

If population inversion by pumping has established a density N (per cm^3) in the upper laser level, the optical gain G in a laser medium of length l is given by the expression

$$G = \exp\{\sigma(\lambda)Nl\} \tag{5.4}$$

Thus, a larger cross section will give a larger optical gain.

Cross Section

A larger cross section also allows more efficient energy extraction. The typical cross section is about 10^{-21} cm^2.

In the absence of laser action, the population of the upper laser level decays at a rate inversely proportional to the lifetime τ. This time is important in determining how much pumping power is required to generate the necessary population of ions (in a solid-state laser, the active agent is usually the ions, so we use ions instead of atoms or electrons as originally stated) for laser action. Fluorescence lifetimes allow net optical gains to be achieved. Typitcal lifetimes in solid-state lasers are in the range 10^{-6} to 10^{-2} s.

Most solid-state laser ions have optical absorption over a wide spectrum of wavelengths leading to reasonable efficiency with sources that are broadband thermal radiators.

The Host Crystals

The host crystal of a solid-state laser plays a different role as substrate in semiconductor lasers. Lasing action in solid-state lasers depends on the interaction between host ions and impurity doping ions. A solid-state host crystal provides not only the containment of the laser-active ions, but also creates a favorable enviroment for the interaction between electromagnetic radiation and electronic states for stimulated emission. It should also provide a sink to dissipate heat energy produced by non-radiative transitions or any excess pumping energy.

A very large number of crystalline and glass hosts have been reported for laser action. Only a limited number of samples are listed here. The development of new host material depends on the demand for it and the price you are willing to pay. The following list is limited to those commercially available.

Sapphire (Al_2O_3). Sapphire is a hard, strong material. When doped with chromium (Cr^{3+}), it is a ruby laser [1]. When doped with titanium, it becomes a Ti:sapphire laser [3] with wide wavelength tunability.

YAG ($Y_3Al_5O_{12}$). Yttrium aluminum garnet is a hard durable material. When doped with neodymium (Nd^{3+}), it can result in an efficient high-gain laser [4].

YALO ($YAlO_3$). Yttrium aluminate is a popular host crystal for neodymium-doped lasers [5].

YLF ($LiYF_4$). Lithium yttrium fluoride doped with neodymium emits 1053 nm, which matches the peak gain of phosphate laser glasses [6].

Many glass materials can be used as host crystals. Infinite combinations are possible. Most of the well-developed glasses are silicates or phosphates. The commercial availability of host materials depends on the demands which promote research and fabrication of new products.

Interactions between Doping and Host Ions

Laser action occurs when the host crystal is doped with impurity atoms, with a divalent or trivalent transition-metal ion substituting for an ion of the same valence state in the host structure. In general, the energy transitions possible are summarized in Fig. 5.1. Three types of transitions—the three-level, two types of four-level, and vibrational transitions—are shown in Figs. 5.1A, B, C, and D, respectively. The difference between b and c are the number of sublevels at level 3.

Figure 5.2 shows the energy-level diagram of the rare-earth ions [7]. The complexity of the energy levels is clear. In general, the energy level of the particular ion determines the laser emission. The host matrix plays a much less important role in the emission process.

Three-Level Energy State

The energy state shown in Fig. 5.1A is a three-level state. The highest level (No. 2) is the pump level. The next high level (No. 3) is the metastable level, and Number 1 is ground level. The No. 3 level is very important for the stimulated emission process because electrons at this level have a long mean lifetime. When the ion is pumped to

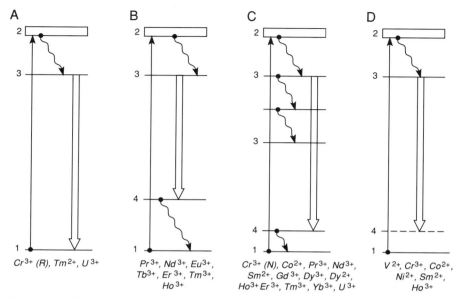

Figure 5.1 Simplified energy-level diagrams of solid-state lasers. (A) Three-level lasers. (B) Four-level lasers. (C) Multiple four-ievel lasers. (D) Vibrational or vibronic lasers. After A. A. Kaminski (1981). "Laser Crystal," Springer-Verlag.

an upper level (No. 2) and nonradiatively relaxes immediately to level 3, the long lifetime helps the state to reach population inversion. Laser action occurs when the population at level 3 exceeds that at the ground state. As the ground state (No. 1) is usually highly populated, the required population inversion ($N_3 \gg N_1$) demands a rather high rate of pumping.

Most of the transitions in solid-state lasers take place in the $3d$ states of the energy-level diagram. Cr^{3+} doping belongs to the three-level type; when ions are pumped to a higher level (No. 2), they immediately relax to level 3 where the mean lifetime is long, making population inversion possible. When stimulated in this state, the ion drops to the ground level, and photons are emitted. The transition is not efficient. Cr:sapphire lasers belong in this group. They emit at $\approx$690 nm with a reasonable quantum efficiency.

Four-Level Energy State

In a four-level energy system, shown in Figs. 5.1B and C, when ions are pumped to a higher energy level, radiative transition takes place between levels 3 and 4. In group B, the third level is a single level, whereas in group C, the third level is divided into sublevels as shown. Normally, as level 4 is not the ground state, it is sparsely populated so that the transition efficiency can be high.

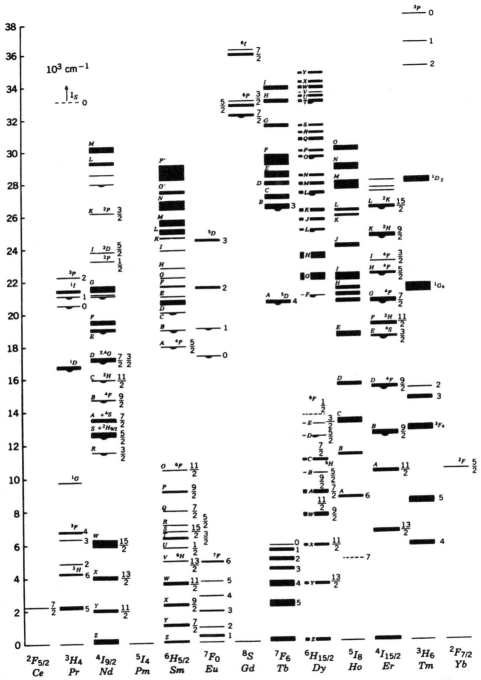

Figure 5.2 Energy levels of the rare-earth ions. Those exhibiting fluorescence are denoted by a semicircle. After [7].

Transitional States

Some transitional energy may also be converted into vibrational and/or rotational motion of the atoms and in phonon interactions. These phonon interactions are non-radiative. Instead, heat is produced that must be removed from the system in operation; however, some do modify the nature of light emission. The existence of a large number of vibrational states, shown in Fig. 5.1D, broadens the energy into a band of energies that also contributes to the tunability of solid-state lasers. Lasers based on this transition are sometimes called the vibronic lasers.

Codoping

From Fig. 5.2, one observes that the rare-earth ions have a number of levels at about the same energy above the ground state. This suggests the possibility of codoping the host with more than one type of impurity. Codoping is also done to increase the absorption of pumping energy and enhance transfer efficiency. One example is the Ho:YAG laser at 2050 nm. When codoped with Er^{3+} and Tm^{3+}, much more efficient laser action can be obtained [8]. Codoping may also serve to effect better lattice matching between different atoms.

Figure 5.2 also shows the complex structures of the energy-level diagrams of rare-earth ions. Note that Cr and Nd ions are both complex, whereas ytterbium is the simplest.

Solid-State Laser Combinations

The goal in preparing a dopant–host combination is to find doping ions that have the same valence state in the host structure and for which a strong absorption band is available. Impurity ions will substitute for a few host ions and promote the interaction when pumped, although almost any material can be made to lase with a proper doping companion. Table 5.1 is a partial list of the most common solid-state laser combinations in use at present.

From this table, one notes that solid-state lasers with a variety of wavelengths and energy or power capability are available. Some lasers cover a band of wavelengths that can be used as tunable lasers. Some require cryogenic cooling. But most combinations are selected to operate at room temperature. There are basically two groups, divalent and trivalent transition metal groups. Cobalt belongs to the divalent group, whereas the Cr, Er, Nd, Tm, Ho, and Sm belong to the trivalent group. Most solid-state laser operations require external optical pumping.

The applications of solid-state lasers are expanding into the industrial and medical fields continuously and into instrumentation and consumer products more recently. In the research and development area, tunable solid-state lasers and ultrafast lasers both have undergone great technological advances in recent years. Coupled with the availability of new laser crystals such as Ho:YAG and Tm, Ho:YAG, new laser sys-

TABLE 5.1
Solid-State Laser Combinations

Chemical formula	Wavelength (μm)	Energy level	Power (W)	Efficiency (%)	$\Delta\nu$ frequency
$Cr^{3+}:Al_2O_3$	0.69	3	5–7 J P	0.1	40 GHz
$Ti^{3+}:Al_2O_3$	0.68–1.13		10 CW	0.01	
$Cr^{3+}:Al_2BeO_4$	0.72–0.79				
$Nd^{3+}:YAG$ ($Nd_xY_{3-x}Al_5O_{12}$)	1.06–1.32	4	60 CW	0.5	120 GHz
$Nd^{3+}:YLF$	1.047–1.6	4	20 CW		5000
$Er^{3+}:glass$	1.06–1.54	3	50 J P	1.0	4000
$Cr^{4+}Mg_2SiO_4$	1.167–1.345				
$Nd^{3+}:Mg_2SiO_4$	1.35–1.61		P		90 fs
$Cr^{4+}:YAG$	1.35–1.45				
$Ni^{2+}:MgF_2$	1.51–2.28 (cryogenic cooling)		170 mJ		150 ns
$Co:MgF_2$	1.75–2.5				
Tm:YAG	2.01				
Ho:YAG	2.10		29 W		20 Hz
$Er^{3+}:LiYF_4$	2.8				
Er:YAG	2.94				

tems can be designed for particular applications. Combinations of technologies such as nonlinear frequency doubling of diode-pumped solid-state lasers further diversify laser choices.

Excitation of Solid-State Lasers

Solid-state lasers use ions that can absorb energy at various wavelengths, from the broadband thermal radiators to ultraviolet emissions. The fluorescence lifetime characteristics of solid-state lasers, typically 10^{-6} to 10^{-2} s, allow optical gain to be achieved with a variety of continuous-wave or pulsed pump sources. The emission cross section of these ions, typically 10^{-19} to 10^{-21} cm^2, permits efficient extraction of energy from either CW or pulsed sources. In principle, solid-state lasers can be excited by many mechanisms, optical sources, either incoherent or coherent light, and diode lasers, in either CW or pulse mode. Optical pumping by flashlamp was used almost exclusively at first because powerful Edgerton flashing strobe lights were already in existence, and they are inexpensive and powerful. Flashlamp pumping is still in use in some applications. The development of diode laser arrays with powerful and efficient pumping, however, has changed the picture of solid-state laser pumping.

Many extensive uses of solid-state lasers are derived from the use of diode laser pumping technique, using diode lasers to pump lasers.

Flashlamp Pumping

A powerful xenon flashlamp for exciting the metastable ions in solid-state lasers was introduced when the first ruby laser was demonstrated. A helical flashlamp was used to surround the solid-state laser rod and pulses were initiated by discharging previously charged condenser banks through electric circuitry. Exhaustive experiments were carried out to study the characteristics for the use of these lamps for pumping solid-state lasers. Powerful as they may be, the efficiency of energy transfer is low. Moreover, for a pulse storage laser, where the laser ions essentially integrate the pumping radiation, the peak stored energy density will clearly be dependent on the ratio of the pulse duration of the flashlamp and the fluorescence lifetime. Flashlamp operation involves many problems, not just those dealing with the change in the ratio of time constants. As the flashlamp pulse is shortened, the current density, lamp spectrum, and opacity of the lamp all change. Xenon flashlamps also cause electrical loading problem.

The repulsive force between adjacent coil turns of a helical flashlamp may mechanically flex the glass tube to crack the glass tube and cause an explosion, thus limiting the input energy as well the pulse width that can be used for a specific laser [9].

To date, for small lamps such as are typically used with Nd : YAG lasers, an overall radiation efficiency of 50 to 60% of capacity has been reported [10]. Xenon flashlamps require heavy current for operation. Yet only a small portion of the electrical energy is used to pump the ions. The excess heat may be so intense that the laser rod needs to be cooled. Laser rods can suffer another type of damage, catastrophic failure. The facet coatings at both ends of the laser rod may be suddenly burned out at high output power. This type of damage is irreversible. Although this situation may not be directly related to the flashlamp, it happens very often in flashlamp-pumped lasers. Catastrophic failure is the result of localized melting at the facet during the initial 1- to 10-μs interval.

Reliable triggering of xenon flashlamps is also very important. Many circuitry problems may be involved.

Arclamps and other filament lamps have also been used to pump solid-state lasers both at room and at reduced temperatures for a laser host operating in CW mode. Generally, the efficiency is very low, typically 0.1%. With continuous pump sources, however, stable long-lived operation can be achieved. Laser gain is much smaller than with pulsed operation, and spontaneous emission and power extraction may become a problem.

Diode Laser Pumping

Semiconductor diode lasers, such as GaAlAs laser diodes which emit near 800 nm, have been used to pump solid-state lasers. One could anticipate that a diode laser pump would have a higher quantum efficiency. Also, as semiconductor diode lasers are very compact, the overall system could be more compact as well. A small amount of thermal heating is to be expected from the diode laser pumping scheme. Earlier attempts to use GaAlAs laser diodes to pump Nd lasers were not cost effective. The GaAlAs diode was too expensive to be used for pumping. These diodes are specially designed for communication applications that require many special features such as single-frequency, single-mode, and narrow-linewidth operations. For general pumping purposes, many of these features may not be required. Special semiconductor diodes for pumping solid-state lasers have since been developed [11]. And as the unit price of the diode laser becomes lower, acceptance of the use of diode lasers for pumping solid-state lasers may become routine.

Pump duty cycle is an issue we have to consider when the diode laser is used for pumping. For neodymium ions in most host materials the fluorescence lifetime is 200 to 250 μs. To pump the laser efficiently, the pumping time should be limited to this period. Pumping longer would not increase the population inversion.

Diode lasers are usually designed for a particular wavelength. To match the output wavelength to the absorption band of the solid-state laser, one has to select the right wavelength. Selecting a diode laser from those available for solid-state lasers, for matching the absorption and extracting energy bands for efficient transfer of energy, is often not very easy. This is because solid-state lasers often require a range of wavelengths for efficient pumping.

Pumping Schemes: End-Pumped and Side-Pumped Lasers

The simplest way to excite solid-state lasers using semiconductor diodes is end pumping. The diode is coupled to the solid-state laser from one end only, and only one pump can be used at any time. The power used for pumping is thus limited. A side pump can be used to increase the pumping power. An array of diode lasers, in the form of bars up to a centimeter in width, and able to deliver up to 10 W of power, are available at a reasonable price. The largest arrays currently available commercially have 40 elements and produce 500 mW CW [12].

Single quantum wells with separated confinement heterostructures (SQW-SCH) have been designed and used to pump Nd:YAG lasers. With laser arrays and bar structures, high-optical-power solid-state lasers have been reported. The arrays reported to date do not in general have optimal spatial or spectral output. An injection locking technique has been used to concentrate energy. Research is ongoing in this area. It is predicted that powerful diode lasers may one day be used to pump solid-state lasers, replacing other gas and dye lasers of high power.

Other rare-earth lasers, such as holmium (Ho, 2050 nm) and thullium (Tm, 2020 nm) lasers, are attractive candidates for diode laser pumping. They have long upper-state lifetimes, more than 10 ms with the use of laser diode radiation. When those hosts are codoped with erbium or thulium, highly efficient CW laser operation with heavy-duty cycles has been reported [13].

Side-pumping technology has been developed steadily. Side pumping removes the restriction of a single power source, thus increasing the input power available for pumping. For this purpose, surface-emitting diodes were developed. At the XIIth IEEE International Semiconductor Laser Conference held in Davos, Switzerland, in September 1990, Geels and Coldren from the University of California, Santa Barbara, reported a vertical-cavity surface-emitting (VCSE) submilliampere laser, shown in Fig. 5.3A. The quantum-well structure is sandwiched between two sets of graded-barrier mirrors. Light propagates perpendicularly to the active layer. Feedback for lasing is obtained from the multilayer semiconductor-stack reflectors shown in the figure. The single-quantum-well structure emits at 980 nm. In CW operation, a threshold current of 0.7 mA is used. The light output versus current characteristic is shown in Fig. 5.3B.

Recently, a verical-cavity diode laser was reported at the CLEO '92 meeting that

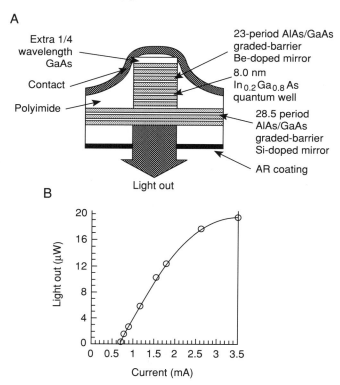

Figure 5.3 Planarized vertical-cavity surface-emitting (VCSE) laser designed by Geels and Coldren of the University of California, Santa Barbara. Light is emitted through the substrate. (A) Sectional structure. (B) Optical output versus threshold current. A threshold current of 0.7 mA is indicated. Courtesy of R. Geels and L. Coldren, UC-Santa Barbara. © PennWell Publishing Co.

delivers 8.5 W of output power in the blue-green wavelength region. It was found that blue and green diode ZeSe lasers can be used efficiently to pump the Cr-doped solid-state laser [14]. Four 20-nm-wide quantum-well regions of ZnSe in a hetero-structure form the active medium of these diode lasers.

Tunability of Solid-State Lasers

An attractive feature of solid-state lasers is the tunability in wavelengths. Until re-cently, the most widely used tunable laser system was the dye laser. Dye lasers are inherently toxic and lack long-term reliability. Their tuning range is narrow com-pared with that of the solid-state lasers. Thus, solid-state lasers have become increas-ingly significant and tend to overtake dye lasers in many applications. Solid-state lasers offer a broader tuning range and extend to longer infrared wavelengths than dye lasers. In addition, they pose no danger and provide long lifetimes, in operation and on the shelf. They have the capability to store higher energy and thus generate high peak powers via a Q-switching arrangement. We shall elaborate on these points later.

Technology is advancing in the direction of using semiconductor diode pumping as an economically feasible, technologically efficient, and physically reliable pump source for solid-state lasers.

Excited-State Absorption

In solid-state lasers, we have investigated the energy transfer mechanism leading to the emission of photons. But there are also possibilities of energy transfer from the upper laser level to higher levels of the energy-level diagram. We call this process excited-state absorption (ESA). The difference between these two counteracting transfer mechanisms determines the net energy left for emission or absorption. Ab-sorbed energy causes attenuation of the stimulated emission as energy is being wasted through ESA. The net optical gain in the system is proportional to the differ-ence between the cross section for stimulated emission and that for ESA. The process of ESA therefore affects the operation of solid-state lasers profoundly. In tunable solid-state lasers, the effect of ESA is felt more severely. The cross section of ESA is difficult to estimate. No direct measurement of ESA is possible. Indirectly, one can estimate the cross section by measuring the optical gain (or loss) in a pumped crystal as a function of wavelength. The ESA effect is particularly detrimental when Q-switch operation is desired, as the upper laser level population can rise to high levels before the Q-switch action is effected.

As the application of solid-state lasers broadens, the demand for good beam qual-ity, higher output power, and wide wavelength availability also increases. A high-quality beam, usually called a diffraction-limited beam (DLB), can be collimated and focused to small spots using relatively simple optics. Much ongoing research is devoted to improving beam quality.

Typical Solid-State Diode Lasers

Solid-state lasers have been studied intensively, particularly with respect to laser fusion technology. Ongoing research projects are so numerous and the number of publications so vast that only persons specialized in the field can follow them. In this book, we limit our attention to devices of much lower power-handling capability. Those include industrial, commercial, and telecommunication applications.

Let us divide the various solid-state lasers listed in Table 5.1 into groups according to their applicable wavelengths. For example, for telecommunication fiber cable applications at 1550 and 1300 nm, Er-doped YAG and Cr-doped glass lasers are most interesting. For medical applications, Ho:YAG lasers operating at 2100 nm and ultraviolet wavelengths, and even at visible wavelengths, are useful. For industry, where high power is required for heating or cutting, Nd:YAF lasers are useful. Other timely topics are tunable solid-state lasers and ultrafast lasers.

Er:YAG and Cr:Glass Lasers for Optical Fibers

In 1988, the announcement of the introduction of Er^{3+}-doped fiber amplifiers to fiber-optic transmission systems operating at 1550 nm electrified the world. The attractive features of the Er-doped fiber amplifiers are high optical gain, low insertion loss, and compatibility with optical fiber systems. Long-haul telecommunications over several thousand kilometers or more without regeneration of repeater stations is now possible. Those amplifiers are pumped by semiconductor diode lasers at either 1480 or 980 nm. Excellent results have been reported [15].

To make use of dispersion-free silica optical fibers at around 1310 nm, Cr^{3+}-doped fiber amplifiers have been introduced that perform similarly to the Er-doped fiber amplifiers at 1550 nm [16].

Ho:YAG and Nd:YAG Lasers for
Medical Applications

Medical applications of solid-state lasers, especially the Ho:YAG at 2100 nm, are gaining popularity [17–20]. The holmium wavelength is absorbed strongly by the water component of body tisse, resulting in shallow tissue penetration and good surgical precision in most types of soft and hard tissues. Unlike CO_2 lasers, holmium laser output can be delivered reliably through silica fiber [17]. New products claim to provide more than 30 W of average power at pulse rates of 10 to 15 Hz.

Pulsed Nd:YAG lasers, operating in a free-running pulsed mode, are used to treat soft as well as hard tissue dental problems. Operating at about 1064 nm, pulse energies of 100 to 700 mJ and maximum average power of 3 to 35 W and duration 150 to 800 μs are employed. Also in dentistry, an Er:YAG laser with output at 2900 nm is expected to replace the conventional dental drills.

Nd:YAG lasers have been used for many dental services such as bleaching stained teeth, guiding tissue regeneration, assisting wound healing, sealing teeth cracks, and desensitizing teeth.

In urology, Nd:YAG lasers have been used to crack kidney stones. General surgery also employs frequency-doubled Nd:YAG surgical lasers to promote soft tissue surgeries including lumbar diskectomy and the treatment of vascular skin lesions. At 532 nm, a power capability up to 36 W is available.

In ophthalmology, Nd:YAG CW lasers equipped with sapphire-tipped probes have been used to reshape the lens structure to achieve better vision in human eyes.

Frequency-tripled Nd:YAG lasers at 355 nm are currently undergoing clinical evaluation for coronary laser angioplasty applications, where they might compete with 308-nm excimer lasers and holmium lasers.

Nd:YAG Lasers for Industrial Applications

It was the CO_2 lasers with high power capability (in the multikilowatt region) that dominated the industrial laser market a few years ago. Now, at the lower end of the power scale, say, of the kilowatt class, Nd:YAG lasers are replacing CO_2 lasers gradually. (See the *Industrial Laser Review*, September 1991.)

The ninth European Machine Tool exhibition, held June 1990 in Paris, France, introduced many interesting laser tools for industrial use. For example, a 250-W pulsed Nd:YAG laser marking system for deep engraving of gears in the automobile industry was demonstrated. This machine is fully automated (made by Carl Baasel Lasertech, Starnberg, FRG). Quantel (Les Ulis, France) demonstrated a range of Nd:YAG modular lasers for industrial applications ranging from 300- to 1200-W cutting and welding machines at 1060 nm. Other exhibits included a 500-W Nd:YAG laser fitted into a six-axis Yaskawa robot that can cut a fender for a Renault sports car (Rofin-Sinar, Hamburg, FRG) [21].

Tunable Lasers

Solid-state lasers can be made wavelength tunable by choosing proper host crystal and doping ion combinations (Table 5.2). With a host crystal, we choose either divalent or trivalent transition metal ions for doping. The doping impurity ions replace a few host ions of the same energy level. It is the vibronic interaction between these ions that gives the solid-state laser a wide range of tunability. A closeup of a vibronic energy-level diagram is shown in Fig. 5.4. Although this graph applies to a Ti:sapphire laser, all vibronic lasers have a similar structure, the bell-shaped energy band levels. Optical energy is absorbed by the crystal, which excites the electron from the

TABLE 5.2
Types of Tunable Vibronic Solid-State Lasers

Laser material	Pump source	Operation	Wavelength range (nm)
Alexandrite	Arc lamp	CW	730–810
Alexandrite	Flashlamp	Pulsed	701–858[a]
Ce:YLF	KrF excimer laser	Pulsed	309–325
Co:MgF$_2$	1320-nm Nd:YAG	Pulsed[b]	1750–2500[c]
Cr:LiCaAlF$_6$	Laser or lamp	Pulsed or CW	720–840+
Cr:LiSrAlF$_6$	Laser or lamp	Pulsed or CW	760–920+
Cr:emerald	Laser	Pulsed or CW	720–842
Cr:forsterite	Laser	Pulsed or CW	1167–1345
Thulium:YAG	Laser	CW	1870–2160
Ti:sapphire	Usually laser	Pulsed or CW	660–1180

[a] Wavelengths longer than 826 nm possible only at elevated temperatures.

[b] At room temperature, CW operation is possible at cryogenic temperatures.

[c] At room temperature, wavelengths as short as 1500 nm are possible at cryogenic temperatures.

[d] Data adapted from J. Hecht,"Tunability makes vibronic lasers a versatile tool," Laser Focus World, p. 93 (1992).

ground state (A) to the excited (B) level; the electron then relaxes to the upper laser level (C). The laser transition is to a vibrationally excited sublevel of the ground state (D), which relaxes back to the ground state (A).

We limit the discussion to only three groups of dopants: the alexandrite group, the Ti:sapphire group, and the forsterite group.

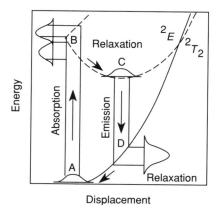

Figure 5.4 Closeup of the energy-level diagram of a vibronic laser: Ti:sapphire laser. The ground state is at A. Optical energy absorbed from the pump excites the atoms to B, which immediately relax to the upper laser state C nonradiatively. Laser action takes place between C and the vibronic sublevels in D, leading to the broad tunability of the laser. After [36].

Alexandrite Group

Alexandrite lasers make use of Cr-doped host crystals such as $BeAl_2O_4$ and LiSAF [22]. As can be seen from Fig. 5.2, their energy-level diagrams are similar to that for ruby. Laser action on the vibronic transition is possible, thus resulting in a wider tuning range provided that the internal crystal field of the host crystal is low. Low crystal field allows vibronic interactions to take place between two excited levels of the Cr ions and the host ions because their energy difference is now small enough.

A unique feature of alexandrite lasers is that they have a lower threshold than other solid-state lasers, and the threshold decreases with increasing crystal temperature. This is because as the crystal temperature increases, so do the gain cross section and thus the optical gain [Eq. (5.4)].

The presence of ESA in alexandrite reduces the gain at both the long- and short-wavelength extremes of the tuning range. But its effect is weak at the peak gain region. A host crystal that exhibits a small ESA effect has been identified; it is LiSAF [23]. But the fluoride materials in general do not have the mechanical strength of alexandrite. Thus its use is being limited to low-power-output applications; however, its broad tuning range and high room temperature gain encourage further laser development.

The major advantage of the alexandrite laser is the possibility for tunable operation with conventional flashlamp pumping and with the crystal at room temperature. The absorption band of the alexandrite group in the visible wavelength range is very strong. Tuning from 730 to 790 nm and beyond has been reported. Good beam quality and narrow laser linewidth have been obtained with diffraction-limited Q-switched alexandrite oscillators that can generate 50 to 200 mJ [24].

Ti:Sapphire Group

The search for a combination of ions and host crystal that minimizes the problem of ESA in solid-state lasers was successful. Ti:sapphire or $Ti:Al_2O_3$, under laser pumping conditions, has demonstrated that the ESA effect is practically nonexistent. The Ti^{3+} ion has only one valence electron, and exhibits the simplest energy-level structure of all the laser-active transition metal ions. Higher energy states of the Ti ion require promotion of the single electron out of the $3d$ shell. With sappphire as the host crystal, the energy required to achieve the transition above the first excited state is so high that neither laser nor pump wavelengths could be absorbed by transitions originating from the upper laser level [25].

The laser gain cross section for Ti:sapphire is large, around 3.5×10^{-19} cm^2, higher than that for other divalent and trivalent $3d$ transition metal tunable lasers by at least an order of magnitude. High-power laser output can be achieved. The highest-energy Nd-pumped Ti:sapphire oscillator to date has produced 430 mJ of energy in a single, 10-ns pulse. At 800 nm, a pulse rate of 10 Hz has been achieved with a pump energy of 1.0 J [26].

The tuning curve (solid line) of a typical Ti:sapphire laser pumped by a 300-mJ-output, Q-switched, frequency-doubled Nd:YAG laser is shown in Fig. 5.5A. A tun-

A

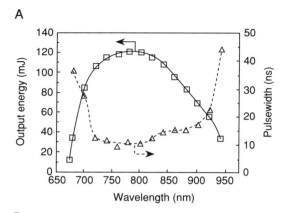

B

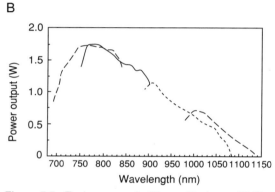

Figure 5.5 Tuning curve of a Ti:sapphire laser. (A) The solid curve represents laser pulse energy versus wavelength for a Ti:sapphire laser pumped by a 300-mJ, Q-switched, frequency-doubled Nd:YAG laser. The dotted curve represents Ti:sapphire laser pulse width as a function of wavelength. (B) Power output versus wavelength for a CW Ti:sapphire laser pumped by 7 W of power from an all-line argon ion laser. Curves shown are for four different mirror sets. After [26].

ing range from 675 to 940 nm is recorded. The dotted line shows the pulse width as a function of wavelength. Within the tuning range the pulse width is maintained at about 12 to 20 ns. In Fig. 5.5B, power output versus wavelength for a CW Ti:sapphire laser pumped by 7 W of power from an all-lines argon ion laser. Note that the tuning range is achieved by using four different mirror sets.

With respect to CW lasers, it has been reported that 43 W of diffraction-limited power was generated from a Ti:sapphire laser pumped by a laser array consisting of four ion lasers with a total pump power of 103 W [27].

Ti:sapphire has a large cross section and thus can be used in pulsed, high-gain amplifiers. The Ti:sapphire laser has found wide applications in basic scientific research as a source for the high-resolution spectroscopy, Raman scattering, and frequency standards. In applied research, it has found use in pumping fiber amplifiers and solid-state lasers. With the mode-locking technique and 100-fs high-energy pulses, Ti:sapphire lasers are expected to be used in atmospheric study projects.

It is expected that when high-quality materials become available, process variables such as optical quality, titanium concentration on optical losses, and postgrowth annealing on optical quality will be optimized to make even better lasers.

Chromium-Doped Forsterite Lasers

A new material, Cr-doped forsterite ($Cr^{4+}:Mg_2SiO_4$), has been added to the family of tunable solid-state lasers in recent years. These lasers can be pumped with 532-, 629-, and 1064-nm wavelengths, and have a tuning range in the near infrared from 1167 to 1345 nm. This wavelength range coincides with the region of minimal dispersion in silica optical fibers, thus increasing their technological importance in applications involving these wavelengths. Table 5.3 shows the spectroscopic and laser properties of the Cr-doped forsterite laser. The communication industry, which uses silica optical fibers, is particularly drawn to the wavelength ranges covered by Cr^{4+}-doped lasers.

The most distinguishing feature of forsterite lasers is that the active ion in lasing is not trivalent chromium as in the case of other chromium-based lasers, but it is believed to be the tetravalent chromium (Cr^{4+}) [28]. Tetrahedral Cr^{4+} has three broad spin-allowed absorption bands that can be pumped efficiently for laser operation. These are the bands at 850–1200, 600–850, and 350–550 nm. The center lasing wavelength is at 1235 nm for pulsed and at 1244 nm for CW operations, respectively. CW laser operation was reported using 1064-nm radiation from a CW Nd:YAG laser

TABLE 5.3
Spectroscopic and Laser Properties of $Cr:Mg_2SiO_4$

Property	Value
Major pump bands	850–1200 nm
	600–850 nm
	350–550 nm
Fluorescence band	680–1400 nm
Room temperature fluorescence lifetime	$\approx 3\ \mu s$
Lasing wavelength (center)	1235 nm (pulsed)
	1244 nm (CW)
Spectral bandwidth	30 nm (pulsed)
	12 nm (CW)
Slope efficiency	23% (pulsed)
	38% (CW)
Tuning range	1167–1345 nm
Gain cross section	$\approx 1.45 \times 10^{-19}\ cm^2$

*Data adapted from [29].

as the pumping source and an output power of 110 mW at room temperatures. The measured slope efficiency was 6.8% and the bandwidth was 12 nm. By cooling the crystal in liquid nitrogen, a power output of 1.5 W for a Nd:YAG pump power at 6 W was demonstrated. For pulsed operation, both 1064- and 532-nm pumping were used. An output power of 1.3 mJ pumped by Nd:glass at 2.3 J and with a peaked spectrum of radiation at 1235 nm and a linewidth of 30 and 27 nm, respectively, was reported [29].

Chromium-doped forsterite laser covers a tuning range in the near-infrared spectral range, which is not covered by any other tunable solid-state lasers. Its applications in medical practice are expected. As the absorption spectrum of water changes significantly within the tuning range of the Cr-doped forsterite laser, the penetration of laser radiation into the body tissue can be cantrolled for specific purposes.

When the Cr-doped forsterite laser is used in conjunction with a titanium-doped sapphire laser, it may cover nearly the whole visible spectrum (including the use of second-harmonic generation) and the near-infrared region up to 1350 nm. Through the technique of mode locking, ultrashort pulses may be generated.

Because large Cr-doped forsterite crystals can be easily grown, their potential use as an amplifier medium to produce high-power lasers can be anticipated; however, the problem of ESA may reduce the gain at longer wavelengths.

In general, the tuning ranges for common solid-state lasers are shown in Fig. 5.6. The total range covered by different lasers can amount to a broad spectrum, although each laser covers only a small section of the range.

The gain of a vibronic laser is strongly dependent on pump power. Figure 5.7 shows the dependence of laser output power versus wavelength for three pump power levels for a Ti:sapphire laser, indicating the effect of pump power level. Different cavities are used to tune the wavelength range as indicated by the arrowheads.

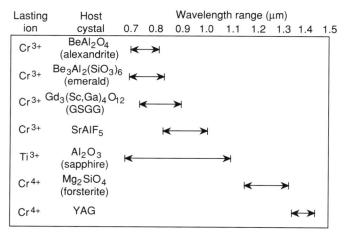

Figure 5.6 Tunable ranges for common tunable solid-state laser materials over a broad spectrum. After [29].

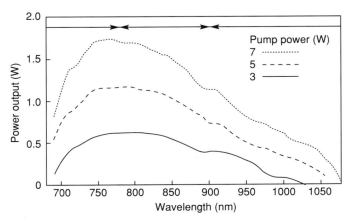

Figure 5.7 Tuning curve showing the effect of various pumping powers on laser power output for a Ti:sapphire laser pumped by a CW argon ion pump. Several cavities are used to extend the tuning range. After [36].

The emission of a vibronic laser is highly temperature dependent. Figure 5.8 shows laser gain versus wavelength for different temperatures. High temperature increases the gain but restricts the tuning range. High temperature populates higher vibronic sublevels of the round electronic state, preventing the laser from oscillating at shorter wavelengths.

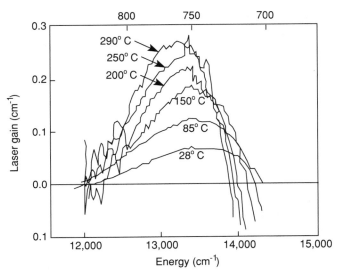

Figure 5.8 Temperature effect on vibronic lasers. In an alexandrite laser, laser gain increases and shifts to slightly lower energies as the temperature is raised. After [36].

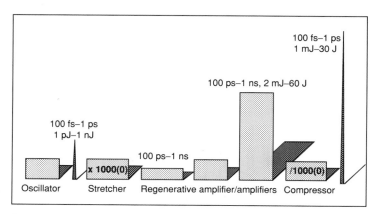

Figure 5.9   In chirped-pulse amplification, a short pulse is stretched, amplified, and compressed. This has been demonstrated with Ti:sapphire, alexandrite, and Cr:LiSAF lasers. J. Squier and G. Neurow (1992). Laser Focus World. © PennWell Publishing Co.

Ultrafast Solid-State Lasers

Solid-state lasers can be forced to create ultrashort pulses by a technique known as chirped-pulse amplification [30]. In normal pulsed laser design, if one tries to extract energy from a high-peak-power short pulse, the intensity developed in the amplifier chain may be so high that it perturbs the amplifier and destroys the laser catastrophically. The Chirped-pulse technique has been designed to counteract this limitation (Fig. 5.9). The pulse is being temporally stretched, or chirped, say, to 10,000 times the original width by standard optical conponents. The stretched pulse is then amplified. The broadened pulse can extract more power from the amplifier without risking a breakdown. After amplification, the pulse is recompressed back to its Fourier transform limit. This technique has been applied to Nd:glass, alexandrite, Ti:YAG, and other solid-state lasers with reported success. To date, pulses as short as 25 fs have been produced in Ti:YAG lasers [31].

Unstable Resonators

Resonators play an important role in laser operation. Stable resonators were instrumental in the development of lasers. A stable cavity in a laser, however, because of its small volume, often limits the output power capability of a laser.

Not all gas, dye, and semiconductor lasers use stable resonators. Most solid-state lasers do use stable resonators. But any asymmetry in the structure of mirrors, such as different input and output mirror sizes and even different mirror curvatures, could cause the resonator to operate in the unstable mode. Sometimes, unstable resonators are purposely introduced to boost the output of the laser. A pair of convex mirrors can form an unstable cavity for laser operation. The advantage of an unstable cavity in a high-power laser structure is that a greater portion of the gain medium is available (compared with a stable cavity) to contribute to the laser output because of its increased volume. Investigation of the electromagnetic field distribution at the edges of a stable resonator by Siegman [32] led to the suggestion of using unstable resonators for laser operation. Fox and Li carried out computer programs to analyze the unstable resonators consisting of different mirror structures, such as strip mirrors and confocal and convex mirrors, to acertain the reduced cavity loss in using unstable resonators [33]. In the 12th Conference on Lasers and Electro-optics held on May 10–15, 1992, in Anaheim, California, DeFreez reported on a high-power InGaAs/GaAlAs SQW semiconductor laser with an unstable resonator [34]. This laser produced 6 W of output (two-facet) at 870 nm.

Summary

In this chapter, we reviewed the working principle underlying solid-state lasers. Then, the differences between semiconductor diode lasers and solid-state lasers were addressed. The importance of developing solid-state lasers in power generation and wavelength tunable lasers was emphasized. The trend in research into new materials for host and doping ion combinations has led to high power capability and wide wavelength tunability in new solid-state lasers, replacing the dye and gas lasers in many applications. Extending the wavelength tuning range of the laser to cover the entire electromagnetic spectrum from microwave to infrared is now possible. Multiple pumping schemes, such as double pumping and triple pumping, to boost the laser output power capability, have been the subject of experiments. Harmonic generation using nonlinear crystals provides the means to generate high-frequency lasers. Extremely short pulse widths in the femtosecond range can be generated with high optical power. It seems that the field of solid-state lasers has no limit. Perhaps the words of Professor Edward Teller, "it's no laser unless it is a solid state laser," are true.

A special issue on quantum electronics contains articles on new developments in solid-state lasers [35].

References

1. T. H. Maiman, Stimulated optical radiation in ruby masers. *Nature* **187,** 493 (1960).
2. A. S. Grove, *Physics and Technology of Semiconductor Devices,* pp. 258–336, Wiley, New York, 1962.
3. P. F. Moulton, Spectroscopic and laser characteristics of Ti: Al_2O_3. *J. Opt. Soc.* **B3,** 125–133 (1986).
4. J. E. Geusic, H. M. Marcos, and L. B. Van Uitert, "Laser oscillations in Nd-doped yttrium aluminum yttrium." *Appl. Phys. Lett.* **4,** 182 (1964).
5. M. Birnbaum and A. W. Tucker, Nd: YALO oscillator at 930 nm. *IEEE J. Quantum Electron.* **QE-9,** 46 (1973).
6. A. L. Harmer, A. Linz, and D. R. Gabbe, "Fluorescence of Nd^{3+} in lithium yttrium fluoride," *J. Phys. Chem. Solids* **30,** 1483 (1969).

7. G. H. Dicke, *Advances in Quantum Electronics,* p. 170, Columbia Univ. Press, New York, 1961.

8. L. F. Johnson, J. E. Geusic, and L. G. Van Uitert, "Coherent oscillations from Tm^{3+}, Ho^{3+}, Yb^{3+}, and Er^{3+} ions." *Appl. Phys. Lett.* **7,** 127 (1965).

9. J. M. McMahon, J. L. Emmett, In *Proceedings, 11th Symposium on Electronegative Ion Laser Beam Technology, Boulder, Colorado, 1971.*

10. J. H. Gomez and P. B. Newell, *J. Opt. Soc. Am.* **56,** 87 (1966).

11. W. Streifer, D. R. Scifres, G. L. Harnagel, D. F. Welch, J. Berger, and M. Sakamoto, Advances in diode laser pumps. *IEEE J. Quantum Electron.* **24,** 883–894 (1988).

12. L. Goldberg and J. F. Wellwe, Injection locking and single-mode fiber coupling of a 40-element laser diode array. *Appl. Phys. Lett.* **50,** 1713 (1987).

13. R. Allen, L. Esterowitz, J. E. Weller, and M. Storm, Diode-pumped 2 μm holmium laser. *Electron. Lett.* **22,** 947 (1986).

14. L. R. Marshall, Blue-green lasers plumb the mysteries of the deep. *Laser Focus World* **29,** 185–197 (1993).

15. J. T. Whitley, Laser diode pumped operation of Er^{3+} fiber amplifier. *Electron. Lett.* **24,** 1537–1539 (1988).

16. Y. Ohishi, T. Kanamori, T. Nishi, and S. Takahashi, A high gain, high output saturation power Pr^{3+}-doped fluoride fiber. *IEEE Photon. Technol. Lett.* **3,** 175 (1991).

17. R. Allen, L. Esterowitz, L. Goldberg, J. R. Weller, and M. Storm, Diode pumped 2 μm holmium laser. *Electron. Lett.* **22,** 947 (1986).

18. J. Manni, Solid-state lasers in medicine today. *Lasers Optronics,* pp. 17–20 (April 1992).

19. P. K. Cheo (Ed.), *Handbook of Solid-State Lasers,* Marcel Dekker, New York, 1988.

20. J. M. McMahon, Solid-state lasers. In *Handbook of Microwave and Optical Components* (K. Chang, Ed.), Vol. 3, Ch. 7, Wiley, New York, 1990.

21. D. A. Belforte, Industrial lasers move up to the power curve. *Laser Focus World* **27,** 69–84 (1991).

22. P. F. Moulton, Tunable solid-state lasers. *Proc. IEEE* **80,** 348–364 (1992).

23. L. K. Smith, S. A. Payne, L. L. Chase, W. L. Kway, and B. H. T. Chai, Cr:$SrGaF_6$—A new laser material of the colquitrite structure. In *Conference on Lasers and Electro-optics,* pp. 388–390. Optical Society of America, Washington, DC, 1991.

24. W. R. Rapaport, J. W. Kuper, J. S. Krasinski, and T. Chin, High-brightness alexandrite laser. In *OSA Proceedings on Advanced Solid State Lasers* (H. P. Jensen and George Dube, Eds.), Vol. 6, pp. 170–173, Optical Society of America, Washington, DC, 1991.

25. P. F. Moulton, Ti:Al_2O_3—*A New Tunable Solid State Laser.* Solid State Research Rep. DTIC AD-A124305/4 (1982:3), pp. 15–21. Lincoln Laboratory, MIT, Lexington, MA, 1982.

26. G. A. Rines and P. F. Moulton, Performance of gain-switched Ti:Al_2O_3 unstable-resonator lasers. *Opt. Lett.* **15,** 434–436 (1990).

27. G. Erbert, I. Bass, R. Hackel, S. Jenkins, K. Kanz, and J. Paisner, 43-W CW Ti:sapphire laser. In *Conference on Lasers and Electro-optics,* pp. 390–391, Optical Society of America, Washington, DC, 1991.

28. V. Petricevic, S. K. Gayen, and R. R. Alfano, Laser action in chromium-activated forsterite for near-infrared excitation: Is Cr^{4+} the lasing ion? *Appl. Phys. Lett.* **53,** 2593–2595 (1988).

29. V. Petricevic, A. Seas, and R. R. Alfano, Slope efficiency measurements of a chromium-doped forsterite laser. *Opt. Lett.* **16,** 811–813 (1991).

30. D. Strickland and G. Mourou, Compression of amplified chirped optical pulses. *Opt. Commun.* **56,** 219 (1985).

31. C. Huang *et al., Opt. lett.* **17,** 139 (1992).

32. A. E. Sigman, Unstable optical resonator for laser applications. *Proc. IEEE* **53,** 277–287 (1965).

33. A. G. Fox and T. Li, Resonant modes in a maser interferometer. *Bell Syst. Tech. J.* **40,** 453–488 (1961).

34. R. K. DeFreez, paper CWN3, CLEO/QELS '92 Convention Record.

35. Special issue on quantum electronics. *Proc. IEEE* **80,** No. 3 (1992).

36. J. Hecht (1992). "Tanability makes vibronic lasers a versatile tool." P. 93. Laser Focus World.

CHAPTER 6

Other Laser Sources

Introduction

Since the invention of the solid-state ruby laser in 1960 [1] at a wavelength of 694.3 nm (4.32×10^{14} Hz), laser applications have spread to virtually all areas of science and technology. Semiconductor injection-type lasers discussed in Chapter 3 for applications to telecommunication systems represent only a small portion of the laser family. Developments in laser technology appear in the news daily. These include gas lasers, chemical lasers, dye lasers, excimer lasers, X-ray lasers, and free-electron lasers, covering a wide frequency spectrum and a variety of applications. At present, we have available lasers that can deliver power varying from milliwatts to gigawatts, and wavelengths from soft X-ray and ultraviolet to far-infrared in all sizes. As some scientists put it, anything can lase if it is hit hard enough. In this chapter we intend to describe briefly these lasers to give readers an idea of what they are all about.

In all lasers, the basic principle of operation is the same: (1) The atomic structure of the medium has a number of energy levels. (2) Pump energy is provided to create population inversion so that the high energy level is momentarily heavily populated. (3) Stimulated emission occurs in the presence of a photon that releses the population of high-energy electrons to the ground level concomitant with emitted photons. (4) A resonant cavity reinforces the feedback oscillation. The frequency of oscillation is proportional to the energy difference between the levels, as $h\nu = E_2 - E_1$. Here, h is Planck's constant, ν is the frequency of the ocillation, and E_2 and E_1 are the respective upper and lower energy levels of the atom. The details of the energy-level structure (two, three, or four levels), the method of pumping, and the structure of the resonator may differ. Also, lasers may differ in the material medium used (gas, liquid, etc.), the excitation method (dc discharge, rf discharge, or others), the power-handling capacity (low and high power), and, of course, the application. We describe the different lasers in the following sections.

Gaseous Discharge Lasers

Why are we interested in gaseous discharge lasers? One can choose among the many gas lasers those that have characteristics desired for laser operation: a gas that is very homogeneous and produces a nearly perfect Gaussian beam that can be analyzed and predicated; a gas laser that is simple to understand in operation; a gas laser that will perform efficiently; and an operation that can be scaled to a larger volume to accommodate larger power capability.

The disadvantage of a gas discharge laser is its huge size. The voltage and current required for excitation and operation are extremely large compared with semiconductor lasers. But gas lasers, in particular, CO_2 lasers, are among the few lasers that can handle higher continuous-wave (CW) power with good efficiency.

The first demonstration of gaseous discharge laser from a mixture of helium and neon at a wavelength of 1552.3 nm occurred in 1961 [2], right after the invention of the ruby laser. Since then, thousands of different reports have appeared in the literature covering different gas mixtures and varying wavelengths. We restrict this discussion to only a few, the helium–neon laser, the argon ion laser, and the CO_2 laser.

Helium–Neon Laser

Gases like helium and neon have well-known energy-level diagrams. Simplified energy-level diagrams of these two gases are arranged side by side in Fig. 6.1. The reason for combining these diagrams will become obvious shortly. Simplified spectroscopic notation is used freely. In the helium atom, the 2^3s_1 and 2^1s_0 levels are metastable. When these atoms are efficiently pumped to higher levels via electrons in the discharge and quickly relax to these metastable levels and stay there for a longer time. Note that the 2^3s_1 level and the 2^1s_0 level of helium lie close to the $2s_2$ and $3s_2$ of neon atoms, respectively. This makes the selective energy transfer from helium to neon very easy. This is the principal reason a He–Ne gas mixture is used. As shown in the diagram, many transitions are possible, including the visible red 632.8-nm and the near infrared 1150- and 3390-nm lines.

Note that at least four $3s$, ten $3p$, four $2s$, ten $2p$, and four $1s$ neon levels and two helium metastable states are to be related to the discharge. Not all transitions are interesting although many have been identified. Actually, many transitions between $3s$ and $2p_{1-8}$ have been observed to have lased, but the most intense one is the "red" laser indicated by the $3s_2$ transition to $2p_4$.

The atomic pumping is provided by electrical discharge, with a discharge current typically in the range 10 to 50 mA and a voltage of 1 to 5 kV. With this low current, we do not expect a He–Ne laser to deliver large power. He–Ne lasers are designed

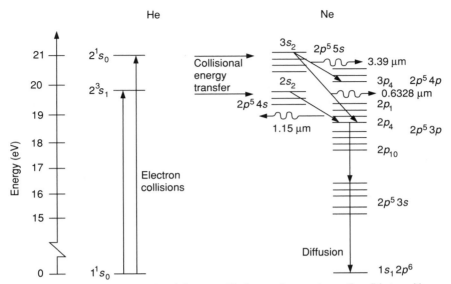

He　　　　　　　　　　　　　　Ne

Figure 6.1 Simplified energy-level diagram of helium and neon atoms. Possible transitions are indicated. Adapted with permission from Lehecka *et al.* (1990). Gas Lasers, in *Handbook of Microwave and Optical Components* (Kai Chang, Ed.). © John Wiley, 1990.

to carry power of less than 50 mW. Besides, as the transfer efficiency for conversion to optical power is quite low ($\approx 0.1\%$), the heat dissipation of the device is large. A massive amount of helium, usually in the proportion of 10:1 to neon gas, is used in the mixture to transport the heat away to the walls. This is a second reason for including helium gas in the laser.

A He–Ne laser consists of a discharge tube made of glass 1 to 2 mm in diameter and 10 to 30 cm long. Helium and neon gases are filled in a proportion about 6:1, which gives a gas pressure of about 3.5 torr. Brewster angle windows and external mirrors are used to form a resonant cavity.

A simple analysis of the rate equations will show that the number density of neon at high energy levels saturates at a high discharge current, and this may destroy the population inversion property of the laser. The He–Ne laser is thereby doomed to small-power operations; however, the usefulness of this laser is still very broad. He–Ne lasers are being used in accurate distance-measuring systems [3], in laser printing [4], and in medical applications [5].

The advantages of He–Ne lasers are the excellent beam quality, visible spectrum, and long lifetime. But as laser technology progresses, the challenge is to compete with the semiconductor diode lasers of similar wavelength ranges now available. Semiconductor lasers are by far more efficient and convenient to use than gas lasers. At present, He–Ne lasers are still popular for laboratory applications.

Ion Lasers

The excited ions of many elements, such as Cd, Se, Ge, Hg, Au, and Pb, can contribute to lasing if properly ionized or brought to the temperature for ionization. Some require extreme temperature or high voltage to reach this stage. At present, only a few of the ions are of potential interest in laser applications. Among these the heavy noble gases argon and krypton are the most common. We describe the argon ion laser as an example.

Figure 6.2 is the energy-level diagram of an argon ion. Although the ionization potential of argon is 15.75 V, the laser transitions occur at much higher energies. One can obtain nearly 50 different transitions in the excited argon ion in the wavelength range 700 to 3000 nm; only a few transitions within the visible spectrum are shown in the figure. These transitions occur between the high levels as shown. A total energy of 35.43 V is needed to reach the upper level of the 4880-Å line. As the mean discharge electron energy is about 20 V, a two-step excitation is usually required to pump from the ground state to a level for lasing. The quantum efficiency of this type of laser is very low, amounting to only 0.72%. The overall efficiency of the argon ion laser is poor, typically in the range 0.05 to 0.2%. The discharge is, however, very intense and the current of discharge is heavy, usually in the range of tens of amperes. As much as 80 kW of electrical power is required to produce approximately 20 W of laser output. The excess heat generated by the discharge requires a massive cooling system for its removal. The discharge tube is constructed with high-temperature-refractory materials such as silica, alumina, and beryllia.

A typical argon laser is about 1.5 m long and has a 2.5- to 4 mm-diameter discharge tube. Brewster angle windows are employed to allow external tuning of the laser. Argon gas is filled at a pressure of 0.1 to 0.5 torr. A discharge current of 25 to 40 A is required with a voltage of 300 to 500 V. The system is water cooled with a

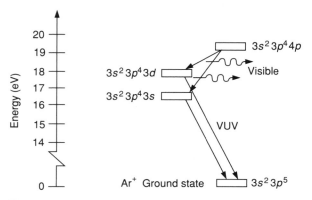

Figure 6.2 Energy-level diagram of an argon ion. The ionization potential of argon is 15.75 V; transitions occur at much higher energies. Adapted with permission from Lehecka *et al.* (1990). Gas Lasers, in *Handbook of Microwave and Optical Components* (Kai Chang, Ed.). © John Wiley, 1990.

flow rate of 5 to 10 liters per minute. The output of the laser is about 10 W. To help confine the argon ion beam, an axial magnetic field of approximately 0.1 T is used.

From the preceding description, we know that an argon ion laser is a massive device. Why do we want to use this laser? At present, it is the only laser suitable for many applications. Argon ion lasers are widely used in atomic physics and chemistry as pumps to tunable dye lasers [6] and displays [7]. They also have potential widespread use in medicine [8]. For example, NEC Electronics brought out a 5-mW air-cooled argon ion laser that can be used to separate colors in photoplotting and in diagnostic medical and clinical systems.

As markets mature for ion lasers, users are concerned more with the improved reliability and safety of the devices. Argon ion lasers have a good safety record.

Carbon Dioxide Laser

The carbon dioxide laser is a spectacular laser. It provides the characteristics of high power output and high efficiency, in both CW and pulsed operation at wavelengths from 900 to 1100 nm, or in the midinfrared range. CW outputs of tens of kilowatts and pulsed output energies up to 10 J are routinely available. By using a master oscillator/power amplifier combination scheme, terawatts of pulse power can be achieved. The overall efficiency of the CO_2 laser is in the 20% range.

The carbon dioxide gas molecule possesses linear symmetric bonds in equilibrium. In the excited state, three vibrational states become possible: the symmetric stretch (SS), the asymmetric stretch (AS), and the bending of the atomic bonds. Each vibrational state is capable of contributing to population inversion when properly pumped. Thus many transitions are possible. Two of these transitions are indicated in Fig. 6.3 in the energy-level diagram. More than 200 vibrational–rotational transitions at 8 to 18 μm are possible. The most important lasing transitions are the 940- and 1040-nm transitions. Lasings take place between the odd rotational levels of the 001 state and the even rotational states of either the 100 or 020 state, as shown in Fig. 6.3. The former emits at 1040 nm and the latter at 940 nm. In the same figure is also shown the energy-level diagram of nitrogen molecules. Nitrogen gas is added to CO_2 to enhance the transfer of excitational energy from the pump, as the vibrational energy of the diatomic nitrogen molecule is so close to the 001 state of the CO_2 molecule that energy transfer from N_2 to CO_2 is very efficient. (This is shown in Fig. 6.3 as the excited nitrogen level $v = 1$, which is at about the same level as the 100 level of CO_2.) Helium gas is also added to enhance the collisional deexcitation of CO_2 molecles from the lower lasing level to the ground state. The high thermal velocity of helium molecules also increases the deexcitation rate of nitrogen molecules; however, their primary purpose is to control the E/p and the electron temperature of the discharge. The gas mixture of a CO_2 laser is usually in the proportion $1CO_2 : 2N_2 : 3He$.

CO_2 lasers are generally divided into three groups, according to the power rating. Low-power lasers (<100 W) units are usually of the sealed-beam type. The electrical discharge tube is about 20 to 50 cm long and typically 5 to 10 mm in diameter. The

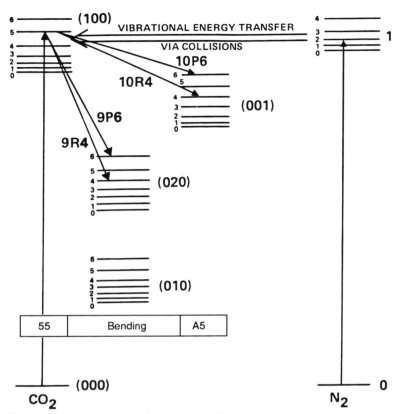

Figure 6.3   Energy-level diagram of carbon dioxide and nitrogen molecules. Lasing between SS (100) and AS (001) produces 10.6–10.4 μm. Lasing between SS (100) and Bending produces 9.6–9.4 μm. Adapted with permission from Lehecka *et al.* (1990). Gas Lasers, in *Handbook of Microwave and Optical Components* (Kai Chang, Ed.). © John Wiley, 1990.

composition of $CO_2:N_2:$He is about $1:4:15$. The discharge tube is sometimes contained in a waveguide structure. In this structure, a radio frequency source is usually used for excitation. The lifetime of these lasers is limited to less than 500 hours, because of contamination of the gas mixture. The CO_2 gas tends to degenerate into CO, thus contaminating the mixture.

Medium-power CO_2 lasers carry a power of 100 to 1000 W. A typical medium-power laser has a discharge tube 1.5 m long and 10 mm in diameter. The electrical discharge is maintained by a voltage of about 10 kV and current of 100 to 200 mA. The discharge tube is vacuum sealed. Brewster angle windows are provided. To facilitate wavelength tuning, a diffraction grating is used in place of a mirror. A number of discrete wavelengths in the range of 920 to 1080 nm can be selected. The discharge tube is provided with a water-cooling jacket.

The high-power CO_2 laser carries power in excess of 1000 W. The design of medium-power lasers must be improved to remove heat generated in the discharge by increasing gas flow rate. For laser applications requiring high peak power, the longitudinal-type discharge tube designed for low- and medium-power lasers that carries a relatively low pressure (10–20 torr) may limit the discharge energy. A new type of transverse excitation atmospheric pressure (TEA) laser has been developed [9]. Construction of high-power CO_2 lasers is much too complex and is beyond the scope of this book.

Applications of high-power CO_2 lasers include their use as pump sources for high-power optically pumped far-infrared lasers for fusion plasma diagnostics, laser–matter interaction studies, laser–plasma accelerators for particle physics, material cutting and drilling for industries, and surgery in medical practice. The range of application is still increasing.

CO_2 lasers are being targeted for very specific industrial and scientific applications. Higher-power sealed CO_2 lasers are one popular trend. Industrial use of CO_2 lasers has increased since 1975. In the material processing industry, particularly in the automobile industry, high-power lasers have been adapted increasingly for cutting and welding. CO_2 lasers provide adequate output power (up to 10 kW now), have good beam focusability, and can be used in combination with automatic control of these parameters. They can provide a very stable beam for operation. Low-price ($3000), low-power (up to 2 W) portable CO_2 lasers are commercially available. On the upper end of the power scale, 5- to 10-kW units are also available. Future development centers on a new resonator design to further reduce unit size without compromising beam quality and performance.

Other Gaseous Lasers

Chemical mixtures of common molecular hydrogen and fluorine may also be initiated to lase by an energetic electron beam to give ultraviolet radiation. Pulse energies up to 1.7 kJ in a 60-ns pulse have been reported [10].

Dye Lasers

Dye lasers usually come in liquid form. The active medium of a dye laser is a solution of organic dye compound in alcohol, ethylene glycol, or water. Stimulated emission from organic dyes was first observed in 1966 [11]; this was followed by the appearance of the first high-gain room temperature laser using trivalent neodymium in the inorganic liquid selenium oxychloride [12]. The first CW liquid laser, the rhodamine 6G laser, was put into use in 1970 [13].

Many dye compounds have been identified that can lase at wavelengths of interest. The following is a partial list [14]:

Dye	Solvent	Wavelength (nm)	Color
Acridine	EtOH	600–630	Red
Puronin	MeOH, H_2O	570–610	Yellow
Rhodamine 6G	EtOH	570–610	Yellow
Na-fluorescein	H_2O	530–560	Green
7-Hydroxycoumarin	H_2O	450–470	Blue
Cyanine		520–1150	

The broad spectrum of the dye laser makes it an extremely useful tool, especially in laser spectroscopy. These lasers are pumped by either CW argon or krypton ion lasers and are capable of generating several watts of output power.

The outstanding advantage of the dye laser is its broad tunability. The problem with these lasers is the need to control and tune the output precisely. Stabilized oscillation can be established by a complex optical and electronic control system [15].

It is not clear how the dye laser works in terms of our understanding of the energy-level concept of atomic structure of simple elements. An organic dye has a complex molecular structure. But experimentally, the system lases quite spectacularly. To explain the experimental results, a first-order approximation to the energy-level diagram was proposed [16]. The suggested energy-level diagram of a dye complex is shown in Fig. 6.4. The excited dye molecules appear to occupy two groups of energy levels, one by the singlets (S) and the other by the triplets (T). The vibrational (thinner lines) and rotational (thicker lines) states alternate in each group at different lev-

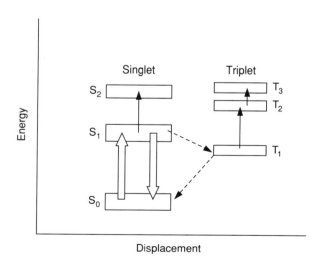

Displacement

Figure 6.4 Energy-level diagram of a typical dye laser. There are two different groups: the singlets (S) and the triplets (T). Within each energy box (S_0–T_2), there exists a number of sublevels due to vibrational or rotational motions, usually indicated by a heavy or light line (not shown in this graph).

els. The S groups are identified as S_0, S_1, and S_2. and the T groups as T_1, T_2, and T_3, respectively. The S and T groups appear separately on the plot as shown. Singlet levels can be excited by optical pumping. The dye molecules absorb the energy and are raised to an upper level. On relaxing to the lower state, they emit a photon. Excited atoms in the upper S_1 state may also transfer to the T_1 state and dissipate energy as heat. To avoid this conversion, the pumping must be intense and fast so that the S_0 state will not be thermally overpopulated, which would terminate the lasing action. Thus, most dye lasers are pulse operated. Flashlamp-pumped dye lasers are excited with pulse widths on the order of 100 to 1000 ns to avoid the buildup of the triplet T_1 population. For CW operation, the dye solution must be refreshed by recirculation.

With intense and rapid pumping, a dye laser can have high power. The pulse can be short as the dye laser is self-quenched by the triplet buildup.

Although tunable solid-state lasers are now available, it is not anticipated that they will replace dye lasers soon. This is because the dye laser remains a cost-effective source of tunable visible and ultraviolet photon sources when compared with the solid-state alternative. The dye lasers for these wavelength regions are more efficient.

Excimer Lasers

The excimer laser belongs to the rare-gas-halide gas laser family which emits powerful ultraviolet pulses of short duration. It is so called because it works only when the molecules of the gas are bound in excited states.

The development of excimer laser has been very rapid since its first demonstration in 1975 [17]. At present, it has become the workhorse of scientific, medical, and industrial applications. Excimer lasers are now standard laboratory tools.

The most important excimer laser media are the rare-gas halides:

Argon fluoride (ArF) which emits about 193 nm
Krypton fluoride (KrF) which emits about 248 nm
Xenon fluoride (XeF) which emits about 308 nm
Xenon chloride (XeCl) which emits about 350 nm

Average power on common excimer lasers ranges from less than 1 W to 100 W. Strong KrF and XeCl lasers can go well beyond 100 W. Excimer lasers with much higher pulse energies and average powers are being developed in military and fusion research.

General Considerations

Rare gases, such as argon, krypton, and xenon, are chemically inert. For the most part, they exist as single monatomic gases and do not form compounds with other elements. But in the excited state, one of the electrons that had otherwise filled the outermost shell of the atom comes loose and is promoted to the next shell, making the excited atom active. We label these excited rare gases R_g^*. An excited Ar* acts as

K, Ne* as Na, Kr* as Ru, and Xe* as Cs, and so on. These gases are vigorous and can react with halogens (F, Cl, Br, etc.) readily to become gas–halide salts.

Energy-Level Diagram of Excimer Gas

The excited gas–halide salt has an energy-level diagram related to the initial states of the rare gas. A typical example of (ArF)* is shown in Fig. 6.5. Here, curve B is the potential energy diagram of the excited gas, and the ground state of the gas is labeled X. Note that while curve X is a monotonously decreasing function with increasing interatomic spacing, curve B undergoes a dip in potential at a special spacing. The shape of this curve dominates the action of the excimer laser as follows: The excitation mechanism follows the expression

$$\text{Ar} + \text{energetic electron} = \text{Ar}^+ + e^-$$

Note that the argon becomes ionized. Next,

$$2\text{Ar}^+ + \text{F}_2 = 2(\text{Ar}^+\text{F}^-)^*$$

The active excimer gas is now $(\text{Ar}^+\text{F}^-)^*$. The argon and fluorine atoms are now oppositely charged and attract each other. The potential energy of $(\text{Ar}^+\text{F}^-)^*$ is shown as curve B. It has an attractive well between $r = $ infinity and $r = 2.3$ Å because of Coulombic attraction between Ar^+ and F^-. The potential minimum appears at an atomic spacing of $r = 2.3$ Å because below this distance, other repulsive forces come into play to prevent the atoms from getting closer to each other. The electrostatic force at that spacing is about 6.54 eV ($V = q/(4\pi\epsilon_0 r_{min})$). The ground

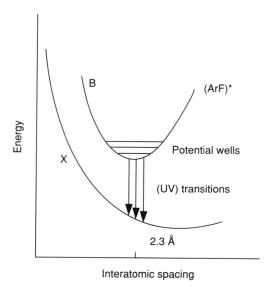

Figure 6.5   Energy-level diagram of excimer gas (ArF)*.

state X lies at about 0.29 eV and is always depopulated. The near-zero populated ground level makes it an ideal state for the radiative emission to relax to.

Energy can be deposited to the laser gas by electrical discharge, microwave excitation, and electron beams. The first two methods are suitable for low-power lasers; the last method is reserved for high-power applications.

Discharge-Pumped Excimer Laser

Discharge-pumped lasers are the most attractive and most frequently used excimer lasers for applications involving ultraviolet spectroscopy. They provide moderate output energy ($<$1J) per pulse at an inexpensive price.

The discharge tube is usually about 1 m long and 1 to 2 cm in diameter. Discharge pulses are directed either longitudinally or perpendicularly to the discharge tube. Several kilovolts and several hundred milliamperes of current are required to maintain the discharge.

The laser gas degrades easily during use. Once the gas is excited and the laser generated, the volume of gas cannot be used for the next pulse. In smaller excimer lasers, the gas is circulated by a pumping and cooling system to coutinuously replenish the supply. Again, there are limits on the energy and the laser aperture that can be pumped by transverse discharge.

The efficiency of discharge excitation is very low. Only about 5% of the electrical energy goes to the laser, with an overall efficiency of less than 2.5%.

Waveguide

A waveguide can be used to carry the laser gas and at the same time to support the discharge, using radio frequency excitation, usually across the waveguide. This excimer gas is housed in a small bore of a dielectric waveguide, about 1 mm across. The gas is allowed to flow slowly through the waveguide at 10 to 20 cm^3/min, at standard temperature and pressure. The cavity needs a little reflective coating on the output mirror. Usually, 4 to 10% reflectivity or an uncoated optical surface can provide enough feedback to ensure laser operation.

Electron-Beam-Pumped Excimer Lasers

This excimer laser requires the excitation of a large amount of excited ions in the gas mixture. It usually takes a great deal of power to produce this excitation. Electron beam-pumped excimer lasers, which are among the highest energy-pulsed lasers in the ultraviolet wavelength range, may require nearly relativistic electron beams to excite the gas. Usually several hundred kilovolts and hundreds of amperes per square centimeter are required. For example, a 500-keV electron beam will generate about $5 \times 10^5/26$, or about 19,000, secondary electron pairs. The divisor, 26 eV is the energy required to be deposited in a neutral gas to create one electron–ion pair of argon. As the ionization potential of argon gas is only 12.75 eV, the energy difference is used to create some excited states that decay readily to the metastable level and become poised for lasing action.

Electron beam-pumped excimer lasers belong to a class of high-energy devices in the ultraviolet wavelength region. Reported energies up to 10 kJ have been obtained with KrF [18], and 5 kJ has been obtained with XeCl [19]. The engineering problems encountered in designing and operating these gigantic experiments are complex and beyond the scope of this book.

Pulse Energy, Repetition Rate, and Average Power

Pulse energy and average power of excimer lasers are functions of the repetition rate. At low repetition rates, pulse energies normally are constant, but at higher rates, they drop. The average power increases with the repetition rate at first and then decreases at further higher rates. The exact point at which the average power reaches the maximum depends on the type of excimer gas used. Moreover, maximum average power is not equal to the product of peak repetition rate and peak pulse energy. Thus, lasers that produce low-energy pulses at a high repetition rate may not be interchangeable with lasers that have a low repetition rate and high-energy pulse, although they may have the same average power.

Applications

Applications of excimer lasers can be divided into two categories: low power and high power. First, we have to recognize that the excimer laser is used chiefly for ultraviolet wavelength operations. Ultraviolet light can be absorbed in very shallow depths and, in combination with its short-duration (nanosecond) pulses, can result in very localized heating and melting. These properties can be applied to material processing in factories. Focusing on the order of a hundred millijoules of quite incoherent excimer laser on the surface of any material will generate a very high temperature, causing the material to melt or to vaporize. This principle has been used to vaporize and sputter highly refractive materials [20]. Excimer pulses can ablate ceramic materials, producing thin films in the fabrication of high-temperature superconductors.

The broad bandwidth and highly multimode output of excimer lasers can result in highly uniform heat to anneal the surface of semiconductor wafers, far better than using other types of lasers [21].

The shorter wavelength of the excimer laser can also be used in photolithography to overcome the chromatic aberration problem of ordinary optical lenses in photolithographic processes in the microelectronics industry.

In medical practice, XeCl and ArF lasers are frequently used. For example, the output of a XeCl laser coupled through an ultraviolet fiber bundle is used to clear human arteries. To propagate maximum energy per pulse through the optical fiber, longer-duration pulses are used [22].

The ArF laser operates at 193 nm. This short-wavelength radiation will make cleaner cuts because it breaks the molecular bond directly. It is possible to operate on the surface of the lens of an eye and to preserve its optical property by reshaping the curvature of the lens. Longer-wavelength lasers may damage the tissue more extensively [23].

High-power excimer lasers are the only lasers that have demonstrated a capacity for direct production of short-wavelength laser light at high energy. The broad bandwidth of the excimer laser has fueled interest in their use in producing very short duration (<1 ps), very high power ($\geq 10^{12}$ W) laser pulses for research and military applications [24].

Excimer lasers are used to pump dye lasers to produce tunable short-wavelength light for research, remote sensing, and other applications. Because excimer lasers can generate high peak powers and their short wavelengths are strongly scattered in the atmosphere, they are used widely in atmospheric research. Excimer lasers have been used to measure ozone concentrations in the upper atmosphere and to monitor atmospheric pollution.

Precautions

Excimer lasers are powerful tools that can be used in many fields. But caution must be exercised in using these tools.

1. All excimer lasers use a high-voltage power supply. Ten kilovolts for discharge is common. Care must be used in handling high-voltage devices.

2. Often, an X-ray preionizing scheme is used to ensure ionization states for the gas mixture. This poses a radiological hazard problem to the worker. Proper shielding must be provided.

3. The output of an excimer laser is in the ultraviolet. Ultraviolet lights are hazardous to human skin and particularly to the eye.

4. The laser gas itself poses another hazard to the operator. The inert gas can displace oxygen and cause suffocation if inhaled. Halogen gases are very active and attack anything in their way. Even the container of these gases and the waveguide structures have to be coated with Teflon to prevent reaction.

5. The fluorine content of air should be limited to less than 0.1 ppm, and chlorine content to less than 1 ppm. Fluorine gas should always be diluted to 5% with other inert gases such as neon and helium before it is released into the air.

The spotlight is now on the application of excimer lasers to remote sensing and photolithography.

Free-Electron Lasers

The free-electron laser (FEL) is a new addition to the laser family. The concept of producing coherent radiation is, however, different from that for conventional lasers. A free-electron laser uses an electron beam of near-relativistic velocity in a strong magnetic field to obtain energy gain. It provides a powerful and tunable source of radiation, covering a wide spectrum, from millimeters in the microwave through the visible and into the ultraviolet and soft X-ray range. Accordingly, many applications can be derived for this invention.

The concept of the free-electron laser is an outgrowth of research on linear accelerators and synchrotron radiation in atomic physics. Madey is credited with turning this concept into an optical device [25].

The free-electron laser makes use of the interaction between an energetic electron beam and an electromagnetic (EM) wave. The beam and wave interchange energy when they travel together long enough and at velocities in synchrony with each other. The electron beam may gain energy from the EM field if the phase relation between beam velocity and field velocity is reversed. Or, conversely, the EM field may gain energy from the beam as an amplifier. In the process of energy exchange, both energy and momentum must be conserved. A familiar example is the microwave traveling-wave tube (TWT). Here an electron beam is led along the axis of a helical coil housing the EM field. As the beam velocity is very low, corresponding to several hundred electron-volts, the EM field velocity has to be brought down by a slow wave structure such as a helical coil. When the phase relations are adjusted correctly by adjusting the beam velocity, energy can be gained by the EM field, and its signal is amplified.

In the free-electron laser, electrons are forced to wiggle through an alternating magnetic field with periodicity W, while the EM field remains unaltered. The beam velocity must therefore be brought up to a near-relativistic value to match their phases.

Structure

An artist's view of a free-electron laser is provided in Fig. 6.6 [26].

An electron beam, accelerated in a linear accelerator to a near-relativistic velocity of, say, 500 keV is directed into a structure consisting of uniformly spaced strong magnetic field pieces of opposite polarity known as a "wiggler." As electrons accelerate in a plane transverse to the magnetic field, they execute a wiggling oscillatory motion. Thus, electron acceleration by the wiggler magnets leads to EM radiation. Of the overall broad spectrum of light emission, there is one component of the wiggler-induced radiation that has the exact phase accompanying the electron motion. This component is reinforced and becomes the FEL wavelength. Placing reflectors for this wavelength can further enhance the coherence of the beam. Laser light is finally extracted from the output mirror. Without feedback mirrors, this scheme can also be used as an amplifier. Peak currents as high as 10,000 A have been reported.

An Approximate Equation

Consider the exchange of power between the moving electron beam and the traveling EM wave with an electric field E_x. Assume that electrons with near-relativistic velocity are forced to wiggle in the transverse direction (x or y) along the traveling

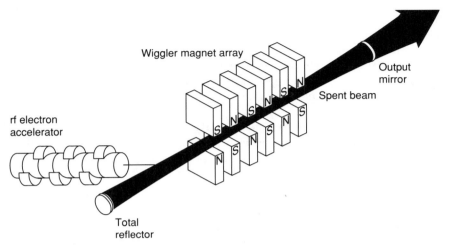

Figure 6.6 Artist's view of a free-electron laser. In a free-electron laser, electrons wiggling in the magnetic field of the wiggler emit light, which amplifies the laser beam entering the device from the left. If mirrors are used as shown, the light reflects back and forth between them, getting stronger on each pass through the wiggler. The light is then extracted as a laser beam through the output coupling mirror. In this configuration, the laser is referred to as an oscillator. After C. A. Brau [27].

wave in the z direction. Under synchrony condition, power $E_x v_x$ does not change sign, as reversal of the sign of v_x occurs at the same time as reversal of the sign of E_x. Depending on the phase relietionship between E_x and v_x, power transfer may be positive or negative. By positive power transfer we mean the electrons lose energy to the EM field. By negative power transfer, power is gained by the beam.

The interaction between the electron and the optical beams intended to extract energy from the electron beam requires that the beam have a high current. Thus, a relativistic beam must be generated with as many electrons (current) as possible, and these electrons must be periodically accelerated as they traverse along the direction of propagation.

Given that the current is high enough, the magnetic wiggler field is strong, and mirror feedback is provided, the coherent radiative output will have a free-space wavelength approximately expressed by

$$\lambda = W\left(\frac{1 + a_w^2}{2\Gamma^2}\right) \tag{6.1}$$

where $\Gamma = 1/(1 - \beta^2)^{1/2}$ is a relavistic factor, $\beta = v/c$, W = the wiggle period $\approx \lambda_0$,

$$a_w = \frac{qB_0^2\lambda_0}{2\pi mc^2} \tag{6.2}$$

Γ is the energy of the electrons in units of their rest energy, which is 0.511×10^6 eV. Thus, for 100-million-volt electrons, $\Gamma = 200$. The free-space wavelength can then be tuned by varying the factors contained in a_w, for example, by varying the magnetic field strength B_0, the wiggler field periodicity W, and also the electron beam velocity v. If the wiggler period is 2 cm, the free electron radiation will have a wavelength of about 0.5 μm in the green light region.

The wavelength is inversely proportional to the square of Γ, the relativistic factor, because of a "double Doppler shift effect," and this results in a much higher frequency or shorter wavelength.

Equation (6.2) enables us to tune the laser. By the use of different energies, it is possible to obtain wavelength from the microwave region (mm), through the visible portion of the optical spectrum (0.5 μm), and up to the ultraviolet and soft X-rays.

Advantages and Shortcomings

Use of the free-electron laser has four advantages:

1. *Tunable wavelength:* By adjusting the electron beam energy and the period of the magnetic wiggler, various wavelength ranges can be achieved.

2. *High output power:* Peak power greater than 10^{10} W has been reported [27]. Unlike excimer lasers, the removal of heat in a free-electron laser poses little problem. The exit electron beam carries the excess energy away. Advanced microwave tube technology has facilitated recovery of the beam energy in this laser.

3. *Good quality of the optical beam:* The long and slender cavity supports only the desirable mode of osillation.

4. *High overall efficiency:* Although the efficiency in extracting light from the electrons is only 2%, the efficiency of the high-current beam accelerator is high, about 50%. Research is underway to improve this efficiency.

The disadvantage is that as the electron beam loses energy to the optical beam at the output, it slows down such that the oscillation predicted by Eq. (6.1) no longer holds. This causes the lasing action to stop. It is also the chief reason why the efficiency of the device is so low. New designs with a tapered wiggler help to adjust the wiggler period and to compensate for this effect and raise the efficiency. Besides the tapered wiggler, the wiggler used in a FEL is usually designed according to the beam quality and energy requirements for the particular application. In one exotic system, a robotic computer program has been written to adjust each magnet to keep the beam in perfect alignment with the fields continuously.

For further information on FEL, see Brau [28].

Applications

The high power and broad wavelength range of the free-electron laser have led to many applications in science and industry.

In materials science, free-electron lasers in the range 100 to 400 μm are used for local heating, abrasing, and annealing.

In chemistry, the shorter-ultraviolet-wavelength laser with tunable capability is important in photochemical processing applications. Some wavelength ranges are inaccessible by any other type of laser.

The high-power capability has been exploited by the military to build strategic defensive weapons, as in "Star Wars."

New applications are expected in the future.

X-Ray Lasers

Lasing action at X-ray frequencies can be acheived by the same principle of population inversion and resonant cavity feedback but with difficulty. Threshold population inversion density is inversely proportional to the square of the wavelength, making it difficult to attain threshold as the wavelength decreases. Simulated emission is proportional both to the cross section for the interaction of photons with excited atoms and to the density of the excited states in the upper level of a radiative transition. As the wavelength corresponding to a transition decreases, the lifetime of the upper level rapidly decreases. The cross section of photon interaction also decreases with wavelength. As a result, the power required to sustain an inversion increases greatly. A 10-nm laser will need 10^{10} times as much pump power as a 1-μm laser. Inefficiencies in couping may make the pump power required for X-ray lasers even greater. There is also appreciable difficulty in fabricating suitable mirrors for X-ray resonators. High-reflectivity multilayer coatings—both the materials and the methods for fabrication—for extreme-ultraviolet and soft X-ray applications are the subject of ongoing research programs.

X-Ray laser action was first observed at the Lawrence Livermore National Laboratory (LLNL) in 1980. Nuclear detonation was involved in the generation of the laser pulse. Controlled experiments were carried out at the Princeton Physics Laboratory (PPL) in New Jersey. Suckewer and De Meo [29] reported on the generation of a 20-ns pulse of soft X-ray ASE with a power of 100 kW, an energy of 2 mJ, and a divergence of 5 mrad. At LLNL, the gigantic NOVA 1.06-μm Nd^{3+} : glass laser system was used to produce 250-ps X-ray laser pulses at wavelengths as short as 4.3 nm.

High-resolution X-ray microscopy, which has achieved 2-μm spatial resolution, will have many applications in the life and physical sciences. X-Ray microlithography for producing the next generation of densely packed semiconductor chips is in big demand. The dynamic imaging and holography of individual cellular structures have many applications in biology and medical science.

Extreme-Ultraviolet Lasers

The demand for the development of X-ray lasers arises from the need to microscopically image very small structures, such as living cells and other biological material. X-Ray lithography will also benefit from the increase in power and brightness of X-ray lasers. Precision can be gained by moving toward extreme-ultraviolet (XUV) lasers. Although XUV lasers with wavelengths as short as 3 nm have been tested in the laboratory, commercial XUV lasers are not yet available. Intensive research on XUV lasers was reactivated in 1985 [30, 31]. Since then, many technological advancements have been reported on XUV lasers. The initial difficulty was the stringent scaling law for operation at very short wavelengths.

At X-ray wavelengths, mirror reflectors do not work as efficiently as those for visible light.

Several pumping schemes for making short-wavelength lasers have been demonstrated: electron collisional pumping, recombination, and photopumping. Readers are referred to the current literature for details [32 – 34].

Commercial Non-Solid-State Lasers

The lasers described in this chapter have been in existence for a long time. For example, the first gas laser, the He–Ne laser, was demonstrated in 1961. CO_2 laser technology emerged in the mid-1960s. Scientific, medical, and industrial applications of these lasers have long been established. Their popularity has been challenged only recently by the development of solid-state lasers; however, new solid-state lasers are very expensive and have their share of shortcomings. This leaves the old-fashioned lasers very much in demand. It is therefore interesting to show some of these laser designs for reference.

Helium–Neon Lasers

Figure 6.7 is a cutaway view of a He–Ne gas laser. A long glass tube (10 cm to 2 m long) is filled with gas, usually a mixture of helium and neon. An electric discharge passing between electrodes placed at opposite ends of the tube, known as the cathode and the anode, is maintained by a high-voltage power supply. The discharge is confined to a center bore with mirrors at each end to form a resonator. Light emerges from one mirror to yield a wavelength at 632.8 nm red line with a few milliwatts of power.

A diode laser that generates 1 -mW beam can be obtained much more cheaply and works more efficiently and in a smalleer package without the need for high voltage. But He–Ne lasers have their advantages, including better coherence, better beam quality, and cost effectiveness for some applications, such as high-speed laser printers and holograms.

Ion Lasers

The most important commercial ion laser is the argon ion laser as described earlier under Gaseous Discharge Lasers. Figure 6.8 is a cutaway view of an ion laser that delivers power at visible and ultraviolet wavelengths. An electrical discharge is maintained between the cathode and the anode of a ceramic tube. Within the tube ion

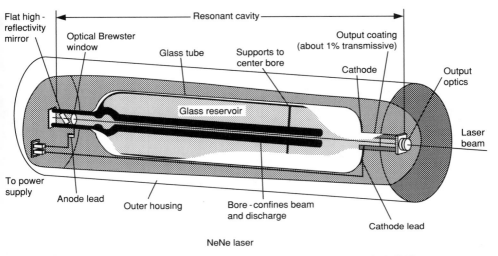

Figure 6.7   Industrial helium–neon laser. Reprinted with permission from Hecht [35].

shields are installed along the discharge to separate the active ions and the gas reservoir with a series of copper cups and tungsten disks. The ceramic tube is then surrounded by a water-cooling jacket. Outside, a magnet array supplies the magnetic field to focus the beam along the axis. The resonator is confined between a pair of Brewster windows and mirrors. Pure argon or an argon–krypton mixture at 1 torr of pressure is used as the gas. A few thousand volts is used to break down the gas; the voltage then drops to 70 to 400 V while the current jumps to 10 to 70 A. Laser action takes place on transitions far above the ground level of the ions. Two principal visible argon lines, 514.5 nm (green) and 488 nm (blue), can be obtained. The output power is in the range 1 to 10 W. Higher discharge currents can produce doubly ionized argon which has ultraviolet laser lines at 275, 300–306, 334, 351, and 364 nm. The efficiency is very low (0.01%).

From Fig. 6.8, this laser is seen to have a complex structure. The aim is to replace these ion lasers with diode-pumped neodymium lasers. On the other hand, other solid-state laser advances have helped create a new market for ion lasers—as pumps for tunable Ti:sapphire lasers.

Carbon Dioxide Laser

Industrial CO_2 technology emerged in the mid-1960s. Development has been rapid. Today, industry uses lasers for cutting, welding, and drilling, whose power ranges from 1000 to 3000 W. The typical CO_2 laser used to be large and was inconvenient for operation. Today, a sealed CO_2 laser, one-tenth the size of conventional CO_2 lasers, offers new opportunities. The basic change in the design is the shape of the discharge tube, shown in Fig. 6.9. The discharge is contained in large electrodes that

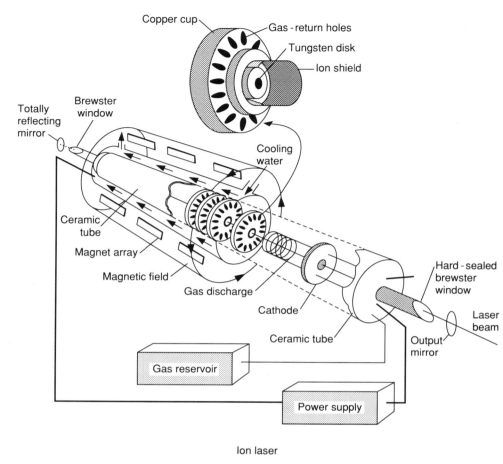

Ion laser

Figure 6.8 Ion laser for visible and ultraviolet wavelengths. Reprinted with permission from Hecht [36].

provide the basis for the discharge. They are also used to cool the mixture and form part of the optical resonator. The whole laser is sealed, thus eliminating the gas supply. The resonator is factory-filled with gas at 80 to 100 torr pressure. It has long operating life. Lifetimes exceeding 10,000 hours before gas refill are expected. The device uses a rf power supply requiring only a 48-V dc input and a 20-A operating current. Stable operation over a wide range of operating temperatures is achievable; 75- to 150-W power output units are commercially available.

Excimer Lasers

Excimer lasers contain rare-gas halides such as argon fluoride (ArF), krypton fluoride (KrF), and xenon chloride (XeCl), which emit at wavelengths of 193, 248, and

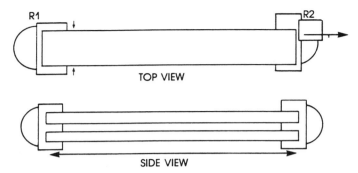

Figure 6.9 Sealed carbon dioxide laser. Reprinted with permission from Shiner [37].

350 nm, respectively. Excimer lasers were first demonstrated in the mid-1970s. Initially, they were designed for medical and industrial uses. They have become standard laboratory tools. Their design is complicated by the corrosive nature of halogen; however, the ability to generate high-energy ultraviolet pulses at a sufficient repetition rate for average powers that can exceed 100 W has created a steady demand for excimer lasers in science, industry, and medicine. The typical design of an excimer laser is shown in Fig. 6.10. The setup is very impressive. First, the gas that supports the lasing action needs to be replenished through the gas-handling module, with an exhaust vacuum pump. A gas reservoir is provided to supply fresh gas. The

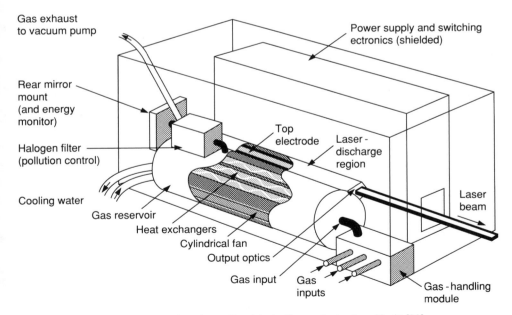

Figure 6.10 Sketch of an excimer laser. Reprinted with permission from Hecht [38].

chamber is water cooled. The laser resonator is about 1 m long. The internal gain is very high. The optics for monitoring the operation are provided. Finally, the power supply and switching electronics are provided and are well shielded. At 193 nm, an ArF excimer laser can produce pulses at a 300-pulse repetition rate to yield an average power of about 50 W.

Lasers made with non-solid-state materials are in great demand at present. Scientists are working hard to replace these with solid-state lasers.

Summary

In this chapter, we have described lasers that are totally different, with respect to size, power-handling capacity, and operating principles, from the semiconductor diode and solid-state lasers. The applications of these lasers are also different. We have included these lasers to open the door to the broad laser field and to show the extent

TABLE 6.1
Characteristics and Parameters for Well-Known Gas (g), Solid (s),
Liquid (l), and Plasma (p) Laser Transitions[a]

Laser	Transition wavelength	Mode	Efficiency	Output power or energy
C^{6+}	18.2 nm	Pulsed	10^{-5}	2 mJ
ArFexc (g)	193 nm	Pulsed	1.	500 mJ
KrFexc (g)	248 nm	Pulsed	1.	500 mJ
He–Cd (g)	442 nm	CW	0.1	10 mW
Ar^+ (g)	513 nm	CW	0.05	10 W
Rhodamine 6G dye (l)	560–640 nm	CW	0.005	100 mW
He–Ne (g)	633 nm	CW	0.5	1 nW
Kr^+ (g)	647 nm	CW	0.01	500 mW
Ruby (s)	694 nm	Pulsed	0.1	5 J
$Ti:Al_2O_3$	0.66–1.18 μm	CW	0.01	10 W
Nd^{3+} glass	1.06 μm	Pulsed	1.	50 J
Nd^{3+} YAG (s)	1.064 μm	CW	0.5	10 W
KF (s)	1.25–1.45 μm	CW	0.005	500 mW
He–Ne (g)	3.39 μm	CW	0.05	1 mW
FEL	9–40 μm	Pulsed	0.5	1 mJ
CO_2 (g)	10.6 μm	CW	10	100 W
H_2O (g)	118.7 μm	CW	0.001	10 μW
HCN (g)	336.8 μm	CW	0.001	1 mW

[a] Adapted with permission from Suckewer and DeMeo [29].

and general interest of lasers for different applications. The descriptions are far from complete. Interested readers should consult the current literature to learn about advances in the field.

To conclude this chapter, we provide in Table 6.1 a list of selected laser transitions to supplement the contents; many of these have not been described in the text.

References

1. T. H. Maiman, Stimulated optical radiation in ruby masers. *Nature* **187,** 493 (1960).
2. A. Javan, W. R. Bennett, Jr., and D. R. Herritt, Population inversion and continuous optical maser oscillation in a gas discharge containing He–Ne mixtures. *Phys. Rev.,* **122** (6) p. 106 (1961).
3. J. F. Ready, *Industry Applications of Lasers,* Academic Press, New York, 1978.
4. G. K. Starkweather, In High-speed laser printing system. *Laser Applications* (J. W. Goodman and M. Ross, Eds.), Vol. 4, Academic Press, New York, 1980.
5. I. Goldman (Ed.), *The Biomedical Laser: Technology and Clinical Applications,* Springer-Verlag, New York, 1981.
6. W. O. N. Guimaraces, C. T. Lin, and A. Mooradian (Eds.), *Laser and Applications,* Springer-Verlag, New York, 1981.
7. R. Iscoff, Ultrafast dye laser. *Laser Optronics* **7,** 46 (1988).
8. J. A. S. Carruth and A. L. McKenzie, *Medical Lasers, Science, and Clinical Practice,* Adam Hilger, Bristol, England, 1986.
9. A. J. Beaulieu, Transverse excited atmospheric pressure CO_2 laser. *Appl. Phys. Lett.* **6,** 504 (1970).
10. G. N. Hays, J. M. Hoffman, and G. C. Tisone, Phoenix 2: Sandia's 1 kJ HF laser system. Paper WE 1, Topical Meeting on Inertial Confinement Fusion, San Diego, California, February 26–28, 1980.
11. F. P. Schafer, W. Schmidt, and I. Voise, Organic dye solution lasers. *Appl. Phys. Lett.* **9,** pp. 306–309 (1966).
12. A. A. Heller, A high gain room temperature liquid laser: Trivalent neodymium selenium oxychloride. *Appl. Phys. Lett.* **9,** 106 (1966).
13. O. G. Peterson, S. A. Tuccio, and B. B. Snavely, CW operations of organic lasers. *Appl. Phys. Lett.* **17,** 245 (1970).
14. B. B. Snavely, Flash lamp pumped dye lasers. *Proc. IEEE* **57,** 1374 (1969).
15. B. Peuse, New developments in CW dye laser. In *Physics of New Laser Sources* (N. B. Abraham, F. T. Arecchi, A. Mooradian, and A. Sona, Eds.), NATO ASI Series, Series B: Physics, Vol. 132, pp. 69–78. Plenum, New York, 1984.
16. M. Bass, T. F. Deutsch, and M. J. Weber, Dye lasers. In *Lasers* (A. K. Levine and A. De Maria, Eds.), Vol. 3, p. 275. Marcel Dekker, New York, 1971.
17. S. K. Searles and G. A. Hart, Stimulated emission at 281.8 nm from XeBr. *Appl. Phys. Lett.* **27,** 243 (1975).
18. L. A. Rosocha, J. A. Hanlon, J. McLeod, M. Kang, B. L. Kortegaard, M. D. Burrows, and P. S. Bowling, Aurora multijoule KrF laser system prototype for inertial confinement fusion studies. *Fusion Technol.* **11,** 497 (1987).
19. J. R. Oldenettel and K. Y. Tang, Multi-kilojoule narrowband XeCl laser. *Proc. Soc. Photo-Opt. Instrum. Eng.* **710,** 117 (1986).
20. D. Dijkkamp, T. Venkatesan, X. D. Wu, S. A. Shahenn, N. Jisravi, Y. H. Min-Lee, W. L. McLean, and M. Croft, Epitaxial ordering of oxide super conductor thin films on (100) $SiJiO_3$ prepared by pulsed laser evaporation. *Appl. Phys. Lett.* **51,** 861 (1987).
21. T. J. McKee and J. Nilson, Excimer applications. *Laser Focus,* **19,** p. 51 (June 1982).
22. J. B. Laundenslager, Excimer lasers adapt to angioplasty. *Laser Focus* **23,** 57 (1988).
23. S. L. Trokel, R. Srinivasan, and B. Brarea, Exciter lasers, surgery of the cornea. *Am. J. Ophthalmol.* **96,** 710 (1983).

24. J. P. Roberts, A. J. Taylor, P. H. Y. Lee, and R. B. Gibson, High-irradiance 248 nm laser system. *Opt. Lett.* **13**, 734 (1987).

25. J. M. Madey, Stimulated emission in Bremsstrahlung in a periodic magnetic field. *J. Appl. Phys.* **42**, 1906 (1971).

26. J. A. Pasour, Free electron lasers. *IEEE Circuits Devices Mag.*, pp. 55–63 (Jan. 1987).

27. C. A. Brau, Free-electron laser experiments. In *Physics of New Laser Sources* (N. B. Abraham, F. T. Arecchi, A. Mooradian, and A. Sona, Eds.), Nato ASI Series, Series B: Physics, Vol. 132, Plenum, New York, 1984.

28. C. A. Brau, *Free-Electron Lasers,* Academic Press, San Diego, 1990.

29. S. Suckewer and A. R. De Meo, Jr. X-ray laser microscope development at Princeton. *Princeton Plasma Phys. Lab. Dig.* (May 1989).

30. D. L. Matthews, P. L. Hagelstein, M. D. Rosen, M. J. Eckan, N. M. Ceglio, A. U. Hazi, H. Medecki, B. J. MacGowan, J. E. Trebes, B. L. Whitten, E. M., Campbell, C. W. Hatcher, A. M. Hawryluk, R. L. Kauuman, L. D. Pleasance, G. Rambach, J. H. Scoffield, G. Stone, and T. A. Waver, Demonstration of a soft X-ray amplifier. *Phys. Rev. Lett.* **54**, 110–113 (1985).

31. S. Sucwer, C. H. Skinner, H. Milchberg, C. Keane, and D. Voorhees, Amplification of stimulated soft X-ray emission in a confined plasma column. *Phys. Rev.* **55**, 1753–1756 (1985).

32. M. D. Rosen, J. E. Trebes, B. J. MacGowan, P. L. Hagelstein, R. A. London, D. L. Mathews, D. J. Nieson, T. W. Phillips, D. A. Whelan, G. Charatis, G. E. Busch, and C. L. Shepard, Dynamics of collisional excitation X-ray lasers. *Phys. Rev. Lett.* **59**, 2283–2286 (1987).

33. C. Chenais-Popovics, R. Corbett, C. J. Hooker, M. H. Key, G. P. Kiehn, C. L. S. Lewis, G. J. Pert, C. Regan, S. J. Rose, S. Saddaat, R. Smith, T. Tomie, and O. Willi, Laser amplification in recombining plasma from a laser irradiated carbon fiber. *Phys. Rev. Lett.* **59**, 2161–2164 (1987).

34. G. Y. Yin, C. P. J. Barty, D. A. King, D. J. Walker, H. E. Harris, and J. F. Young, Low energy pumping of a 108.9 nm xenon Auger laser. *Opt. Lett.* **12**, 331–333 (1987).

35. J. Hecht, Helium neon lasers flourish in face of diode-laser competition. *Laser Focus World,* pp. 99–108 (Nov. 1992). PennWell Publishing Co., Westford, Massachusetts.

36. J. Hecht, Ion lasers deliver power at visible and UV wavelengths, *Laser Focus World,* pp. 97–105 (Dec. 1992). PennWell Publishing Co., Westford, Massachusetts.

37. W. H. Shiner, Sealed CO_2 laser opens door to new industrial uses. *Photonics Spectra,* pp. 109–114 (Sept. 1992). Laurin Publishing Co., Inc. Pittsfield, Massachusetts.

38. J. Hecht, Excimer lasers produce powerful ultraviolet pulses. *Laser Focus World,* pp. 63–72 (June 1992). PennWell Publishing Co., Westford, Massachusetts.

CHAPTER 7

Photonic Detection

Introduction

Light detectors have been in existence for many years, long before the invention of lasers. The detection of electromagnetic waves has always been an important problem in scientific and engineering fields; however, recent advances in communication systems employing laser and fiber technology have spurred renewed interest in making better detectors. We devote this chapter to these advancements.

Photonic detection is the reverse process of photonic generation. It is used not merely to detect the presence of radiation, but to measure its intensity quantitatively as well. If the light source carries information, that is, a modulated signal, detection could include the process of demodulation. Thus, besides the high sensitivity required for detecting radiation, a fast response, faithful reproduction of the signal, and many other desired characteristics to be specified later in the chapter are required.

For optical detection, we rule out the use of thermal detectors. Thermal detectors convert optical energy into heat first and are usually inefficient and relatively slow in response. In this chapter, only photoelectric detectors using semiconductor materials are discussed.

The operation of photoelectric detectors is based on the photoelectric effect, which converts the absorbed photon energy of materials to generate free carriers (electrons and holes) of electricity. The motion of free charges, electrons and holes, in response to the applied electric field constitutes the electric current flow in the external circuit. This current is used for quantitative measurement. The operation may also be called *quantum detection.*

There are two forms of photoelectric effect: photoemission and the photoconductive. Both can be used for photodetectors, but their principles of operation are quite different. Photoemission, in which the photogenerated electrons escape from the material as free electrons, is also called the *external photoeffect*. A vacuum photomultiplier, a diode consisting a photoemission cathode and many secondary emission dynodes connected to progressively increasing potentials, has been in use for photon detection for many years. It is still in use for some applications, but we do not discuss these devices here. Our concern is confined to semiconductor varieties.

In the photoconductive effect, the generated carriers remain within the material, but the conductivity of the bulk material is changed. The current response to applied electric field, due to the presence of these incremental free carriers, changes considerably. This is called the internal photo effect. In detectors such as *p–i–n* and avalanche photodiodes, although both involve the generation of electron hole pairs, their motion is confined within the semiconductor. These are classified as *internal photoeffect detectors*. Both forms of photoeffect can be used to build photodetectors.

To qualify a semiconductor diode as a good detector, it must be sensitive and fast in response, low in noise, high in efficiency, simple in structure, and low in cost. Not all requirements can be met in one design.

Properties of Semiconductor Photodetectors

A photodetector converts an input photon flux to an output electric current by using semiconductor materials. One needs to find those semiconductor materials that can absorb the incident photon energy and generate electron hole pairs. Suitable materials for this purpose are shown in Fig. 7.1 [1]. The coefficient of absorption of semiconductor materials is given as a function of wavelength, indicating the wavelength-sensitive property of the absorption coefficients of most semiconductor materials. A useful semiconductor must have an absorption coefficient (α value) greater than 10^4/cm. When α equals this value, $1/\alpha = 1$ μm; most photons are absorbed near the surface of the semiconductor device, where the recombination time is quite short. The wavelength corresponding to this value of α becomes the short-wavelength limit of the material. The long-wavelength limit is determined by the expression $\lambda = hc_0/E_g$, where E_g is the bandgap energy and c_0 is the speed of light *in vacuo*. For longer wavelengths, the photon energy is insufficient to overcome the bandgap energy. One can see that a given material is useful for only a limited wavelength range.

Quantum Efficiency

When photon flux strikes a semiconductor material, because of the probabilistic nature of the absorption process, not all photon energy is absorbed. Of the total number of electron hole pairs generated, not all pairs are available to be converted into electric current. The effectiveness of the generation of electron hole pairs in response to incident photon energy is defined by the quantum efficiency η_q:

$$\eta_q = (1 - R)\beta[i - \exp(-\alpha d)] \qquad (7.1)$$

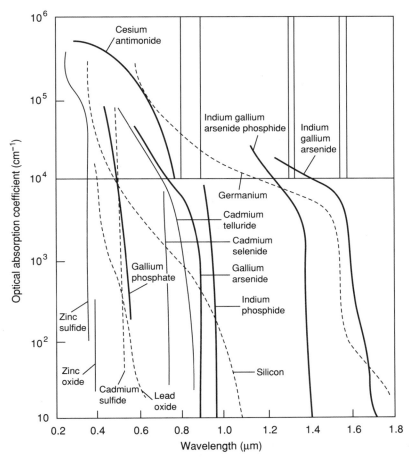

Figure 7.1 Absorption coefficients of semiconductor materials as a function of wavelength. Semiconductor materials: ──── ; III−V compounds, -----; Elemental semiconductors, ──── ; II− VI compounds. Reprinted with permission from Melchior [1].

where R is surface reflectivity, β is the fraction of electron hole pairs that contribute successfully to the photo current, α is the absorption coefficient, and d is the width of the absorption layer.

Quantum efficiency can be increased by (1) reducing the surface reflectivity R, (2)increasing β by selecting material growth property, (3) selecting a large product of αd so that $\exp(-\alpha d) \leq 1$. For a ready-made detector, quantum efficiency can also be defined as the artio of electric current (carriers per second) to the number of photons per second absorbed, or

$$\eta_q = \left(\frac{i}{q}\right)\left(\frac{P_{opt}}{hv}\right) \tag{7.2}$$

where i is current flow, q is electric charge, h is Planck's constant, and v is the frequency of the optical source.

Responsivity

The sensitivity of a detector can be expressed as the amount of electrical current the detector can produce for a watt of input optical power, or

$$I_{ph} = \frac{\eta q \mathbf{P} \lambda}{h c_0} \tag{7.3}$$

where η_q is quantum efficiency (coupling efficiency can be included in this notation by using η instead of η_q), $\mathbf{P}$ is the average optical power at wavelength λ, and $hv = hc_0 \lambda$ is the photon energy. The responsivity r is defined as the I_{ph} per unit power, or

$$r = \frac{I_{ph}}{\mathbf{P}} = \frac{\eta_q \lambda}{h c_0} \tag{7.4}$$

For an ideal detector, $\eta = 1$, and the responsivity is directly proportional to the wavelength. This is shown as the dotted straight line in Fig. 7.2, where responsivity is plotted against wavelength λ [2]. The responsivities of a silicon detector, a Ge diode, and an InGaAs diode are also shown as solid curves in Fig. 7.2.

Silicon diode, because its absorption coefficient decreases very rapidly at longer wavelengths, soon loses its responsivity above $\lambda > 1$ μm. At these wavelengths, the absorption layer d can never be wide enough to satisfy the relation $\alpha d \gg 1$, and its quantum efficiency also drops drastically.

For long-wavelength detection (1.0–1.7 μm), Ge and InGaAs p–i–n diodes have been developed. Their response curves are very attractive, as can be seen from Fig. 7.2, but both are troubled by the large dark current caused by low bandgap energy [3]. Other materials such as InGaAsP/InP and AlGaAsSb have been developed.

Response Time

The flow of detector current does not occur instantly in response to incident optical power. This time delay affects the operational characteristic of the detector immensely. Mechanisms responsible for the time delay can be attributed to the following causes: (1) the diffusion time, the time for carriers generated within a diffusion length from a junction to diffuse through the depletion region, (2) the drift time for the carriers to drift through the depletion regions, and (3) the effect of junction capacitance.

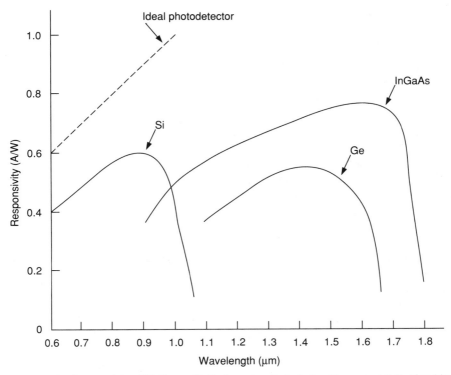

Figure 7.2 Responsivity of Si, Ge, and InGaAs *p–i–n* photodiodes. The responsivity of an ideal photodiode is represented by the dashed line. After C. Yeh, *Handbook of Fiber Optics,* reprinted by permission. © 1990 Academic Press.

Diffusion Time

If d is the depth of photon absorption, the diffusion time can be expressed approximately as $t_{\text{diff}} = d^2/2D$, where D is the diffusion constant of the minority carriers. (In junction diodes, only the minority carrier density changes significantly.) As the absorption coefficient is strongly dependent on the wavelength, so is the diffusion time. To ensure that as few carriers as possible are generated outside of the depletion region, we require that at a desired wavelength, the depletion width $w/(w \approx d)$ be larger than $1/\alpha$. If this criterion cannot be met, which may be the case with low reverse bias, diffusion time may become a limiting factor in the response time.

Drift Time

Within the depletion width, carriers drift across this region with a velocity v_s, known as the saturated velocity of the semiconductor material. For a narrow depletion width, even at moderately low-reverse-bias voltage, the electric field within this region is sufficiently high that only saturated velocity needs to be considered. Thus, $t_{drift} = w/v_s$. For $v_s = 10^7$ cm/s, the drift time is about 10 ps/µm.

Junction Capacitance Effect

Junction capacitance of a $p-n$ diode is expressed as

$$C_j = \frac{A}{2}\left[\frac{2q\epsilon}{\phi_0 + V} \cdot \frac{N_d N_a}{N_a + N_d}\right]^{1/2} \tag{7.5}$$

where A is the junction area, ϕ_0 is the built-in potential, V is the reverse bias voltage, and N_a and N_d are the carrier densities of the p and n sides, respectively. Usually, $V \gg \phi_0$, and $N_a \gg N_d$, so that

$$C_j \sim \left(\frac{A}{2}\right)[2q\epsilon N_d]^{1/2} \; V^{-1/2} \tag{7.6}$$

The junction capacitance shunts across an input resistance R_L of the amplifier. The time constant $R_L C_j$ gives rise to a bandwidth B such that $B = 1/2\pi R_L C_j$. The combination of R_L and C_j acts as a low-pass filter to limit the frequency response of the detector. Frequency response can, therefore, be improved by reducing C_j, which is accomplished by reducing the diode area A, reducing the carrier density N_d, or increasing the reverse bias voltage V. Usually, there is a limit to which each item could be adjusted. Compromises must be used in actual diode design.

The sum of these times represents the total time delay from diode alone.

Noise Considerations

The detection of weak signals is limited by the presence of background noise. Noise comprises spurious fluctuations that are not wanted. The hiss noise we hear in radio when the receiver is not in tune to a station is considered as noise in analog systems. Even if the receiver is in tune, weaker stations are usually masked by noise that make the listening difficult or unpleasant. In digital system, noise presents itself as bit error, in which "0" and "1" can no longer be clearly distinguished.

Noise in photodiodes can be considered as random fluctuations of photon flux that generate electric current. It is statistical in nature [2]. After we identify the noise sources, we discuss the statistical noise estimation for each type of detector separately.

Sources of Noise in Photodetectors

The inherent noise of photodetectors is of four types: quantum noise, background noise, dark current noise, and surface leakage current noise. In case the detector involves again mechanism, as in an avalanche photodiode (APD), gain noise should also be counted. When the detector is connected to a receiver, various components in the electrical circuitry also contribute to the receiver as circuit noise.

Quantum Noise

When an optical signal enters the detector, the photoelectrons carry a quantum noise component [4]. This arises from intrinsic fluctuations of photon excitation of carriers. It is fundamental in nature and sets the ultimate limit for detector sensitivity. Both the number of electron hole pairs generated and the generation rate are statistically random quantities that satisfy Poisson statistics.

Recent studies in quantum optics have revealed that by means of nonlinear optical processes, squeezed states can be generated that have sub-Poissonian distribution and are called squeezed light. When used with matching detectors in optical receivers, the squeezed states can lead to significant improvement in the signal-to-noise radio. In fact, it allows all optical measurements with a precision beyond the limit set by the zero-point or vacuum fluctuations of the optical field [5]. Reduction of quantum noise of 24% below the shot noise limit has been reported [6].

Background Noise

Background noise is the photon noise associated with light reaching the detector from extraneous sources other than the signal of interest, such as sunlight and starlight. It is particularly detrimental in middle- and far-infrared detection system because objects at room temperature emit copious thermal radiation in this region.

Dark Current Noise

Photodetectors also generate dark current noise, even in the absence of light [7]. Dark current noise results from random electron hole pairs generated thermally or by tunneling. Noise caused by dark current is sensitive to temperature changes, as the intrinsic carrier density is proportional to $\exp(-E_g/2kT)$. In silicon, for example, a temperature $50°$ C higher than room temperature will result in an increase in the dark current by a factor of 20.

Surface Leakage Noise

Defects on the surface of semiconductor may lead to surface leakage current and cause surface leakage noise. In a carefully prepared semiconductor detector, this noise is usually negligible. The noise contribution caused by photon current I_{ph} and dark current I_d, neglecting the background and leakage current, can be treated as shot current noise whose mean square value is given by

$$\langle i^2_{shot} \rangle = 2q[I_{ph} + I_d]B \tag{7.7}$$

where B is the bandwidth. Shot noise is a signal-dependent noise.

Circuit Noise

In the case where the detector output is followed by an amplifier, the noise from the amplifier, which is dominated by Johnson noise, must be considered. The noise contribution from thermal noise at the output of a preamplifier is

$$\langle i_{th}^2 \rangle = \frac{4kTB}{R_L} \tag{7.8}$$

where R_L is the load resistance of the detector or the resistance of the amplifier, k is Boltzmann's constant, and T is the temperature in degrees Kelvin. Thermal noise may be reduced by increasing R_L at the expense of reducing the maximum available bandwidth B and $B = 1/\{2\pi R_L C_d$, where C_d is the detector capacitance. Circuit noise is signal independent.

Noise in Detectors with Magnification

In the case of an avalanche photodiode (see next section), where photocarriers are multiplied in the avalanche process, the expression for the shot noise is modified by two factors: the multiplication factor M and an excess noise factor F. That is,

$$\langle i_{shot}^2 \rangle = 2qFM^2IB \tag{7.9}$$

where I is the total current. The avalanche multiplication factor M is

$$M = \frac{1}{[1 - (V/V_{br})^2]} \tag{7.10}$$

where V_{br} is the avalanche breakdown voltage, and

$$F(M) = M\{1 - (1 - k)[(M - 1)M]^2\} \tag{7.11}$$

Note that $k = \gamma_n/\gamma_n$, the ratio of the hole ionization coefficient to that of the electron. If $k = 1$, F has the maximum value, as does the shot noise. Thus, to minimize shot noise, k should be chosen as small or as large as possible [8].

Noise in Photoconductors

The noise in photoconductor comes from the thermal effect. The finite dark current, when the device is not illuminated, gives rise to a random fluctuation known as Johnson noise current, the mean square of which is

$$\langle i_j^2 \rangle = \frac{4kTB}{R_D} \tag{7.12}$$

where R_D is the dark resistance of the photoconductor. The effect of Johnson noise in the accompanying amplifier should also be considered.

Noise in Receivers

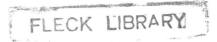

Noise in detectors contributes to the performance of a receiver. Our concern in a receiver with respect to noise is expressed in a slightly different way. We are interested in (1) the signal-to-noise ratio, or (2) the bit error rate, and (3) the minimum detectable power.

Signal-to-Noise Ratio

The signal-to-noise ratio is defined as the ratio of the mean square of the signal current to the sum of the mean square currents from all noise sources:

$$\frac{S}{N} = \left\langle \frac{i_{ph}^2}{i_n^2} \right\rangle \tag{7.13}$$

where i_n is the sum of the noise contributing fluctuation currents. These are

$$\langle i_n^2 \rangle = \langle (I_{ph})^2 \rangle = \langle (i_Q)^2 \rangle + \langle (i_d)^2 \rangle + \langle (i_L)^2 \rangle + \langle (i_{th})^2 \rangle$$

where $\langle (i_Q)^2 \rangle = 2qI_{ph}B$ = quantum noise, $\langle (i_d)^2 \rangle = 2_qI_dB$ = dark current noise, $\langle (i_L)^2 \rangle = 2qI_LB$ = leakage current noise, and $\langle (i_{th})^2 \rangle = 4kTB/R_L$ = thermal noise. If the signal current is expressed in terms of signal power as in Eq. (7.2), then the signal-to-noise ratio can be written as

$$\frac{S}{N} = \frac{\left[\dfrac{P_0 \eta q}{h\nu} \right]^2}{\left[\dfrac{2q^2 P_{th} \eta}{h\nu + 8\pi TBC_d} \right] B} \tag{7.15}$$

In this equation, the noise contribution from dark current and leakage current has been neglected for simplicity as they are small in well-designed detectors. Power P_0 is the signal power including the sideband powers.

Bit Error Rate

In digital modulated systems that require only identification of the "1" or "0" state in each sampling interval, one is interested in knowing the probability of making a false identification [9].

The arrival of a pulse may be represented by a collection of photons. On average, there will be 10 to 20 photons per pulse. As photon shots are random, there is a probability that a pulse may fail to contain a single photon. This indicates a false identification or error. How often this kind of error occurs in a system actually limits the speed of operation or the bit rate. If $p(n, \Omega)$ is the probability of detecting n

photons per unit time interval, and Ω is the average rate of photon per pulse, then by Poisson statistics [10],

$$p(n, \Omega) = \frac{\Omega_n e^{-\Omega}}{n!} \tag{7.16}$$

If we let $\Omega = 20$ and $n = 0$, representing the case where 20 photons are registered per pulse, we call this a "1". If there is no photon in a pulse, we call this a "0." Using our numbers, the probability is

$$p(0, 20) = \frac{(20)^0 e^{-20}}{0!} = 2 \times 10^{-9} \tag{7.17}$$

If we assume a system containing equal numbers of zeros and ones, the preceding probability gives rise to a bit error of $(\frac{1}{2})(2 \times 10^{-9}) = 10^{-9}$. This number is reasonable and has been universally adopted as the standard bit error rate (BER) for digital systems. In fact, many authors derive the sensitivity of a detector using the minimum detectable power as a function of the bandwidth or bit rate for a specified BER (10^{-9}). If a Gaussian distribution is assumed, the S/N ratio can be related to the bit error rate as

$$\text{BER} = (\tfrac{1}{2}) \frac{\text{erfc}\left(\dfrac{S}{N}\right)}{\sqrt{2}} = (\tfrac{1}{2})\text{erfc}(Q) \tag{7.18}$$

where $Q = (S/N)/\sqrt{2}$ and erfc is the complementary error function defined as

$$\text{erfc}(x) = 1 - \left(\frac{2}{\sqrt{\pi}}\right) \int_0^x \exp(-t^2)dt \tag{7.19}$$

For an error rate of 10^9, $Q = 6$, and the corresponding $S/N = 6\sqrt{2}$, or about 9.3 dB.

Minimum Detectable Power

The minimum power P_{min} required to obtain a reliable identification of the signal for a given bandwidth is obtained by assuming $S/N = 1$. A plot of P_{min} against bit error rate gives useful information about detector performance. One such plot is shown in Fig. 7.3. First, we calculate the power required to identify a signal for a bit error rate of 10^{-9} for a finite number of photon energies. These are marked as lines corresponding to 10, 100, and 1000 photons per bit, respectively. The 10-photon line is accepted as the limiting case for detection. Then the minimum detectable receiver power using $p-i-n$ and avalanche photodiodes is estimated separately from the power equations.

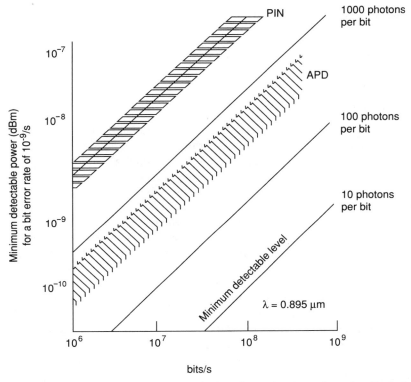

Figure 7.3 Minimum signal power required to give a signal-to-noise ratio of unity in a photo-detector. After C. Yeh, *Handbook of Fiber Optics,* reprinted with permission. © 1990 Academic Press.

When these values are plotted on the graph, they appear as lines or bands for each type of detector as shown. Calculation of detector power is discussed in a later section.

Photodetectors

As stated in the introduction, semiconductor materials are used exclusively for the optical detectors discussed in this chapter. Bulk semiconductor bars are used as photoconductors to detect optical power, especially in the far-infrared and longer-wavelength regions. Most other detectors are built by virtue of $p-n$ junctions formed by dissimilar semiconductor materials as used in fabricating optical sources.

Detectors based on silicon (Si) are useful in the wavelength range 0.6 to 1.0 μm, whereas Ge and InGaAs work best in at 1.3 to 1.7 μm.

Two types of detectors manufactured by junction techniques are of of interest in optical communication systems. These are $p–i–n$ photodiodes (PINs) and avalanche photodiodes (APDs). We describe these separately.

PIN Photodiodes

PIN Diode Structure

The PIN is the most popular diode used in fiberoptics because it is compatible in size with fibers, is simple in structure, gives good performance in operation, has long life, and is reliable [11]. The basic structure of a PIN is shown in Fig. 7.4. A thick i layer (intrinsic layer) is grown epitaxially and sandwiched between a thin p^+ layer and a thick n substrate, forming a basic $p–i–n$ diode. A metallic contact with an opening to admit light serves as one contact. The other contact is on the substrate. When a reverse bias is applied, the intrinsic region is fully depleted. The incident light penetrates the thin p^+ layer and reaches the i layer. There, absorbed photon energy generates electron hole pairs (EHPs). The high electric field in this region rapidly sweeps the electrons and holes to the n and p sides of the diode, respectively, causing current to flow in the external circuit in proportion to the illumination level. Any photons absorbed by the top $^+$ layer and the bottom n layer also produce EHPs. But these carriers will have to diffuse back into the depletion region before they can be collected. As the diffusion process takes time (10^{-9} s), it affects the response time of the diode. In comparison, the transit time of carriers across the depletion region is very short ($\approx 10^{-12}$ s). A good diode should, therefore, be designed to limit the absorption of photons to only the i layer.

Another factor that limits the response time of a diode is the shunt capacitance associated with the junction. The junction capacitance can be kept small by (1) limiting the diode area and (2) reducing the doping density of the i region. But the small diode area restricts the power-handling capability. Reducing doping density may widen the depletion layer thickness which would increase the transit time for the carriers, thus spoiling the frequence response. Carefully prepared epitaxial processing purifies the material which results in less doping density.

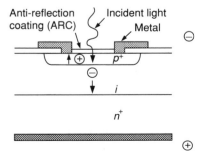

Figure 7.4 $p–i–n$ photodiode structure. Adapted with permission from Suematsu [11].

Several methods have been improvised to minimize the absorption of photons in the top p layer. One way is to use higher-bandgap p-type GaAlAs for the top layer, which allows light to pass through the p layer without absorption because high-bandgap GaAlAs is transparent to optical radiation.

Another way is to use side or bottom illumination [12]. In a top-illuminated diode, the metallic contact on the p^+ layer usually has a hole to admit light to the p layer. Large area is necessary for this structure to provide more photons into action. This causes an increase in diode junction capacitance which decreases the frequency response. By turning the diode upside down, that is, to admit light from the substrate, which can be made with transparent n-type semiconductor material, the whole substrate can be exposed to illumination. This increases the quantum efficiency of the diode without increasing the junction capacitance. Similar results can be achieved by using side illumination. In this case, illumination enters the i layer from the sides.

PIN Diode Performance

For the wavelength range 0.8 to 0.9 μm, a silicon PIN can easily attain a quantum efficiency of 85% from dc to 1 Gbit/s. The noise source is primarily shot noise, as the dark current is sufficiently low and causes no additional noise. Generally, the signal output from a detector requires amplification. This makes the amplifier noise very important for the receiving system. Because sensitivity decreases with increasing noise and noise increases with bandwidth, the sensitivity of PINs decreases at high bit rates.

The speed, or bandwidth, of PINs is very high, up to 15 or 16 GHz. But any parasitic capacitances added by the external circuitry will lower the speed correspondingly.

The signal-to-noise ratio of a PIN can be expressed as

$$\frac{S}{N} = \frac{\langle I_{ph}^2 \rangle}{\left\{ 2q(I_{ph} + I_d)B + \dfrac{4kTB}{R_L} \right\}} \tag{7.20}$$

Using Eq. (7.8) to eliminate R_L, we have

$$\frac{S}{N} = \frac{\langle I_{ph}^2 \rangle}{\{2q(I_{ph} + I_d)B + 8\pi C_d kTB\}} \tag{7.21}$$

Let us define a minimum signal current from which we can calculate the minimum detectable power. For $S/N = 1$, Eq. (7.21) yields a signal current

$$(I_{ph})_{min} = \{2Bq(I_{ph} + I_d) + 8\pi C_d kTB^2\}^{1/2} \tag{7.22}$$

and

$$P_{min} = \left(\frac{2hc}{\eta q\lambda}\right)\{2Bq(I_{ph} + I_d) + 8\pi C_d kTB^2\}^{1/2} \tag{7.23}$$

In the case the receiver is amplifier-noise-limited, that is,

$$8\pi C_d kTB^2 > Bq(I_{ph} + I_d)$$

then

$$P_{min} = \left(\frac{2hc}{\eta q\lambda}\right)B(2\pi kTC_d)^{1/2} \tag{7.24}$$

The minimum detectable power is then directly proportional to the bandwidth.

Avalanche Photodiodes

Structure

In the PIN-type photodiode just discussed, at most one EHP is generated for each photon absorbed. The process therefore possesses no gain mechanism; however, for a wide depletion width operating at a high reverse bias, the field within the region may reach a value sufficiently high ($E > 10^5$ V/m) to produce avalanche breakdown. The mechanism of avalanche multiplication is schematically shown in Fig. 7.5. In this case, electrons and holes traversing under such a high field can acquire sufficient kinetic energy to produce additional EHPs through inelastic collisions. This process builds internal gain within the width of the depletion layer. Gain coefficients of 55 are common.

A silicon APD has a structure similar to that of a PIN, as shown in Fig. 7.4. The design criterion regarding optimization of the absorption region also applies. The principal difference lies in the bias. In APDs, a reverse-bias voltage as high as 300 V is used, compared with the 5 to 10 V used in PINs.

Avalanche photodiodes are more expensive than PINs and, because of the higher

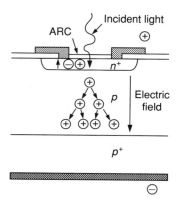

Figure 7.5 Structure of an avalanche photodiode showing the avalanche multiplication mechanism. Adapted with permission from Suematsu [11].

voltage levels, are probably less reliable. The biggest trade-off for APDs is noise. Excess noise is created by the avalanche phenomenon and sets an upper limit to the bandwidth of these devices at about 10 GHz.

Responsivity

Avalanche photodiodes are more sensitive, as more EHPs are available to participate in the current production when illuminated. The avalanche is measured by a multiplication factor M defined as

$$M = \frac{1}{\left\{ 1 - \int_0^w \gamma(x)dx \right\}} \tag{7.25}$$

and the breakdown criterion is

$$\int_0^w \gamma(x)dx = 1 \tag{7.26}$$

Here $\gamma(x)$ is an ionization coefficient defined as

$$\gamma = A \exp\left(\frac{-B}{E}\right) \tag{7.27}$$

where A and B are material constants, and E is the electric field intensity. The ionization coefficients for electrons and holes may be different. The multiplication factor M therefore includes the factor $k = \gamma_h/\gamma_n$, which is the ratio of the ionization coefficient of holes to that of electrons [13]. Thus,

$$M = \frac{(1 - k)}{\{\exp(-[1-k]\delta) - k\}} \tag{7.28}$$

$$\delta \equiv \int_0^w \gamma(x)dx \tag{7.29}$$

In Eq. (7.28), it can be seen that if $k = 1$, the denominator approaches zero and M approaches infinity, or large multiplication factor M can be obtained. In practice, however, other factors, principally the noise factor, increases more rapidly as to make the amplification worthless. But the importance of having divergent ionization coefficients of electrons and holes cannot be overemphasized. The worst-case ratio occurs when $k = 1$. A GaAs APD where $k \approx 1$ is therefore more noisy. Silicon APDs for shorter wavelengths are less noisy because for Si, $k < 1$.

Stability may become a problem for APDs with a high M factor. As M is highly sensitive to temperature, extra stability schemes may be necessary for operation. M should be kept below 50. For longer wavelengths, germanium APDs are used [14].

The responsivity of an APD is that of a simple photodiode modified by a multiplication factor, and the modified Eq. (7.3) becomes

$$r = \frac{I_{ph}}{P} = \frac{(M)(\eta q \lambda)}{hc} \tag{7.30}$$

The noise is modified by a more complicated manner. Excess noise is generated by the avalanche process, and the excess factor F as stated in Eq. (7.8) is also modified.

Similarly, the shot noise and the dark current noise are all modified by $\langle M^2 \rangle F$ as $\langle M^2 \rangle$ is the mean square of M.

$$\langle (i_{sh})^2 \rangle = \langle M^2 \rangle (2qI + ph_{ph}BF)$$
$$\langle (i_d)^2 \rangle = \langle M^2 \rangle (2qI_dBF)$$

and the signal-to-noise ratio, for $k > 1$, becomes

$$\frac{S}{N} = \frac{(P_n q \eta / h)^2 \langle M^2 \rangle}{(2qi_d + 2qP_{ph}q\eta/h)\langle M^2 \rangle F + 4B(kT/R_L)} \tag{7.31}$$

The signal-to-noise ratio thus changes with the change in M in a more complex manner. The response time, because of the speed of the detector, is a function of the time spent by electrons and holes in the depletion layer, and because it takes more time to build up these electron hole pairs through the avalanche process, APDs are not as fast as PINs. The avalanche process is gain bandwidth dependent.

The sensitivity of APDs is very attractive between about 100 Mbit/s and 4 Gbit/s, making them ideal for high-speed, long-haul fiberoptic communications.

Long-Wavelength Avalanche Photodiodes

High-quality APDs for detecting long-wavelength radiation are considerably more difficult to design than those for short wavelengths. At long wavelengths, one needs semiconductor materials with narrow bandgaps, such as Ge and InGaAsP, to absorb low-energy photons. But low-bandgap materials have a large dark current at high-reverse-bias voltage as a result of the tunneling effect. More recent research has focused on solving this problem.

One design, known as the SAM-APD, is an APD with separate absorption and multiplication regions [15] (Fig. 7.6). The SAM-APD diode has a heterostructure. Photons are absorbed in a thin, active layer of InGaAs. The photogenerated holes are swept into a layer of InP, which has a significantly larger bandgap. In such a structure, the electric field never exceeds values that would induce significant leakage current from quantum-mechanical tunneling. The highest electric field occurring at the $p-n$ junction is in the highest field region. In indium phosphate, the holes are more highly ionizing than electrons. The device is thus optimized for the injection of holes by being constructed with n-type material; however, the detector efficiency and response speed are both low, rendering it useless in high-bit-rate applications. This is because the bandgap energy difference between InP and InGaAs introduces an energy step that traps holes generated in the InGaAs active layer either to cause recom-

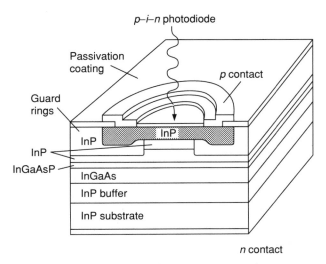

p–i–n photodiode

Passivation coating

p contact

Guard rings

InP

InP

InGaAsP

InP

InGaAs

InP buffer

InP substrate

n contact

Figure 7.6 Structure of a SAM-APD, a separate absorption-and-multiplication avalanche photodetector. Adapted with permission from Susa [15].

bination or to delay their travel to the multiplication region. Both processes slow the response. Considerable research remains to be done before SAM-APDs made for long-wavelength InGaAS/InP can be extensively employed in practical optical fiber systems.

Photoconductors

Photoconductivity

Photoconductivity is a long-recognized fundamental phenomenon in semiconductor material, but its application as a photodetector in optical fibers is relatively new. In fact, renewed investigations, particularly for long-wavelength optical fiber applications, showed that photoconductors may even be advantageous over the use of other detector schemes, such as PINs and APDs.

When a semiconductor bar is uniformly illuminated, if the photon energy is greater than the bandgap energy E_g of the semiconductor, EHPs will be generated. In response to an applied electric field, the conductivity of the semiconductor bar increases with the illumination intensity. If g_{op} is the optical generation rate per unit volume, the equilibrium carrier density is determined by the difference between the generation and recombination rates. Let δp and δn be the excess carrier concentrations of the holes and electrons, respectively. Then the new carrier densities become $n = n_0 + \delta n$, and $p = p_0 + \delta p$. As $\delta n = \delta p$ for the DHPs, then the charge balance relation states that

$$g_T + g_{0p} = \alpha_r np = \alpha_r(n_0 + \delta n)(p_0 + \delta p)$$
$$= \alpha_r[(n_0 + p_0)\delta n + n_0 p_0 + (\delta n)^2] \qquad (7.32)$$

where g_T is the thermal generation rate and is equal to $\alpha n_0 p_0$, and α_r is the recombination coefficient. The optical generation rate (per volume wA) can now be written as

$$g_{0p} = \alpha_r(n_0 + p_0)\delta n = \frac{\delta n}{\tau} \tag{7.33}$$

where τ is the lifetime:

$$\tau = \frac{1}{\alpha_r(n_0 + p_0)} \tag{7.34}$$

Note that we have neglected $(\delta n)^2$ for low excitation. The change in conductivity as a result of illumination can be written as

$$\delta\sigma = qg_{0p}(\tau_n\mu_n + \tau_p\mu_p) \approx qg_{0p}\tau(\mu_p + \mu_n) \tag{7.35}$$

where μ_n and μ_p are the mobilities of electrons and holes, respectively, and $\tau_n \approx \tau_p$ is used for approximation. We use mobility to specify the motion of carriers because electrons and holes in semiconductor are not free to move. They suffer multiple collisions and drift across the device in response to the applied electric field. The mobility of an electron is defined as $\mu_n = \langle v_n \rangle / E$, and $\mu_p = \langle v_p \rangle / E$, respectively. $\langle v \rangle$ is the statistical average velocity of the respective carriers. Equation (7.35) indicates that the conductivity of the semiconductor changes proportionally with the illumination. This phenomenon suggests that it can be used for detecting signals.

If Eq. (7.33) is substituted in Eq. (7.35),

$$\delta\sigma = q\delta n(\mu_n + \mu_p) \tag{7.36}$$

As the current density $J = \delta\sigma E$, and $v_n = \mu_n E$ and $v_p = \mu_p E$, therefore $J = q\delta n(v_n + v_p)$. In semiconductors, $v_n \gg v_p$. For an area A, the total current flow becomes

$$I = AJ \approx \frac{q\delta n \, Aw \, v_n}{w}$$

$$= \frac{qG_{0p}\tau}{\left(\dfrac{w}{v_n}\right)} = qG_{0p}\left(\frac{\tau}{\tau_n}\right) \tag{7.37}$$

where $G_{0p} = g_{0p}wA =$ total light flux, and $\tau_n = w/v_n$ is electron transit time.

Responsivity

The responsivity of a photoconductor is given by Eq. (7.37). It is proportional to the total light flux as expected. The other factor, the ratio of the lifetime to the transit time, is very interesting. Recombination time τ is usually much greater than electron transit time τ_n, $\tau/\tau_n > 1$, indicating a gain. The gain mechanism can be explained physically as follows: Electrons travel faster than holes and the recombination time is very long. As the electrons and holes are transported to opposite sides of the photoconductor, the electron completes its trip sooner than the hole. The external circuit must provide another electron immediately to satisfy the current continuity requirement. Thus, during the lifetime of one hole's trip across the diode, many electrons may be demanded to sustain the current flow. The charge delivered to the circuit becomes $Q = (\tau/\tau_n)q$, and the device exhibits gain.

InGaAs/InP Photoconductor Structure

A long-wavelength photoconductor is shown in Fig. 7.7. A thin layer of indium gallium arsenide, either $n, p,$ or i type, that can absorb radiation having a wavelength as long as 1.65 μm is grown on an i-type InP substrate [16, 17]. For good crystalline match, the composition has been chosen as $In_{53}Ga_{47}As$. Metallic electrodes are deposited in an interdigital pattern on the surface and alternative electrodes are connected to a power supply. In this manner, the carriers generated between the conducting layers have the shortest distance to travel before they are collected.

Even before this photoconductor is illuminated, thermally generated EHPs have been collected on the electrodes: electrons to the anode, and holes to the cathode. When short-circuited, a dark current flows in the external circuit. This dark current causes Johnson noise to become excessive in a photoconductor. When the layer is illuminated, the absorbed photons generate additional EHPs, giving rise to a signal current as a result of illumination.

Why has the photoconductor become important as a long-wavelength detector? One answer is that its structure is simple and can be integrated easily with other

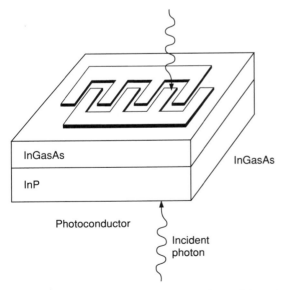

Figure 7.7   Interdigital InGaAs photoconductor. Adapted with permission from Forrest [16].

semiconductor components. But the real reason is that for detecting long-wavelength signals above 3.0 μm, no other types of detector have been found satisfactory yet.

Noise in Photoconductors

Photoconductors have many sources of noise. The dominant noise arises from the dark current. It flows whenever an external voltage is applied. This finite dark conductivity generates a randomly fluctuating noise current known as Johnson noise current. The mean square of the current is

$$\langle (i_{\mathrm{T}})^2 \rangle = \frac{4kTB}{R_c} \tag{7.38}$$

where R_c is the dark resistance of the conducting layer.

The photoconductor signal current S is equal to the gain m multiplied by the primary photogenerated current I_{ph}, which is a function of the time-averaged power of the incident light:

$$S = mI_{ph} = \frac{m\eta qP\lambda}{hc} \tag{7.39}$$

Therefore, the signal-to-noise ratio becomes

$$\frac{S}{N} = \frac{m\eta qP\lambda}{hc} \cdot \frac{R_c}{4\pi kTB} \tag{7.40}$$

Equation (7.40) shows that S/N ratio increases with increasing conductive layer resistance R_c, and gain factor m increases with the increasing ratio between hole lifetime and electron transit time. Unfortunately, hole lifetime is inversely related to bandwidth B. Increasing the gain to improve sensitivity decreases the frequency response. This seems to be the trade-off between sensitivity and frequency response as in all photodetectors.

In contrast to photodiodes, photoconductive detectors require little photon energy to operate. Often the difference potential between the band edge and the impurity level, which is usually a few tenths of a millivolt, is enough for this purpose. Consequently, photoconductors can be used as detectors well into the infrared, out to wavelengths of perhaps 30 μm where the photon energy is in milli-electron volts. This is one major advantage of photoconductors over all other types of photodetectors.

Comparison of Photodetectors

It is informative to compare the three basic types of photodetectors. The basis for comparison is the relative sensitivity or the minimum detectable power (time-averaged signal power) for a prescribed operating condition. For digital optical fiber communication, the sensitivity is usually expressed in terms of the minimum detectable signal power as a function of the bit rate for a BER of 10^{-9}. The bit error rate is

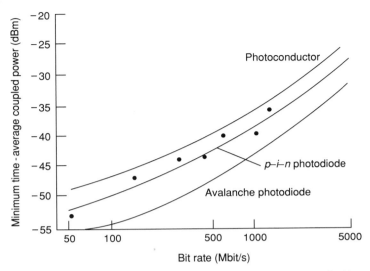

Figure 7.8 Comparison of three types of photodetectors. Adapted with permission from Forrest [17].

a measure of how often a transmitted digit "1" will be mistakenly identified by the receiver as a "0" as described by the probability function in Eq. (7.16). The value 10^{-9} is considered sufficiently low for most applications. Figure 7.8 is such a plot based on the calculated and experimental data [18] for an InGaAs structure operating at 1.3 μm. The minimum time-averaged power, rather than peak power, is specified to yield a value that is independent of the transmitted bit pattern. The unit of power is the number of decibels below 1 mW of power (dBm). Figure 7.8 shows that the avalanche photodiode is the most sensitive and is followed by the PIN, which can be used at even higher bit rates. The photoconductor is the least sensitive. This figure also shows that the sensitivity of all three types of photodiodes decreases with increasing data rate. This is because the noise of all these detectors depends increasingly on the bandwidth.

Although the PIN possesses no gain, the diode can be designed to have low background noise and high quantum efficiency. The bandwidth of an InGaAs/InP PIN may exceed 10 GHz. But it is usually limited by external effects such as the *L-C* time constant of the detector and the noise in the amplifier circuit.

The sensitivity difference between the PIN and the photoconductor is not apparent at low bit rate. There, the high Johnson noise of a photoconductor makes it significantly less attractive than the PIN for applications demanding high sensitivity. At higher bit rate, the detectors appear to be nearly comparable.

Avalanche photodiodes have internal gain, but they also tend to be very "noisy." The excess noise indicated by the factor $F(M)$ in Eq. (7.11) increases with the gain except for very small k. For $k < 1$ (as in silicon), the statistical fluctuations in the collected current are considerably smaller than in materials with $k = 1$ (as in GaAs).

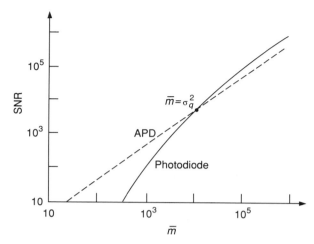

Figure 7.9 Signal-to-noise ratio (SNR) versus photon flux $\bar{m}$ for a photodiode receiver (solid curve) and an APD receiver (dashed curve). For small photon flux (circuit noise-limited case), the APD yields a higher SNR than the photodiode. For larger photon flux (photon noise-limited case), the photodiode receiver is superior to the APD receiver. With permission from Saleh and Teich [18].

At some high value of gain, the avalanche noise of an APD dominates all other noises, including the amplifier noise. A further increase in gain only reduces the signal-to-noise ratio. The trade-off occurs when the noise from an APD equals the noise from the amplifier. This gain is referred to as the optimum avalanche gain. Figure 7.9 compares a photodiode and an APD under similar operating condition. The signal-to-noise ratio is plotted against the optical power for a photodiode (solid curve) and that for an APD (dashed curve). These curves intersect at a certain level of illumination. For smaller photon flux, the APD yields a higher signal-to-noise ratio than the photodiode. For larger photon flux, the photodiode receiver is superior to the APF receiver.

The highest bandwidth (when the gain is unity) of an APD is ultimately limited by the carrier transit time across the depletion region, as in a PIN. With gain, the bandwidth is reduced, because it takes time to build up an avalanche as discussed earlier.

The best combination of bandwidth and sensitivity can be achieved with a PIN at very high bit rates, whereas the APD is useful for moderate rates between 100 Mbit/s and 4 Gbit/s.

Although photoconductors have a gain dependent on the ratio of hole lifetime to electron transit time, this gain is usually insufficient to offset the increase in dark current noise to the extent that it would render its performance superior to that of reverse-biased detectors such as the PIN and the APD. On the other hand, the simplicity of the photoconductor and its potential compatibility with electronic devices enhance its potential as a detector in future monolithic integrated optical electronic circuits. Long-wavelength applications of photoconductors have become increasingly interesting as no other diode has been found to perform as well as they do above a wavelength beyond 3.0 μm.

Extrinsic Semiconductors for Detection

Most p–n photodiodes discussed so far have used intrinsic semiconductor materials. The intrinsic long-wave cutoff of silicon is about 1 μm. Effort has been devoted to extending silicon sensitivity to the infrared, to take advantage of the mature integrated circuit technology of this material. The use of extrinsic photoconductor is one example. By introducing deep-level impurity levels within the bandgap, the absorbed photon energy may liberate electrons from the filled traps to the conduction band or fill the empty impurity level with electrons from the valence level. Thus, the trap center is used as the origin of carrier generation. One disadvantage of doping silicon for infrared area arrays is that low-temperature operation is required [19].

The spectral responses of several extrinsic photoconductor detectors are shown in Fig. 7.10. Responsivity increases approximately linearly with the wavelength, peaks slightly below the long-wavelength limit, and falls off beyond it. Quantum efficiency η_q is a function of wavelength, principally because the absorption coefficient α is dependent on wavelength. For photodetector materials of interest, α is large within a spectral window determined by the characteristics of the material. For sufficiently large λ_g, η_q becomes small because absorption cannot occur when $\lambda \geq \lambda_g = hc_0/E_g$. The photon energy is then insufficient to overcome the bandgap. The wavelength λ_g is the long-wavelength limit of the semiconductor material. Intrinsic materials have been used for detecting photon energies up to about 2 μm. For longer wavelengths, extrinsic photoconductor materials operating on transitions involving forbidden-gap energy levels have been used. Materials used in Fig. 7.10 are typical examples. The long-wavelength limits of these materials are well defined.

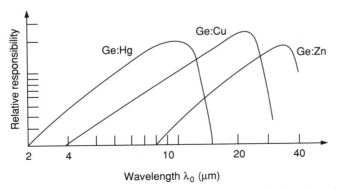

Figure 7.10 Relative responsivity versus wavelength for three doped-Ge extrinsic infrared photoconductor detectors. With permission from Saleh and Teich [18].

Schottky Barrier Photodiodes

In an attempt to improve the photodiode structure to yield better performance and higher quantum efficiency, we run across several problems. Not all semiconductors can be prepared on both the *n* type and *p* type needed to form junctions of a *p–n* photodiode; quantum efficiency of a photodiode will be low if photon-produced EHPs suffer recombination before being collected. Unwanted recombination can occur at the surface of the semiconductor diode when the incident photon energies are well above the bandgap energies and the material has a large absorption coefficient; the response speed of PINs can be slow if the diffusion current associated with photocarriers generated outside of the depletion layer is slow in diffusing across the layer and being collected. One way to improve this is to make one of junction layers very thin. But this will cause an increase in the series resistance, which, combined with the device capacitance, will cause an increase in the time constant and thus deteriorate the response time.

There are reasons to believe that Schottky barrier photodiode may be useful in this respect. Schottky barrier photodiodes are formed from metal–semiconductor heterojunctions. A thin semitransparent metallic film is used in place of the *p*-type (or *n*-type) layer in the *p–n* junction photodiode. The Schottky barrier structure and its energy-band diagram are shown schematically in Figs. 7.11A and B, respectively. The Schottky barrier height is $W - \chi$, where W is the work function of the metal and χ is the electron affinity of the semiconductor. For photon energy $h\nu > W - \chi$, diode current can flow in response to the bias voltage. For a gold film (Au) on *n*-type silicon (Si), the diode responds in the visible wavelengths. For platinum silicide (PtSi) on *p*-type Si, it operates over a range of wavelengths stretching from the near ultraviolet to the infrared [20].

For these semiconductor materials with which it is difficult to form both *n* and *p*

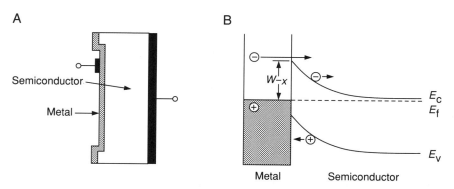

Figure 7.11 Schottky barrier diode. (A) Structure. (B) Energy-band diagram. When the photon energy $h\nu > W - \chi$, the Schottky barrier height, majority carriers flow to produce photocurrent in the external circuit. A platinum silicide (PtSi) on *p*-type silicon diode can operate over a range of wavelengths stretching from the near ultraviolet to the infrared. With permission from Saleh and Teich [18].

types, the use of Schottky barrier structure is of particular interest. The structure is so simple that only a thin layer of metal needs to be deposited. The layer can be made thin enough as to be transparent to photons; thus, the whole area of the diode becomes an effective EPH generator. Surface defects are practically eliminated, reducing the problem of noise caused by surface leakage. A thin metallic layer does not contribute to larger series resistance; thus circuit time constant is not affected.

Schottky barrier photodiodes are majority-carrier devices and therefore have inherently fast responses and a large operating bandwidth. Response time is in the picosecond region, corresponding to a bandwidth of about 100 GHz.

New Approaches for Photodetectors

Development of photodetector technology has two main goals: (1) to develop photodiodes with improved sensitivity for operation at long wavelengths and very large bandwidth and (2) to develop structures with improved performance at low cost. For high sensitivity, the avalanche photodiode is favored, provided the increased noise current at long wavelengths can be reduced.

Multiple-Quantum-Well Photodetectors

One approach is to use a multi-quantum-well structure [21]. In this design, the avalanche region is a sandwiched structure, consisting of narrow-bandgap layers such as GaAs alternating with a layer of wide-bandgap material, such as AlGaAs. The narrow-bandgap layers function as quantum wells in confining carriers. This structure creates a boundary through which the holes and electrons can have markedly different ionization coefficients and thereby reduces the noise. This is because the edges of the valence and conduction bands have been offset at the interface, a feature that enhances the ionization coefficient of the electrons dramatically. The energy-band diagram of an AlGaAs/GaAs MQW APD is shown in Fig. 7.12A under reverse-bias condition. Another advantage of this device is the expanded bandwidth resulting from the concomitant reduction in avalanche buildup time. Figure 7.12B shows a staircase APD for use at longer wavelength under zero bias (top) and reversed bias (bottom) [21].

Unfortunately, this technique does not render an easy solution when InP material is used, where long-wavelength operation is desired. For the range 0.8 to 0.9 μm, silicon works well. Unfortunately, growth of such structures for use at long wavelengths is extremely difficult as molecular beam epitaxial growth of InP-based components is still at an immature stage of development. Research is underway to find new materials and techniques for long-wavelength applications.

Integrated Photodetector Array

To develop a structure with good detector performance and low cost, one returns to the PIN structure. But this time with the help of integrated electronics technology,

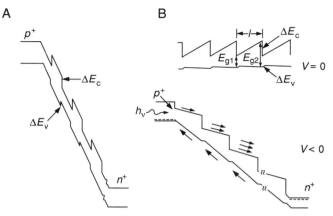

Figure 7.12   Energy diagram of an (A) AlGaAs/GaAs APD under reverse bias and (B) a stair-case APD for use at longer wavelength under zero (top) and reverse bias (bottom). After Chin *et al.* (1980). *Electron. Lett.* **16**, p. 467.

one can incorporate a PIN diode with a metal semiconductor field-effect transistor (MESFET) or transistor [22, 23]. The replacement of a conventional discrete photo-diode and field effect transistor by its integrated counterpart reduces the capacitance of the photodiode, thus increasing the bandwidth. The integrated approach increases sensitivity and reduces cost. For high frequencies, however, the field effect transistor should be replaced by a bipolar transistor. Figure 7.13 shows the schematics of an integrated PIN photodiode and a field effect transistor.

An array of photodiodes have been constructed to increase the bandwidth. A 12-photodiode array is sketched in Fig. 7.14 [24, 25]. Each diode is addressed through an optical fiber. The array may be used in series or in parallel. The effective bandwidth with a 12-element array, each operating at 50 Mbit/s, would give a total bandwidth of 600 Mbit/s. The limiting factor is interchannel cross-talk induced by capacitive coupling between nearby elements. By integrating the entire receiver front end with each photodiode element, this problem can be somewhat alleviated. Again, it is still too early to apply this technology to long-wavelength materials, such as indium phosphate-based materials. Arrays of more than 50 PIN photodiodes have been reported.

Hybrid Photodetectors

A simple, inexpensive yet effective way to improve the performance of photodetec-tors is by means of hybrids. A photodetector hybrid consists of one or more photo-diodes, amplifier transistors, or integrated circuits, all mounted in a single package. The integration of photodiodes and amplifiers minimizes the lead connections be-tween the components, reduces the parasitic capacitances, and thus improves S/N and response speed. The chief difficulty at present seems to be that it is not yet realistic

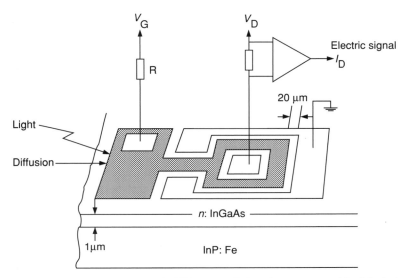

Figure 7.13 A *p–i–n* and MESFET integrated photodetector assembly. Adapted with permission from Leheny *et al.* [21].

to build PINs (usually on Si substrate) and amplifiers (on GaAs substrate) on the same semiconductor substrate. Also, even if silicon amplifiers were used, photodiodes and operational and charge-sensitive amplifiers are each fabricated from very

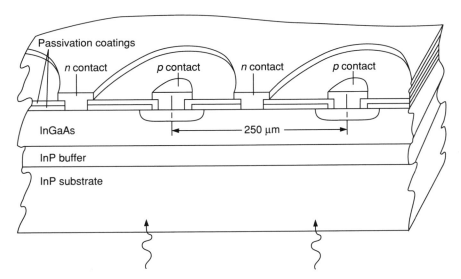

Figure 7.14 Linear array of *p–i–n* photodiodes. Adapted with permission from Ota and Miller [23].

different types of intrinsically doped silicon substrate. Hybrid structure may still be used for practical applications. Many designs of these hybrid photodetectors are on the market.

An example of the use of the high Schottky barrier on AsInAs in an optical device has been reported that has achieved very good performances by using gradual quaternary layers between the Schottky gates and the photoconductive region [26].

Summary

In principle, detector technology has not changed in the past 15 to 20 years. Basically, three types of optical detectors—p–i–n photodiodes, avalanche photodiodes, and photoconductive diodes—are still the mainstream photodetectors in use. For long-wavelength detection, the photoconductive diodes are preferred; however, enormous strides have been made in developing optical detectors for lightwave communications systems providing an increase in sensitivity to weak signals, improving bandwidth and response time, and lowering the cost of production of the photodetectors. New materials and methods of epitaxial growth for long-wavelengths applications rank high in the research effort. Sophistication in the advanced design of photodetectors has turned into multidimensional arrays detectors. Broadening the spectrum range, from ultrashort to long-infrared wavelengths also has become a common goal.

In conclusion, Table 7.1 lists different types of commercially available photodetectors together with their properties. Only quantum detectors and those limited to

TABLE 7.1
Quantum Detectors

Type[a]	Range (μm)	Responsivity (A/W)	NEP (W/$\sqrt{\text{Hz}}$)	Threshold (W/cm^2)	Frequency response at 3 dB	R_s (Ω)	Array size
GaAs (S)	0.4–0.9	0.2		5 mW	To 60 GHz		
Ge (APD)	0.8–1.8	0.8	1×10^{-10}		To 2 GHz		
Ge (PC)	0.5–1.8	1.0	1×10^{-13}	1–10 mW	To 50 MHz	10^6	1×128
InGaAs (PIN)	1–1.7	0.6–0.9	1×10^{-14}	100	To 5 GHz	10^7	1×256
InGaAsP	0.9–1.65	0.6	1×10^{-12}		10 MHz		1×16
InGaAs	0.4–1.6	0.2			25 GHz		
PbSe	1–4.8		1.5×10^{-11}	10^{-3}	0.1–15 kHz		1×256
PbS	1–3	0.1–2	1.5×10^{-11}	10^{-3}	0.2–0.8 kHz		1×256
HgCaTe (PC)	2–20	0.002–5	5×10^{-12}	100	400 MHz	10^5	1×640
Si (APD)	0.4–1.1	120	1×10^{-15}	100	2 GHz		1×32
Si (PIN)	0.2–1.2	0.6	1×10^{-16}	4	2 GHz		1×1000

[a] S, Schottky barrier diode; PC, photoconductor; PIN, p–i–n photodiode; APD, avalanche photodiode.

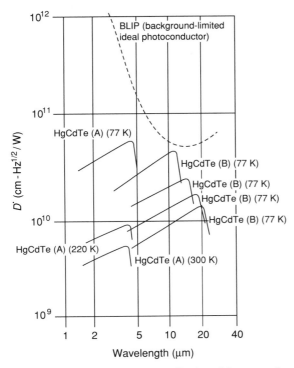

Figure 7.15 Representative specific detectivity curves for various mercury–cadmium telluride detectors versus wavelengths at several temperatures. Adapted from product data from Infrared Industries and New England Photoconductor.

room temperature operations are listed. The effect of cooling in some detectors is shown in graph form in Fig. 7.15.

References

1. H. Melchior, Demodulation and photodetection techniques. In *Laser Handbook* (F. T. Arechi and E. O. Schulz, Eds.), pp. 725–835, North-Holland, Amsterdam, 1972.
2. F. N. H. Robinson, *Noise in Electrical Circuit,* Oxford Univ. Press, London, 1962.
3. S. R. Forrest, M. DiDomanice, J., R. G. Smith, and H. J. Stocker, Evidence for tunneling in reverse-biased III–V photodetector diodes. *Appl. Phys. Lett.* **36,** 580–582 (1980).
4. S. D. Personick, Receiver design. In *Optical Fiber Telecommunications* (S. E. Miller and A. G. Chynoweth, Eds.), pp. 627–651, Academic Press, New York, 1979.
5. H. P. Yuan, Nonclassical light. In *Photons and Quantum Fluctuations* (E. R. Pike, Ed.), pp. 1–9, Adam Hilgen, Bristol/Philadelphia, 1988.
6. A. Heidman, Generation of nonclassical light. In *ECOOSA 90—Quantum Optics* (M. Bertolotti and E. R. Pike, Ed.), Institute of Physics Conference Series Number 115, pp. 53–60, Institute of Physics, Bristol/Philadelphia/New York, 1991.

7. T. P. Persall and M. A. Pollack, Compound semiconductor photodiodes. In *Semiconductors and Semi-metals* (W. T. Tsang, Ed.), Vol. 22, pp. 174–245, Academic Press, Orlando, FL, 1985.

8. R. J. McIntyre, The distribution of gains in uniformly multiplying avalanche photodiode theory. *IEEE Trans. Electron. Devices* **ED-19**, 702–713 (1972).

9. R. G. Smith and S. D. Personick. Receiver design for fiber communication systems. In *Semiconductor Devices for Optical Communication* (H. Kressel, Ed.), Chap. 4, p. 89, Springer-Verlag, Berlin, 1982.

10. G. E. Stillman, L. W. Cook, G. E. Bulman, N. Tabatabaie, R. Chin, and P. D. Dapkins, Long wave-length (1.3 to 1.6 μm detectors for fiber-optical communications. *IEEE Trans. Electron. Devices* **ED-29**, 1355 (1982).

11. Y. Suematsu, Long-wavelength optical fiber communication. *Proc. IEEE* **71**, 692 (1983).

12. T. P. Lee and T. Li, Photodetectors. In *Optical Fiber Telecommunications (S. T. Miller and A. G. Chynoweth, Eds.), pp. 593–626, Academic Press, New York, 1979.*

13. G. E. Stillman and C. M. Wolfe. In *Semiconductors and Semimetals. Vol. 12. Infrared Detectors II* (R. K. Willardson and A. C. Beer, Eds.), p. 291, Academic Press, New York, 1977.

14. T. Takano, T. Iwakami, I. Mito, and Y. Tashiro, Optical Fiber Conference Technical Digest, Paper WB5, 1988.

15. N. Susa, H. Nakagome, D. Mikami, H. Ando, and H. Kanbe, New InGaAs/InP avalanche photodiode structure for the 1.0–1.6 μm wavelength region. *IEEE J. Quantum Electron.* **QE-16**, 864–870 (1980).

16. S. R. Forrest, Optical detectors: Three contenders. *Spectrum* **23**, 76 (May 1986).

17. S. R. Forrest, The sensitivity of photoconductor receiver for long wavelength optical communications. *IEEE J. Lightwave Technol.* **LT-3**, 347–360 (1985).

18. B. E. A. Saleh and M. C. Teich, *Fundamentals of Photonics* (J. W. Goodman, Ed.), Wiley Series in Pure and Applied Opticals, Chap. 17, p. 656, Wiley, New York, 1991.

19. S. M. Sze, *Semiconductor Devices: Physics and Technology,* Wiley, New York, 1985.

20. F. Capasso, In *Semiconductors and Semimetals. Vol. 22D. Lightwave Communication Technology* (W. T. Tsang, Ed.), Chap. 1, p. 166, Academic Press, New York, 1985.

21. R. F. Leheny, R. Nahory, M. Pollack, A. Ballman, E. Beeke, J. DeWinter, and R. Marin, Integrated InGaAs p–i–n FET photoreceiver. *Electron. Lett.* **16**, 353–355 (1980).

22. J. C. Renaud, N. L. Nguyen, M. Allovon, P. Blanconnier, S. Vuye, and A. Scavennec, GaInAs mono-lithic photoreceiver integrating p–i–n/JFET with diffused junctions and a resistor. *IEEE J. Lightwave Technol.* **6**, 1507–1512 (1988).

23. Y. Ota and R. C. Miller, "12-channel PIN and LED arrays and their package for 1.3 μm applications. *Proc. SPIE—Int. Soc. Opt. Eng.* **839**, 143–147 (1987).

24. M. G. Brown, S. R. Forrest, P. H-S. Hu, D. R. Kaplan, D. Koza, Y. Ota, J. R. Potopowicz, C. W. Seaburg, and M. A. Washington, IEDM Technical Digest, Paper 31.5, 727, 1984.

25. Y. Le Bellego, J. P. Praseuth, and A. Scavennec, AlInAs/GaInAs SAGM-APD. In *Proceedings, Third International Conference on InP, Cardiff, April 1991,* Paper WN5.

26. A. Temmar, J. P. Praseuth, and A. Scavennec, AlInAs/GaInAs MSM photodiodes with graded AlInAs transition regions. In *Proceedings, Sixth EURO-MBE Conference, Tampere, Finland, April 1991.*

CHAPTER 8

Optical Amplifiers

Introduction

Electronic amplifiers have been successfully used in telecommunication engineering for many decades. The design technique for such amplifiers has been perfected to take care of almost every detail to satisfy the system requirements. Why the optical amplifier? To answer this question, let us look at a conventional system designed for a long-haul optical fiber repeater station employing wavelength division multiplexing (WDM). (Fig. 8.1A). Here, an input fiber carrying signals from many wavelength channels is led to a wavelength demultiplexer in which each wavelength channel is separated and individually processed as indicated in Fig. 8.1B. After processing, the channels are recombined in a multiplexer and coupled back into the fiber for further transmission. The electronic circuitry required, as shown in Fig. 8.1B, is very complicated. For each channel, the signal at that wavelength is first detected, amplified, and processed. The processing includes pulse slicing, retiming, and reshaping. The output is used to drive a laser of the exact wavelength for that channel. The amplified signal from each laser then rejoins the group in a multiplexer for further transmission. Optoelectrooptic circuitries as well as the laser regenerator are duplicated for each channel.

Would it be better if a single optical amplifier with enough bandwidth could be designed to replace all these circuitry? An optical amplifier is the single device designed to do what has been specified by the preceding question. When it is used in conjunction with the transmission system, no additional optoelectric and eletrooptic conversion devices are needed. Pulse retiming and reshaping are not required. A regenerative laser circuitry is not needed. The same amplifier can be used simultaneously to amplify the signals at different wavelength channels and even with different bit rates. That is, every piece of the devices between and including the demulti-

A

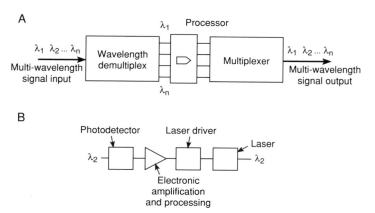

B

Figure 8.1 Setup for a WDM repeater station. (A) Conventional setup. (B) Detail of the processor.

plexer and the multiplexer of Fig. 8.1A can be replaced by a single, simple optical amplifier. It is the purpose of this chapter to describe recent progress on optical amplifiers designed for just this purpose. An optical amplifier can make up the losses suffered by signal transmission in lines and coupling devices; however, it does not solve the problem of signal dispersion, the other factor that limits the distance that can be covered by a useful transmission system. For a long-haul transmission system, a series of optical amplifiers can be connected in tandem to overcome the losses. But signal dispersion is cumulative and must be treated separately.

In this chapter, we describe two general types of optical amplifiers: the semiconductor type and the parametric type.

Semiconductor Optical Amplifier

A schematic diagram of a semiconductor optical amplifier is shown in Fig. 8.2. It is similar to a diode laser structure without the resonator [1–3]. The structure has an epitaxially grown active layer surrounded by layers of high-energy-bandgap semiconductors. The end facets are coated with antireflective materials so that laser oscillation is prevented. Input optical fiber is coupled to the active layer at one end and

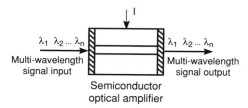

Figure 8.2 Optical amplifier.

output fiber is coupled to the other end. As the diode is forward biased, injected carriers confined to the active region will reach a high density, such that an input signal will induce stimulated emission as in a laser. But unlike laser operation, the active layer now serves as a waveguide for the light to propagate along the device without reflecting back and forth from the end facets. Amplification is the result of a single-pass gain. As the physical dimension of the active layer is large (250–500 μm) compared with the light wavelength (1.3–1.56 μm), light travels along the waveguide many wavelengths and gains energy from each stimulated emission inbetween. Thus, this type of amplifier is also called a traveling-wave amplifier (TWA). In fact, an ordinary laser (or the resonant or Febry–Perot amplifier, FPA) can also be used as an optical amplifier by driving it to just below the threshold current. Simply coating the end facets of an ordinary laser with antireflection coating to prevent oscillation will do the same. But in the latter case the gain–bandwidth product may be relatively small, and the bandwidth will be uncomfortably narrow for many applications [4].

Gain

The achievable gain g of a traveling-wave-type semiconductor optical amplifier is determined by the amplifier material, the structure, and the driving conditions [5]. It has the general form

$$g = \exp\{L[\Gamma g_0(N - N_0) - \Gamma \alpha_1] - \alpha_2 L\} \qquad (8.1)$$

where L is amplifier length; g_0 is the linear gain coefficient, device parameter; N_0 is the carrier density required for transparency; N is the carrier density in the active region, related to the drive current; Γ is the mode confinement factor, determined by the amplifier waveguiding characteristics; α_1 is the optical loss in the input of the waveguiding region; and α_2 is the optical loss in the output of the waveguiding region.

Equation (8.1) indicates that the choice of material for the amplifier affects the gain in two places: g_0 and N_0. Any material suitable for making a laser is also adequate for making an amplifier. Structurally, as in the laser, the importance of the confinement factor Γ is obvious. But in amplifiers, the mode control requirements used in lasers need not be applied as strictly. The length of the structure L affects the time of interaction between the light wave and the stimulating energy source as is expected. In practice, it is difficult to maintain a uniform and waveguiding property of a long structure. Thus, the length of the amplifier structure is usually limited to about 500 μm. The fact that we have to pay a high price in gain for the losses in end facets is unfortunate. Ideal facets present no noise. As we will see later, the facet's reflectivity also affects the noise characteristics of the amplifier.

The dependence of gain on Γ gives rise to another problem important to the operation of an optical amplifier, that is, the dependence of gain on polarization. In a nonsymmetric structure like these used in ordinary injection lasers, light waves of orthogonal polarizations, TE and TM waves, although they have the same propagation constant, react differently with Γ, the confinement factor. It is significantly larger for light polarized in the junction plane (TE) than for light polarized in the orthogonal polarization (TM). Thus, the gains for the TE and TM polarized waves may be different. This problem can be minimized by redesigning the amplifier structure with symmetric waveguides rather than using laser structures. Better antireflection coatings on the facets may also help to improve the polarization dependency problem [6].

In Fabry–Perot amplifiers with low reflective facets, a gain g as high as 30 to 35 dB for a 500-μm-long device has been reported [3]. The avaliable power at the output must account for the power losses in the input and output couplers, which could amount to 3.5 dB each. Permanent couplers are recommended. The small-signal gain spectrum of an optical FPA amplifier is very broad, of the order of several thousand gigahertz. This is because as a fraction of the light is reflected back into the active gain region, it gives rise to a series of peaks corresponding to the longitudinal modes of the amplifier.

In an ideal TWA, because of its single-pass property, the gain spectrum has no resonances and therefore is very smooth. The requirement on the facet reflectivity for achieving traveling-wave operation is also strict. Usually, a reflectivity of less than 6×10^{-5} is required. Moreover, antireflection coatings are optimized for a particular wavelength only. An amplifier will in general operate in the entire range of wavelengths of the gain spectrum.

The gain is linear for low to moderate input light power. At high input light power, the amplifier gain is reduced as a result of nonlinear effects. It has been observed that the gain is saturated and compressed at high powers. Gain saturation is a common phenomenon in laser amplifiers as the rate of stimulated emission increases. This topic is discussed further under Parametric Optical Amplifiers.

Amplifier Noise

Like any other amplifier, semiconductor optical amplifiers generate and amplify noise that appears at the output along with the signal. The noise can be traced to two origins: spontaneous quantum fluctuations and facet reflectivity. Noise resulting from quantum emission is wavelength dependent in favor of longer wavelengths and can mostly be filtered out using optical filters. Facet reflectivity noise can be reduced using special antireflecting coatings.

Each type of noise has its spectral density which depends on the amplifier quality. In coherent detection, the different portions of the noise spectrum may frequently mix with each other to cause additional noise known as *beating* noise [7]. In a specially designed single-pass optical amplifier, quantum noise is usually small compared with that from imperfect end facets. Because of the finite residual reflectivities of the facets, the gain function may become modulated by the Fabry–Perot resonance mode. The depth of modulation is determined by the product of the gain and the root mean of the product of the end facet's reflectivities. To limit the noise contribution from this source, it is often necessary to reduce the reflectivity to less than 2×10^{-4} [8]. Special antireflection coatings have been developed to minimize noise generation of this type.

Signal-to-Noise Ratio

In a communication system, it is the signal-to-noise ratio (SNR) that is more important than absolute noise. The SNR is the power ratio of signal output to that of the noise at the same output. It should include detector noise as well as receiver noise. For this discussion, let us concentrate only on amplifier noise. If the spectral distributions of noise powers are assumed to be statistically independent of each other, then noise powers are additive. Therefore the total noise power must be used to calculate the SNR.

Noise power is usually given in terms of the product of the spectral power density and the noise-equivalent bandwidth of the amplifier. The spectral power distribution of quantum emission noise can be represented by $H_{aq} = Kh\nu$, where K is a material constant and is dependent on operating conditions of the amplifier, and $h\nu$ is photon energy. The noise power resulting from the quantum noise is $P_{aq} = Kh\nu B$, where B is bandwidth. For reflectivity noise, $P_{ar} = H_{ar}B$, where H_{ar} becomes a function of the gain, R_1, and R_2. R_1 and R_2 are the reflectivities of end facets 1 and 2 of the amplifier, respectively. Thus,

$$\text{SNR} = \frac{P_0}{[P_{aq} + P_{ar}]} \tag{8.2}$$

Marshall *et al.* presented an interesting calculation describing the use of a series of optical amplifiers in a long-distance transmission system to replace repeaters between optical fiber sections [9]. Let L_t be the total length of the transmission system and L the distance between repeaters; the number of amplifiers is then L_t/L. If the gain of each amplifier can just compensate for the fiber loss in each section, then the total optical power at the output is $P_0 = P \times 10^{-\alpha L}$, where α is the fiber loss in decibels per kilometer. If an amplifier with perfect end facet reflectivities ($R_1 = R_1 = 1$) is assumed, so that only quantum fluctuation noise exists,

$$\text{SNR} = \frac{P \times 10^{-\alpha L}}{\{[L_t/L]Kh\nu B\}} \tag{8.3}$$

Assume $\alpha = 0.25$ dB/km, and let $P = 1$ mW, $L = 100$ km, $B = 1$ Gb/s, and $K = 4$ ($K = 1$ is an ideal case). Then for a SNR of 13 dB, one can calculate from Eq. (8.3) the total line length L_t. It is about 100,000 km, or four times the earth's circumference! This, of course, is overly optimistic. The noise power caused by end facet reflectivity may prove to be of more importance. Actually, the distance may have been limited by fiber dispersion, a factor not yet considered. The point we wish to bring out here is that, if one-tenth of this optimistic value is realizable, say 10,000 km, it is far enough to span the Pacific ocean without using a single repeater. According to the present rate of progress in this field, it may not be far away.

A study has been made on the use of semiconductor optical amplifiers in conjunction with photonic space-division switching system to enhance the line capacity of a transmission by a thousandfold [10].

Applications

In lightwave communication systems, optical amplifiers are used in three different ways: as in-line amplifiers, as power amplifiers, and as receiver preamplifiers. The optimum design of the amplifier depends on the application.

For in-line applications, the amplifier operates as a simple repeater which provides a broadband gain to compensate for fiber loss between the transmitter and receiver. For long fiber line, a number of amplifiers can be cascaded before noise buildup in the chain becomes unacceptable. Assume that the system is to maintain a 10^{-9} bit error rate as a result of the added amplifier noise and the system's penalty is less than 1 dB. With an amplifier input of -20 dBm, up to 300 amplifiers can be cascaded. For a fiber loss of 0.25 dB/km and a net gain of 15 dB per amplifier, the allowable transmission distance would be 18,000 km. This is overly optimistic, as system dispersion would overtake long before this distance is reached.

For power amplifiers, large-output saturation power is the most important parameter. With multi-quantum-well (MQW) amplifiers, a saturation output power in excess of 50 mW has been reported [11]. Although polarization dependence is a problem in semiconductor optical amplifiers in general, it is not as critical in power amplifiers. The amplifier can be connected to the transmitter with a polarization-preserving pigtail. Mikami has shown that by introducing stress into the active region of a MQW amplifier, the TE and TM gain characteristics can be modified to give polarization-independent gain over some wavelength range [12].

For preamplifier applications, the optical amplifier is used to boost the optical signal immediately before the photodetector. The important parameters are the noise figure and the input coupling efficency. Receiver sensitivity is directly proportional to both. To reduce the effect of thermal and excess noise, the optical preamplifier is usually fitted with isolators and a grating filter to limit the amount of spontaneous emission noise from the receiver.

There are some strong points in favor of semiconductor amplifiers. They can be operated at any wavelength of interest, can be integrated with other semiconductor devices, and may consume less power than other amplifiers [13].

Although development of the Er-doped fiber amplifier (EDFA) may have overshadowed the semiconductor optical amplifier, we will still find applications for the semiconductor optical amplifiers in signal processing, and in other wavelength ranges where erbium-doped fiber amplifiers cannot be used.

Parametric Optical Amplifiers

People working in the microwave field may still remember the day when the parametric amplifier was first announced. A low-noise solid-state amplifier without refrigeration has been successfully exploited as a preamplifier of a microwave receiver in radio astronomy [14]. Unlike other electronic amplifiers whose energy is converted from a dc supply, the parametric amplifier, or paramp, derives its energy from another ac source of slightly different frequency as the signal. Usually, at least three frequencies are involved in a paramp. If the input and output are using the same frequency f_s, a pump frequency f_p and another idler frequency f_i are involved so that

$$f_i = f_p \pm f_s \qquad (8.4)$$

Usually, the difference frequency is preferred.

In general, there are two classes of paramp, the degenerate and nondegenerate. In a degenerate paramp the signal frequency f_s is close to the idler frequency $f_i = f_p - f_s$, so that in practice, the pump frequency is about twice the signal frequency. The output of this class of paramp contains more noise because both signal and idler frequencies contribute to the noise. In a nondegenerate paramp, the signal and idler frequencies are widely separated. This type is more common because, if the idler circuit is properly terminated, it contributes little noise. The total noise is then that from the signal circuit only. The gain of the amplifier is derived from the added pump energy. In the language of electric circuit engineering, with a current flow in the idler circuit, a negative resistance is presented to the output at the signal frequency. Thus, energy is fed into the output to amplify the signal. A more physical explaination of this effect will be given later.

For optical amplifiers, at present we have at least two types of paramps: the Raman scattering paramp and the erbium-doped fiber paramp. Each type is briefly discussed in the following sections.

Stimulated Raman Scattering Paramp

In the discussion on nonlinear effects in laser operations, we have been cautioned that the Raman scattering effect (commonly appears when strongly excited) is undesirable and should be avoided. Nonlinear interactions may cause unwanted effects in optical amplifiers, such as frequency conversions and limitation of power-handling capability of the amplifier. But constructive use of the Raman effect to amplify a signal was studied as early as the 1960s. Stolen and Ippen in 1973 proposed a practical optical amplifier using this effect [15]. To operate the amplifier, a pump of frequency slightly higher than the signal frequency is coupled to the input fiber through a directional coupler. In this case, it becomes imperative that the idler frequency, which is the difference between the pump and signal frequencies, correspond to the vibrational frequency of the molecules of the glass. For an operation at 1.55 μm, the pump must operate at 1.45 μm. The molecular vibrational frequency of silica glass is then centered at this idler frequency. The energy of this frequency difference corresponds to 450 cm^{-1}. This unit is convenient as it corresponds to v/c, where c is the velocity of light *in vacuo*. It represents the frequency difference in units of c. This unit of energy was introduced in Chapter 2.

Raman scattering can be either spontaneous or stimulated. Spontaneous Raman scattering occurs when electrons and holes recombine spontaneously, constrained only by the condition of thermodynamic equilibrium, and is independent of the number of photons that may already be in the mode. In the presence of a photon in the upper energy level of specified frequency, however, the direction of propagation and the defined polarization may stimulate the scattering of a duplicate photon with precisely the same characteristics as the original photon. This is called stimulated Raman scattering. Spontaneous Raman scattering can cause noise problems in an optical amplifier and is discussed later. Only stimulated Raman scattering (SRS) is useful for optical amplification.

There are advantages to SRS: (1) Like semiconductor amplifiers, SRS paramps can be used to replace repeaters in a long-haul transmission system. A SRS paramp uses the optical fiber as a medium. It does not require another device like a semiconductor amplifier. Only a directional coupler is needed to couple a pump onto the fiber. Thus, fewer components are required for the system. (2) The bandwidth of the SRS amplifier is quite broad. When applied to a WDM system in fiberoptic transmission, a single amplifier is enough to cover several channels. In other cases, several pumps, each tuned to a slightly different idler frequency, can be combined to cover the whole wide bandwidth of the system. (3) A SRS paramp is bidirectional; that is, it works in either the forward or backward direction.

The disadvantage of the SRS amplifier is that it usually takes a fair amount of power to operate a Raman pump, at present, on the order of 30 to 100 mW. SRS paramps are also troubled by noise, mainly as the result of the presence of stimulated Brillouin scattering (SBS) and amplified spontaneous emission (ASE). SRS is also sensitive to polarization.

Gain

The gain mechanism of a SRS amplifier involves photon–phonon interaction of the medium with high polarizibility. The theory is beyond the scope of this book. A short physical explaination is given here.

A SRS paramp achieves its gain within the transmission medium. In silica optical fibers, we wish to place the signal at the minimum loss point of the fiber, at 1.55 μm. Now the pump requires a frequency slightly higher than the signal frequency, so that the idler frequency, $f_i = f_p - f_s$, corresponds to the vibrational frequency of silica. For silica fibers, the vibrational frequency corresponds to a differential frequency of about 450 cm^{-1}. A pump at 1.45 μm will satisfy this requirement. The pump acquires energy hf_p. It is then divided into two parts; one part maintains the idler and the other part, hf_s, goes to amplifying the signal, making the signal grow exponentially with length (exp γL). The Raman gain coefficient γ is a function of the optical power, the dielectric constant, and the nonlinear polarizability of the medium.

Noise

The noise sources of a SRS paramp are the spontaneous Raman scattering, the competing stimulated Brillouin scattering, and self-induced phase modulation.

Brillouin scattering occurs when an optical fiber is intensively excited. When the dielectric medium is exposed to strong light, molecular vibrations are excited to a frequency much lower than the pump frequency. The excited modes act like grating for the pump frequency and cause light waves to be diffracted accordingly. If the resultant frequencies fall into the range of the signal, competing with the Raman scattering effect, this could unwanted noise can be produced in the amplifier.

Self-induced modulation is caused by the change in phase resulting from the non-linearity of the refractive index of the fiber, which affects the TE and TM modes of transmission differently and gives rise to polarization of the medium. The resultant electric field contains many frequency components, including the signal frequency and other high-order modulation components. The component at signal frequency then modulates the signal again, producing self-induced phase modulation. The effect of self-modulation is particularly noticeable when the driving pump power is larger than the critical power required for stimulating Raman scattering. Analytically, it has been shown that if fiber losses are included in the wave equation, the effect of self-induced phase modulation becomes evident [16]. Self-induced phase modulation is detrimental if the original signal was phase modulated.

Another kind of noise may arise if a SRS paramp is used to amplify a WDM system containing at least two channels. If the wavelength difference between these channels corresponds to the vibrational frequency of the medium, the signal power of one channel may act as a pump power for the other channel. It also supplies the power consumed in the idler. The interplay of powers increases the output of one channel at the expense of the other. The channel that gains power may increase its output slightly, but the channel whose output is reduced suffers because its bit error rate increases rapidly, thus reducing the SNR. Coupled-mode theory has been developed to analyze this problem [17].

Brillouin scattering effect and self-induced phase modulation noises can be suppressed by increasing the spectral width of the source and by adjusting the proper pump power level. It is spontaneous Raman scattering noise that is more troublesome. Because it is a random process, its noise spectrum is broad. Beside, it is amplified just as the signal. Amplified spontaneous emission may be of comparable magnitude as the signal at low input level.

For pumping a Raman amplifier, a YAG-pumped CW 1.45-μm color center laser is used.

Erbium-Doped Paramp

Rare-earth-doped fiber can serve as useful component of paramps. Rare-earth ions, such as Nd, Er, Yt, Pr, and Sm, are used to dope silica fibers that serve as single-mode waveguides of optical transmission systems. These doped fibers can also be used as amplifiers. They offer polarization-independent gain and low insertion loss. For silica fibers operating near their minimum fiber loss at wavelength $\lambda = 1.55$ μm, erbium (Er^{3+} ions) doping is particularly attractive. This is because the emission and absorption spectra of erbium ions fit the transmission characteristics of silica fibers perfectly. Erbium offers a broad laser transition frequency ($\nu = 4000$ GHz) at room temperature and can be pumped by semiconductor laser diodes. The most interesting pump wavelengths are 0.81, 0.98, and 1.48 μm. When erbium is pumped at 0.81 μm, undesirable excited-state absorption (ESA) can be developed; this requires high pump power to achieve a gain. High pump power is usually a cause of lower lifetime for the pump laser. Other pumping wavelengths have been tried. When the fiber is pumped directly at 1.48 μm by the light from an InGaAsP semiconductor laser, a gain from 30 dB to 45 dB is achieved for a 50-m-long fiber doped with about 300 ppm of Er_2O_3; an optical bandwidth of 30 nm has been reported [18]. Much less pump power is required for this scheme. The width of the emission spectrum, and therefore the bandwidth of the paramp, can be broadened by the addition of Al_2O_3 to the glass [19]. Sometimes, an index-raising codopant with germanium (Ge) is also useful. Later, the use of a 0.98-μm pump wavelength is described.

The rare-earth ion Er^{3+} has a trivalent electronic structures, whereas silica glass has a covalent structure. This makes their alloying in glass very difficult, or, we say that the solubility of erbium ions in silica glass is very low. Addition of Al_2O_3 to silica glass alleviates the problem. It also improves the host-glass environment to enable incorporation of several degrees of doping concentration to favor different applications.

The results reported on erbium-doped fiber paramps excited the communication world even more when it was learned that less power is required to pump a paramp and less noise is produced by this paramp compared with a SRS paramp. In fact, beginning in 1989, erbium-doped fiber amplifiers became the central research and developmental topic of optical amplifiers.

Gain and Saturated Gain

When pumped at 0.98 μm, the Er-doped optical fiber paramp makes use of a three-level pumping scheme (Fig. 8.3). The ground state is level 1 ($E_1 = 0$). Above that, level 2 is long-lived (τ_{21} is large), and level 3 is short-lived (τ_{32} is small). A pump supplies the energy to raise the population from level 1 to level 3 (e.g., by absorbing radiation at the frequency E_{31}/h), which decays immediately to level 2 nonradioactively. As the population N_2 at level 2 builds up because of its long lifetime τ_{21}, population inversion is established; that is, $N = N_2 - N_1 > 0$. Each signal photon of frequency $(E_2 - E_1)/h$ that travels in the z direction will induce a stimulated emission to duplicate itself on relaxing back to level 1. The released energy is added to the original signal photon and amplifies it. As the ground level is normally heavily

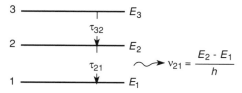

Figure 8.3   Three-level energy scheme.

populated, large pump power is required to achieve the condition of population inversion. When pumped at 1.48 μm, the Er ion is excited directly into the upper laser level (level 2), which means that the amplifier acts as a two-level laser system and that the inverted population density N_2 could be limited to a maximum value determined by the ratio of emission to absorption cross sections. We return to compare these pumping schemes in a later section.

Gain is proportional to the population density $N (= N_2 - N_1)$. Other factors that contribute to gain include the material constants such as the transition cross section, spontaneous lifetime, and lineshape function of the material at that wavelength. The overall gain of the amplifier can again be represented in exponential form as in SRS amplifiers with different γ values. We present the gain characteristics of the amplifiers in graphic form in Figs. 8.4 to 8.6.

The small signal gain, that is, the gain at low pump power, increases as the pump power increases as shown in Fig. 8.4, where unsaturated signal gain is plotted against pump power. For a finite pump power, the gain is almost directly proportional to the fiber length as is expected. The gain for a longer fiber section is larger because the fiber has a longer interaction length to acquire power. The power required to achieve gain for a shorter length of fiber is lower than that for a longer one; however, the gain of the amplifier increases with the amplifier length at first, reaches a maximum, and decreases with further increase in length as shown in Fig. 8.5.

The input/output characteristics of a typical erbium-doped fiber amplifier (EDFA) are shown in Fig. 8.6 [20]. Three basic arrangements are shown: the forward, backward, and bidirectional schemes. Note that the output power is not always directly proportional to the input power, and it shows gradual saturation. Power saturation is inherent to laser amplifying systems. This is because the population inversion density N, which implements the amplifier gain, is not a simple function of the pump energy that increases with it proportionally. If can be shown by solving the rate equations, even in a two-level energy system, that N, after reaching the stimulated emission level, decreases as pump power further increases. We use a two-level system to illustrate this point because it is simple to handle.

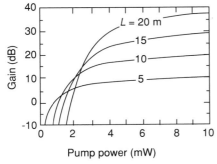

Figure 8.4 Calculated gain versus pump power for various L. Adapted from Giles *et al.* [20], with permission.

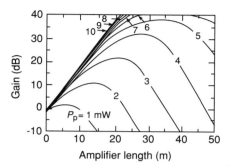

Figure 8.5 Calculated gain versus L for various pump powers. Adapted from Giles *et al.* [20], with permission.

For a two-level system, in the absence of saturation, $N = N_0$, where N_0 is the steady-state population density difference. Once the stimulated emission starts, N is modified as

$$N = \frac{N_0}{(1 + \tau_{21}W_i)} \tag{8.5}$$

where τ_{21} is the lifetime of the transition from level 2 to level 1 and W_i (in s^{-1}) is the probability densities of stimulated emission at this level.

Thus, as pump power increases, output power at first increases. As W_i increases, N becomes smaller. It soon reaches a maximum or saturation and then decreases steadily [21].

Another competing process, known as amplified spontaneous emission (ASE), is always present at the output. Spontaneous emission, resulting from spontaneous recombinations between electrons and holes, is present throughout the active region of

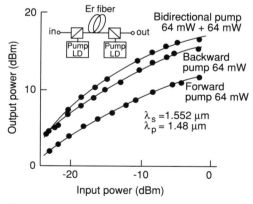

Figure 8.6 Output power versus input power for forward, backward, and bidirectional pumping schemes. After Nakagawa *et al.* [18], with permission.

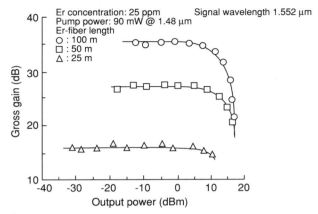

Figure 8.7 Saturated gain versus output power for various fiber L. After Nakagawa et al. [18], with permission.

the amplifier and is amplified along with the signal. An increasing portion of pump power is expended to amplify the ASE and causes further saturation of the power output. As the spectral distribution of ASE is broad and random, it contributes to the noise of the amplifier.

Fiber amplifiers can be pumped forwardly as well as backwardly. In Fig. 8.7, a backwardly pumped amplifier is shown to yield more output power.

The effect of saturation is illustrated in Figs. 8.7 and 8.8. In Figure 8.7 shows the

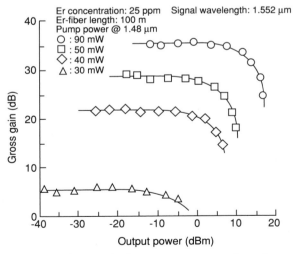

Figure 8.8 Saturated gain versus output power for various pump powers. After Nakagawa et al. [18], with permission.

signal gain saturation characteristics for various EDFA lengths, and Fig. 8.8, those for various pump powers.

In Fig. 8.8, note that saturated output power is proportional to pump power. On each of the constant pump power curves, a 3-dB bandwidth point can be marked that defines the bandwidth of the amplifier at that operating condition. At large gain compression, nearly all pump power is converted to signal power.

Analytical expressions used to calculate and plot the curves in Figs. 8.4 through 8.8 can be found in many publications; for example, see Giles and Desurvire [22].

Noise

The dominant noise sources of erbium-doped optical amplifiers are amplified spontaneous emission, signal–spontaneous (s-sp) beat noise, and spontaneous–spontaneous (sp–sp) beat noise. The resonant medium that provides amplification by the process of stimulated emission also generates spontaneous emission. ASE is broadband, unpolarized, and multidirectional; it contributes to wideband noise. In systems employing cascade amplifiers, ASE is particularly severe as the noise builds up in successive amplifiers. A Broad-bandpass filter may be used to limit the beat noises. To evaluate ASE noise, one has to estimate the ASE power from the rate equations describing the effects of pump power (P_p), signal power (P_s), and ASE power (P_a). On the basis of a three-level energy system, Giles and Deurvire [22] used the rate equations to derive a set of expressions for the maximum gain G (dB) and minimum noise figure F (dB) of an erbium-doped optical amplifier and used numerical simulation to calculate G and F in terms of signal wavelengths for various values of pump powers. The results are plotted in Figs. 8.9 and 8.10, respectively. These figures are calculated for an aluminosilicate glass host erbium-doped optical amplifier with $l = 9$, where l is the amplifier length relative to the reference signal wavelength at 1531 nm. In Fig. 8.9, maximum gain G is plotted as a function of signal wavelength

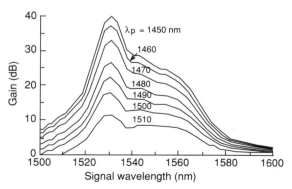

Figure 8.9 Gain versus signal wavelength for various pump wavelengths for an Er-doped fiber amplifier. After Giles and Desurvire [22], with permission.

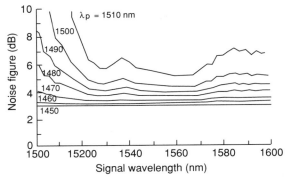

Figure 8.10 Noise figure versus signal wavelength for various pump wavelengths. After Giles and Desurvire [22], with permission.

for various values of pump power. The peak gain is observed is not at 1545 nm, but at 1531 nm. The variation of G with pump power is also clearly shown. Pumping at short wavelength results in larger gains. This is probably due to high population inversion at high pumping frequency. In Fig. 8.10, the minimum noise figure F (dB) is plotted as a function of signal wavelength for the same range of pump wavelengths. Note that the noise figure decreases to about 3 dB at lower pump wavelength. At signal wavelength 1.545 μm, using a pump wavelength of 1.480 μm, $F = 3.5$ dB. The measured value is close to 4.1 dB.

Comparison of the 0.98- and 1.48-μm Pumping Schemes

In comparing the pumping wavelengths for an Er-doped fiber amplifier, we have to distinguish between the uses of the amplifier. When the EDFA is used as an in-line amplifier or preamplifier, the signal is low and the amplifier will be working in the unsaturated region. The design goal is to achieve the highest gain for an available pump power needed on for a specific gain. High pumping efficiency is important. Lower pumping power is also desirable, as a smaller pump power will lead to a longer lifetime for the pumping laser, which is what communication systems long for. When the EDFA is used as a power amplifier, the signal at the input is relatively high, and the amplifier is working in the saturated region. The maximum obtainable signal output power is determined by the quantum efficiency. The design goal is to convert as many pump photons into amplified signal photons, that is, to optimize the ratio of optical power output to maximum obtainable signal power output, known as the *booster efficiency.*

Pedersen *et al.* made a numerical analysis of amplifier performances to answer the question of which pump wavelength to choose for the Er-doped fiber amplifier [23].

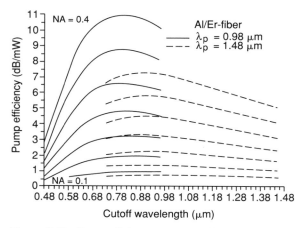

Figure 8.11 Pump efficiency versus cutoff wavelength at various numerical apertures (NA) of an Al/Er fiber for two pump wavelengths, 980 and 1480 nm. After Pedersen *et al.* [23], with permission.

In Fig. 8.11, pump efficiency, that is, the ratio of small signal gain in decibels to pump power in milliwatts, is plotted versus the cutoff wavelength for step-index Al/Er fiber pumped at 0.98 μm (solid curve) and 1.48 μm (dashed curve) for various numerical apertures (NAs). It is shown that for larger-NA fibers, pump efficiency is higher at pump wavelength 0.98 μm than at 1.48 μm. At smaller NAs, the difference is not that great. In Fig. 8.12, booster efficiency, that is, the ratio of the number of

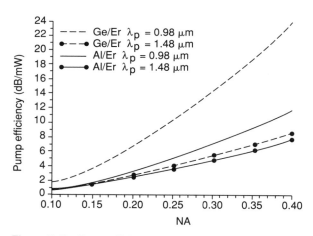

Figure 8.12 Pump efficiency versus numerical aperture (NA) for two pump wavelengths, 980 and 1480 nm, for an Al/Er fiber. After Pedersen *et al.* [23], with permission.

signal output photons to the number of total input photons, is plotted versus pump power in step-index Al/Er fibers pumped at 0.98 μm (solid lines) and 1.48 μm (dashed lines). The difference between them is very small.

From this analysis, it can be concluded that it seems profitable to use the 0.98-μm pump for a preamplifier. For booster amplifiers, the choice depends on which laser is readily available.

One of the most important advantages of the EDFA is that the fiber amplifier is index matched to the transmission fiber, thus removing the restriction of the need for low-reflectivity antireflection coating. Other advantages include the fact that erbium-doped fiber presents no nonlinearities and the gain is inherently polarization independent. The absence of nonlinearity in EDFAs is due to the long lifetime (10 ms) of the upper states in the amplifier; only the average gain can be saturated.

Should Use of the 1.3-μm Window Be Resumed?

Most fiber telecommunication systems developed in the United States and United Kingdom employ the 1310-nm window of silica optical fiber. This is because at this wavelength, the fiber dispersion is zero, making the system easy to design and operate. Since the discovery of Er-doped fiber amplifier in 1989, interest has been shifted toward use of the 1.550-μm window. New fiber installations for longer wavelengths have begun to be developed. We ask the fundamental question: Can we find a similar X-doped system that works at 1.31 μm? The answer is yes and we describe it next.

ZBLAN Fiber Amplifier

$ZrF_4BaF_2LaF_3NaF$ (ZBLAN) is a heavy-metal fluoride nonsilica-based fiberglass. ZBLAN has the potential to be a low-loss fiber, lower than silica fiber, but the technology is not yet available. Standard ZBLAN fiberglass has core diameters of 2.3 and 4 μm and a refractive index difference from 3.6 to 1.13% and is doped with 2000 ppm praseodymium ions. The most desirable feature of this fiber is that it is free from any serious problem of excited-state absorption and free from any competing transitions of higher gain like that suffered by Nd-doped fibers at the 1.06-μm transition. Also significant is that quenching of the lifetime of various energy levels of the dopant via multiphoton emission is greatly reduced. This is desirable for fibers to be used as amplifiers or lasers.

When Pr-doped ZBLAN is used as an amplifier, it is pumped at 1.017 μm with a Ti:sapphire laser. In Fig. 8.13 gain is plotted as a function of pump power and it is shown that high gain is possible (about 40 dB). A peak in gain of 38.2 dB at $\lambda = 1.31$ μm with a pumping power of about 100 mW has been reported [24].

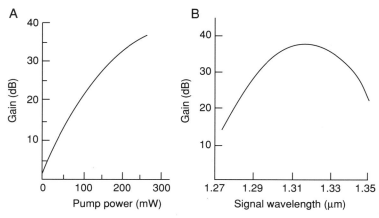

Figure 8.13   (A) Gain versus pump power for a ZBLAN fiberglass amplifier at 1300 nm. (B) Gain versus signal wavelength, showing a 40-dB gain at 1310 nm. After Oshishi *et al.* [24], with permission.

Neodymium-Doped Fluoride Fiber Amplifiers

In the 1300-nm telecommunication window, neodymium-doped fluoride glass can also be employed, using the broadband $^4F_{3/2}-^4I_{13/2}$ transition in Nd^{3+}. Comparatively very little work has been done on neodymium-doped silica glass fiber. This is because in silica glass, the excited-state absorption from the $^4F_{3/2}$ to $^4G_{7/2}$ dominates the transition and renders the amplifiers little gain. When Nd^{3+} is hosted in fluoride glass, however, the ESA spectrum is found to shift both in wavelength and in strength, allowing positive gain in the range 1310 to 1370 nm [11]. It is demonstrated that although the amplifier gain may not be as large as that from a Er^{3+}-doped silica fiber, the saturation outpower in a Nd-doped single-mode fluoride fiberoptic amplifier is very large [13].

Summary

In this chapter, several types of optical amplifiers have been described. Operating principles gain mechanisms, and noise properties of individual amplifiers were discussed. In general, it was recognized that optical amplifiers have a simple structure and wide bandwidth, operate reliably, and are cost effective for most applications. We can expect to see widespread use of reliable optical amplifiers in both telecommunications and the vast broadband local arena in the next few years.

It is instructive to compare the advantages and disadvantages of these amplifiers at this time. To make a fair comparison of these three types of optical amplifiers—semiconductor laser amplifier (SCL), stimulated Raman scattering (SRS) amplifier, and erbium-doped fiber amplifier (EDFA)—let us specify what the amplifier is to be used for.

Consider a telecommunication system employing optical fibers in the wavelength range 1.3 to 1.5 μm. Replacement of a regenerative repeater station with an optical amplifier is required. Which of the three amplifiers would you choose? First, consider the structure. The SCL amplifier has a laser structure less a resonator (that is, with the endfacets antireflectively coated). The design is complex as it involves current confinement and waveguide structure of thin epitaxial semiconductor materials of intricate configuration. In particular, the requirement of perfect antireflective material for the endfacets is very demanding. On the other side, both SRS amplifiers and EDFAs require no specifically designed structures. SRS amplifiers use ordinary Ge-doped silica fiber, and EDFAs use special Er-doped fibers.

Second, from what source does the amplifier derive its power? For SCL amplifiers, a simple injection current source suffices. It is dc pumped and requires only several tens of milliamperes to be powered. Both SRS amplifiers and EDFAs are parametric amplifiers that require an external ac power source. The pump source must be of particular frequency to satisfy the required idler frequency condition. The pump power required to operate a SRS amplifier is large; that required to operate an EDFA is moderate. Pump power can be applied to the fiber via directional couplers.

Third, the gain attainable in SCL amplifiers is moderate and saturates readily at lower values. For SRS amplifiers, only a small gain can be achieved (several to 10 dB). EDFAs have by far the best gain attainable, up to 35 to 40 dB. They also have the largest saturation power at moderate pump power.

Fourth, SCL amplifiers are polarization sensitive (only slightly), whereas EDFAs are not.

Fifth, the noise output for the EDFA is smaller than that for the other two. In EDFAs, the chief noise source is amplified stimulated emission. The best noise figure for EDFA is about 4.5 dB, whereas in SRS amplifiers, ASE becomes a major noise problem. ASE also contributes to the noise in another way. Spontaneous emission will beat with the signal to produce the beat noise or s–sp noise. ASEs of different frequencies will also beat among themselves to create noises, or sp–sp noises. In addition, the competing Bruilloun scattering emission contributes to noise in SRS amplifiers.

From these findings, it can be concluded that the best choice for this application is the EDFA. This explains why in 1987, before which all research efforts were concentrated mostly on SRS, all attention suddenly shifted when the development of a successful EDFA was announced. At present, EDFAs are a hot topic in research in the communication field.

The development of EDFAs had its problems in the early days. Doping of erbium ions with silica fiber is not easy, as the solubility of erbium ions in silica is very low. This problem was solved by adding by intermediate element, Al_2O_3, which helped to link the bonds between atoms. This process made it easy to fabricate erbium-doped fibers of any doping concentration to make the long fibers suitable for distributed fiber amplifiers. The future of EDFAs is unlimited. The spread of this technology of using fiber amplifiers to the local loop arena, increasing its impact in business and in the home, is just begining.

With respect to telecommunications in the 1300-nm window, Nd-doped fluoride glass and ZBLAN fiberglass become interesting subjects of research and development. European countries are anxious to develop this wavelength window because of the existing fiber connections.

References

1. M. J. O'Mahony, Semiconductor laser optical amplifiers for use in future fiber systems. *J. Lightwave Technol.* **6,** 531 (1988).
2. T. Saitoh and T. Mukai, 1.5 μm GaInAsp traveling-wave semiconductor laser amplifier. *IEEE J. Quantum Electron.* **23,** 1010 (1987).
3. J. C. Simon, InGaAsP semiconductor laser amplifiers for single-mode fiber communications. *J. Lightwave Technol.* **5,** 1286 (1987).
4. J. C. Simon, Semiconductor laser amplifier for single mode fiber communication. *J. Opt. Commun.* **4,** 51–62 (1983).
5. T. Mukai, Y. Yamamoto, and T. Kimura, Optical amplification by semiconductor lasers. In *Semiconductors and Semimetals,* Vol. 22E, p. 265, Academic Press, New York, 1985.
6. I. Cha, M. Kitamura, and I. Mito, 1.5 μm band traveling-wave semiconductor optical amplifiers with window facet structure. *Electron. Lett.* **25,** 242 (1989).
7. G. Eisenstein, Theoretical design of single-layer antireflection coating on laser facets. *AT&T Bell Lab. J.* **63,** 113 (1986).
8. Y. Yamamoto, Noise and error rate performance of semiconductor laser amplifiers in PCM optical transmission systems. *IEEE J. Quantum Electron.* **QE-16,** 1073 (1980).
9. J. W. Marshall, M. J. O'Mahony, and P. D. Constantine, Measurements on a 206 km optical transmission system at 1.5 μm using two packaged semiconductor laser amplifiers as repeaters. Technical Digest, 12th Conference on Optical Communications I, Barcelona, pp. 253–256, 1986.
10. C. H. Henry, Theory of spontaneous emission noise in open resonators and its application to laser and optical amplifiers. *J. Lightwave Technol.* **LT-4,** 288–297 (1986).
11. W. J. Miniscalco, L. J. Andrews, B. A. Thompson, R. S. Qimby, L. J. B. Vacha, and M. G. Drexhage, 1.3 μm fluoride fibre laser. *Electron. Lett.* **24,** 28 (1988).
12. Mikami, LiNbO$_3$ coupled-waveguide TE/TM mode splitter, *Appl. Phys. Lett.* **36,** 491–493 (1980).
13. J. E. Pedersen and M. Brierley, High saturation output power from a neodymium-doped fluoride fibre amplifier operating in the 1300 nm telecommunications window. *Electron. Lett.* **27,** 817–818 (1990).
14. R. J. Robinson, Development of parametric amplifiers for radio astronomy. *Proc. IRE* **24,** 119–127 (1968).
15. R. H. Stolen and E. P. Ippen, Raman gain in glass optical waveguides. *Phys. Rev. A* **17,** 276–278 (1973).
16. R. C. Davies, P. Melman, W. H. Nelson, M. L. Dakss, and B. M. Foley, Output moments and photon statistics in fiber Raman amplification. *J. Lighwave Technol.* **LT-5,** 1068–1073 (1987).
17. T. Nakashima, S. Seikai, and M. Nakazawa, Configuration of the optical transmission line using stimulated raman scattering for signal light amplification. *J. Lightwave Technol.* **LT-4,** 569–573 (1986).
18. K. Nakagawa, S. Nishi, K. Aida, and E. Yoneda, Trunk and distribution network application of erbium-doped fiber amplifier. *J. Lightwave Technol.* **LT-9,** 198–208 (1991).
19. W. J. Miniscalco, Erbium-doped glasses for fiber amplifiers at 1500 nm. *J. Lightwave Technol.* **LT-9,** 234–250 (1991).

20. C. R. Giles and E. Desurvire, Modeling erbium-doped fiber amplifiers. *J. Lightwave Technol.* **LT-9,** 271–283 (1991).
21. E. Desurvire and J. R. Simpson, Amplification of spontaneous emission in erbium-doped single-mode fibers. *J. Lightwave Technol.* **LT-7,** 835 (1989).
22. C. R. Giles and E. Desurvire, Propagation of signal and noise in concatenated erbium-doped fiber optical amplifiers. *J. Lightwave Technol.* **LT-9,** 147–154 (1991).
23. B. Pedersen, A. Bjarkiev, J. H. Povlsen, K. Dybdal, and C. C. Larsen, The design of erbium-doped fiber amplifiers. *J. Lightwave Technol.* **9,** 1105–1112 (1991).
24. Y. Oshishi, T. Kanamoki, T. Nishi, and S. Takakoshi, A high-gain, high-output saturated power Pr^{3+}-doped floride fiber. *J. IEEE Photon. Technol. Lett.* **3**(8), 175 (1991).

CHAPTER 9

Solitons in Optical Fiber Telecommunications

Introduction

A soliton is a special pulselike traveling-wave solution, the only stable solution of a dispersive wave equation. A solitary wave, as the name suggests, consists of a single peak moving in isolation. The unique property of a soliton pulse is that it is so stable that its shape and velocity are preserved while traveling along a transmission medium. This means that these light pulses do not spread in optical fiber even after thousands of kilometers, a property we all long for in telecommunication systems. A great deal of research and development have been reported on this subject. Here we intend to describe the properties and the application of solitons to optical fiber telecommunication systems.

The first observation of the existence of a solitary pulse was made by Scott-Russell in describing water waves about 100 years ago [1]. In 1838, Korteweg and deVries [2] provided a simple analytic foundation for the study of a solitary wave in shallow water. They formulated the KdV (after Korteweg and deVries) differential equation, obtained the soliton solution, and defined the properties of a soliton. Interest in solitons soon spread to many branches of science and engineering. Scott *et al.* made a comprehensive survey of the literature concerning the theory and applications of solitons in applied sciences [3]. Application of solitons to communication engineering was first suggested by Hasegawa and Kodama in 1981 [4]. Further investigation of solitons in optical fibers was done by Hasegawa [5, 6] and Mollenauer *et al.* [7, 8]. More than a hundred equations have been found to have solitary waves as solutions. Solitary waves have been observed in a variety of natural realms, such as the atmosphere, the oceans, plasma physics, and possibly the nervous system of living organisms. At present, solitons are taking on an important role in telecommunications, where their persistent shape and immunity to distortions make them ideal carriers of long-distance signals.

The conditions under which a solitary wave can be excited are somewhat unique. For example, in an optical fiber telecommunication system, it takes an exact balance between the effects of nonlinearity and dispersion in an optical fiber to obtain a solution of the wave equation that leads to solitons. Furthermore, it takes a definite amount of power and the correct range of wavelengths to excite a soliton. These conditions are illustrated in the following section.

The Soliton

A qualitative picture of a soliton may be gained by considering the waveform displayed in Fig. 9.1. The envelope of a traveling-wave function $\Phi(z, t) = \Phi(\xi)$ is plotted on a moving frame of reference defined by $\xi = z - vt$, where v is the speed of the traveling wave. Note that the waveform consists of a periodic train of localized humps for which pulsewidth is Δt, the half-power width of the pulse, and the humps are located a distance d apart from each other. A soliton is defined as such a pulse that as d increases to infinity in both directions, Δt remains unchanged. A moving frame of reference is used to graph the traveling wave so that it appears stationary.

There are two types of solitons: temporal solitons and spatial solitons; however, the same differential equation describes the phenomena of both types, with time and space interchanged. In fact, spatial solitons are the transverse analogs of longitudinal temporal solitons. We describe only the temporal soliton in this section. Spatial solitons and their applications in photonics are treated later under The Spatial Soliton.

Starting with a one-dimensional wave equation familiar to electrical engineers, Marcuse first assumed a lossless fiber and neglected the fiber dispersion [9]. By introducing the nonlinearity of the optical fiber in the form of its index of refraction n, he rewrote the wave equation as

$$E_{zz} = \left(\frac{1}{c^2}\right)(n^2 E)_{tt} \tag{9.1}$$

where $E(z, t)$ is the electric field of the light pulse traveling in the positive z direction, c is the speed of light *in vacuo*, and n is the index of refraction. The double-letter subscripts zz and tt indicate the second-order partial differentials of E with respect to z and t, respectively. The nonlinearity in n is specified by $n = n_0 + n_2 E^2$. Here we assume that the nonlinear refractive index consists of the original n_0 (without the electric field) plus $n_2(E)$ due to the electric field. Assume $E(z, t)$ is traveling with a propagation constant β_0 and an angular frequency ω_0; then

$$E(z, t) = \Phi(z, t) \exp i(\omega_0 t - \beta_0 z) \tag{9.2}$$

where Φ is an envelope function. Substituting Eq. (9.2) into (9.1) and performing the necessary differentiations, one obtains a differential equation for $\Phi(z, t)$ that de-

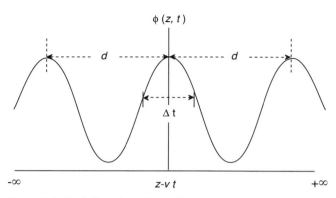

Figure 9.1 Definition of a soliton pulse.

scribes the transmission of an optical wave along an optical fiber in a telecommunication system, with nonlinearity but with no dispersion. The solution of this equation, however, does not yield a soliton solution. Marcuse then considered a pulse train propagating in a linear but dispersive medium in which the propagation constant becomes nonlinear; that is,

$$\beta = \beta_0 + \dot{\beta}_0 \, (\omega - \omega_0) + \left(\frac{1}{2}\right) \ddot{\beta}_0 (\omega - \omega_0)^2$$

where $\dot{\beta}_0 = d\beta_0/d\omega_0$ and $\ddot{\beta}_0 = d^2\beta_0/d\omega_0^2$ for constant z. If each pulse is expressed as $E = A \exp i(\omega t - \beta z)$, and these pulses are superposed over the entire frequency domain, the total pulse can be written as

$$E = \int_{-\infty}^{\infty} A(\omega) \exp i(\omega t - \beta z) d\omega$$

where $A(\omega)$ is the amplitude at frequency ω. Marcuse then derived another differential equation for Φ that can be put in a form similar to that of the dispersionless case. As both equations contain the same derivatives of Φ but with different constraints, they can be added to reflect the effects of both nonlinearity and dispersion. When weak dispersion and small nonlinearity are assumed, the resultant approximate final form of the differential equation becomes

$$\Phi_z + \dot{\beta}_0 \Phi_t = \left(\frac{i}{2}\right) \ddot{\beta}_0 \, \Phi_{tt} + i\left(\frac{n_2}{n_0}\right) |\Phi|^2 \, \Phi \qquad (9.3)$$

where $\Phi(z, t)$ is the envelope function, Φ_z is the partial derivative of Φ with respect to z, and Φ_t and Φ_{tt} are the first and second partial derivatives of Φ with respect to t, respectively. Equation (9.3), when expressed in normalized form (normalized against special soliton units to avoid the appearance of scaling constants in the equation [8]) is also known as the nonlinear Schrodinger equation.

The left-hand side of Eq. (9.3) shows both space and time derivatives of an envelope function. The first term on the right side represents the contribution from dispersion, and the second term is due to nonlinearity.

The simplest solution of the soliton equation (9.3) is known as a soliton of the first order. It is given by

$$\Phi(z, t) = \Phi_0 \text{sech}\left\{\frac{[t - \beta_0 z]}{\tau_0}\right\} \exp(i\alpha z) \tag{9.4}$$

provided that

$$\alpha = \frac{[\ddot{\beta}_0]}{2\tau_0^2} \tag{9.5}$$

where $\beta_0 = n_0 w_0/c$, τ_0 is the time duration of the initial pulse, α has a dimension of the fiber loss, and $1/\alpha = z_0$ is often used as a distance scale. The amplitude of the soliton is then

$$|\Phi|^2 = -\frac{(n_2/n_0)\ddot{\beta}_0}{\beta_0 \; \tau_0^2} \tag{9.6}$$

Here, τ_0 is closely related to the pulse width, measured at half-width, Δt(FDHM), or the full duration half-time width $\Delta t = 1.76\tau_0$.

Discussion of Soliton Results

A Soliton

To see that Eq. (9.4) is a soliton, we note that the amplitude function expressed by $|\Phi|^2$ in Eq. (9.6) is independent of z. This is to say that a soliton traveling in a dispersive medium will not alter its pulse width in time domain. The Fourier transform of Eq. (9.4) reveals that the envelope of the pulse in the frequency domain is also nonspreading.

Amplitude of a Soliton

Equation (9.6) also shows that the amplitude of a soliton is uniquely determined by three factors: n_2, the nonlinearity of the medium; the dispersion, which is proportional to the dispersion factor $D = -(\lambda/c)[d^2n/d\lambda^2]$ at a correspondent wavelength; and the pulse width, specified by τ_0.

Sign of the Amplitude

The negative sign on the right-hand side of Eq. (9.6) is very significant. As the amplitude of the soliton must be positive, and n_0, ω_0, and τ_0 are always positive, only

$\ddot{\beta}_0$ n_2 or $d^2n/d\lambda^2$ can change sign. Thus for a soliton to exist, $\ddot{\beta}_0$ or n_2 or ($\ddot{\beta}_0/n_2$ or $d^2n/d\lambda^2$) must be negative. For fused silica fibers, n_2 is a positive quantity [10]. But dispersion can change sign depending on the operating wavelength. Accordingly, for silica, the dispersion changes sign at 1.27 μm. The fiber must be operating at a wavelength longer than the zero dispersion wavelength, at 1.3 μm or longer.

The Condition

The condition imposed by Eq. (9.5) is also significant. As α has a dimension of the reciprocal of a length, then $z_0 = 1/\alpha$ becomes a characteristic distance of the soliton. Once a soliton has formed, it will be attenuated by the fiber loss as expressed by the exponential decay of Eq. (9.4). For a distance $z < z_0$ or $< 1/\alpha$, the soliton will adjust its width to satisfy Eq. (9.5), and remains a soliton. If $z > z_0$, a soliton ceases to have its unique character and becomes an ordinary pulse instead (z_0 is the distance over which the Kerr effect induced-phase shift equals $\pi/4$).

Starting a Soliton

To start a soliton on silica fibers, one must make sure that the operating wavelength is above the zero dispersion condition, the initial pulse power is high enough, and

$$\Phi_0\tau_0 = \left(\frac{\ddot{\beta}_0}{\beta_0}\right)\left(\frac{n_2}{n_0}\right)^{1/2}$$

High-Order Solitons

Although higher-order solitons do exist as there are an infinite number of solutions of Eq. (9.3) that yield solitons, higher-order solitons may have complicated waveforms, varying pulse widths, periodically changing pulse shape. We do not discuss these cases.

The Energy

To complete the discussion, let us mention that the energy carried by a soliton may be expressed as

$$W_e = \sqrt{\frac{e_0}{\mu_0}}\, A_{\text{eff}}\left(\frac{n_0^2}{n_2}\right)\frac{[\ddot{\beta}_0]}{\beta_0\tau_0} \tag{9.7}$$

and the power at the peak of the pulse is

$$P(\text{max}) = \frac{W_e}{2\tau_0} \tag{9.8}$$

where A_{eff} is the effective area of the mode supported by the fiber. In Eq. (9.7), note that for an optical fiber operating at a given wavelength, the constants are all known.

Thus, we may say that the product of the energy and the pulse width remains a constant. This implies that a shorter soliton pulse requires more energy. This equation will be useful in computing the necessary power required to excite a first-order soliton.

Some Physical Interpretations of Solitons

Exciting a Soliton

The formation of solitons in optical fiber telecommunication systems depends on the balance of two physical factors: the group delay effect of the optical signal in an anomalously dispersive medium and the effect of nonlinearity in optical fibers. Qualitatively, it means that the operating wavelength must be above the zero dispersion wavelength to qualify for anomalous dispersion. For silica fibers, it is in the wavelength range above 1.3 μm. Thus, to operate a soliton wave at 1.55 μm, where the fiber loss is minimum, makes sense.

The Nonlinear Medium

A dielectric medium is nonlinear if its refractive index is a function of light intensity. In an optical fiber, this could lead to frequency chirp, which degrades the transmission. Formation of a soliton in a transmission line depends on the balance between the effects of dispersion and nonlinearity as explained next.

Balance Effects

Figure 9.2 illustrates the balancing of the effects of dispersion and nonlinearity. An original pulse of a specified pulse width and shape is traveling along an anomalously (above the cross-over wavelength) dispersive fiber. The received pulse is illustrated at the end of the line. (Fig. 9.2A). A pulse has a complex waveform. Fourier analysis shows that the sharper the pulse is, the more frequency components it contains, ranging from low to high frequencies. Different frequency components of the pulse will travel at different group velocities and therefore undergo different time delays as a result of fiber dispersion. The higher-frequency waves, which have a lower velocity, begin to fall behind and appear at the back of the main body of the wave. As the wave propagates, the higher-frequency wave falls steadily farther behind, causing frequency spread. Note the spreading in frequency of the trailing end of the wave as shown in Fig. 9.2A. To see how dispersion alone can destroy the signal transmission, let us imagine that further down the line, the amplitude of a pulse, representing a signal, decreases and becomes broader as it propagates. In the end, a pulse that starts out tall and narrow gradually broadens as various frequency components separate. The amplitude diminishes as the wave spreads out over a larger area, the amplitude becomes undetectably small, and the wave disappears even if the fiber is lossless.

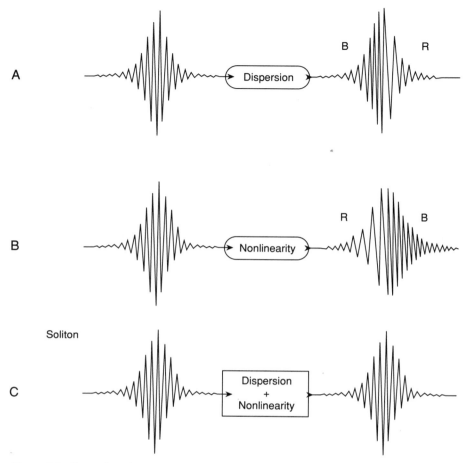

Figure 9.2 Formation of a soliton pulse. (A) Effect of dispersion of the fiber. (B) Effect of non-linearity of the fiber. (C) By combining these effects properly, a soliton is formed when balanced. After T. E. Bell in *Spectrum* **29**, p. 57 (1990). © August, 1990 *IEEE,* with permission.

The nonlinearity of fiber affects transmission differently than does dispersion. Fiber nonlinearity is intensity sensitive. A higher pulse amplitude will cause the wave to travel faster, or the lower-amplitude components of the wave will travel more slowly, allowing the peak of the wave to overtake the lower-amplitude leading edge. This is equivalent to introducing more high-frequency modes, as seen in Fig. 9.2B. Figure 2.B shows the same pulse moving down a nonlinear fiber that is nondispersive. Note that the pulse is also distorted.

The Kerr nonlinearity is expressed as $n = n_0 + n_2 E^2$, where E^2 is proportional to the intensity of the pulse. If the pulse intensity changes, the index of refraction of the medium change. Thus the carrier frequency will change, leading to a phenomenon known as frequency chirp [11]. The change in frequency can be estimated from the phase shift incurred in time. Thus, $\omega = d\phi/dt = \omega_0 - k_0 n_0 \Delta z dE^2/dt$. Here, ϕ is the phase angle, and k_0 is the wavenumber. Carrier frequency may thus increase or decrease depending on the slope of dE^2/dt as k_0, n_0, and Δz are all positive numbers. The envelope of the pulse is positive for the leading edge and a negative slope for the trailing edge. This brings about the phenomenon of frequency expansion or compression depending on the sign of the pulse slope. For positive n, the frequency of the trailing half of the pulse is increased whereas that of the leading half is decreased. This is sometimes called self-induced phase modulation or frequency modulation.

If the chirped wave in Fig. 9.2B is moving down an anomalously dispersed fiber, and if the correct amounts of nonlinearity and dispersion are selected, the resultant pulse, by adding Figs. 9.2A and 9.2B, can become effectively unperturbed as is shown in Fig. 9.2C. This is a soliton. If these two effects were not balanced, pulse compression or expansion might result periodically to destroy the soliton.

The Parameters

The required balancing parameters can be estimated from Eqs. (9.5), (9.6), and (9.7).

An Example

We now use a numerical example to calculate the power required to excite a soliton. The following data are given: $n_0 = 1.46$, $n_2 = 6.1 \times 10^{-19}$ cm²/V², $A(\text{eff}) = 7.85 \times 10^{-7}$ cm², and $\lambda = 1.5$ μm. At this wavelength, the propagation constant of silica fiber is found to be $\beta_0 = 6.12 \times 10^4$/cm. From the dispersion curve, $[\ddot{\beta}_0] = -2 \times 10^{-28}$ s²/cm. For a pulse width of 10ps, the peak power required to excite a soliton can be calculated from Eq. (9.7) to be only 375 mW. If the pulse width is doubled, the peak power required is only one-fourth that much, or 88 mW. New lasers can easily supply this power.

Peak Power

The peak power mentioned in the preceding example is the minimum power required to excite a soliton. Power lower than that will not excite the soliton at all. In such cases, the light pulse will suffer the normal spreading. Too much power will excite solitons of higher orders in addition to the desired soliton. These spurious pulses can give rise to overall pulse distortion.

Excitation

As the power required to excite a soliton is critical, and the soliton in a fiber does not solve the problem of fiber losses, it becomes obvious that some kind of amplifier is required to boost the pulse amplitude repeatedly. The critical distance z_0 becomes a parameter to gauge when a boost in power is required to sustain a soliton. A distributed amplifier may be the best choice but amplifiers at intervals of z_0 distance apart may also be used. Either Raman effect amplifiers [12] or erbium-doped fiber amplifiers [13] can be used. At present, however, Er-doped fiber optical amplifiers have risen to prominence to become the central component of all optical amplifiers [14]. These amplifiers are discussed in Chapter 8.

A Question

From the definition of soliton in Eq. (9.1), it is known that a soliton is a pulse, not a periodic function. But from elementary mathematical theory, any nonsinusoidal function, such as half of a sinusoidal standing alone, can be decomposed into superposition of a number of sinusoidal functions of different frequencies. For a pulse there may be an infinite number of frequencies. As the velocities of propagation of the frequency components are different from one another, which is the cause of dispersion, why not use a soliton pulse? The answer is yes. A pulse is subjected to dispersion. The important difference is that in a soliton pulse, another reflect does a balancing act to compensate for the effect of dispersion. This is the nonlinear effect of the medium where the wave propagates. The compensating effect in solitary waves is a coupling between amplitude and speed. Because of nonlinearity, in the various media, that support solitary waves, waves with larger amplitudes move faster and overtake the lower-amplitude waves (resulting from dispersion). The amplitude effect steepens the wave, which is equivalent to introducing more high-frequency modes. The two competing phenomena reach a point of stable equilibrium, where the waveform remains constant. A recent illustration equating runners on a mattress to soliton formation on optical fiber is very enlightening [14]. The runners create a moving valley on the mattress that pulls along slower runners and retards the faster ones, helping to bunch them together as in solitons.

The Spatial Soliton

Another type of soliton that is mathematically analogous to the temporal soliton is the spatial soliton. Spatial solitons are localized pulses in the transverse plane and travel in a nonlinear medium without altering their spatial distribution. In this case, the soliton property is the result of the balancing between light diffraction (dispersion) and self-induced phase modulation.

If it is assumed that the envelope of the lightwave is $\Phi(x, z)$, now a function of x in the transverse plane with constant y and traveling in the z direction in a nonlinear medium, the differential equation of wave motion can be written as

$$\Phi_{xx} + \left(\frac{n_2}{n_0}\right) k \left|\Phi\right|^2 \Phi = 2ik\Phi_z \qquad (9.9)$$

Note that Eq. (9.9) is similar to Eq. (9.3), with the variable t and its derivatives being

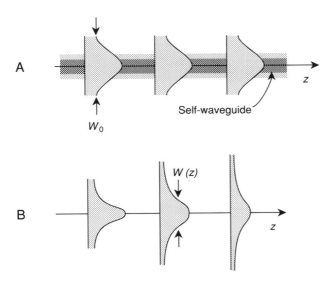

A

W_0

Self-waveguide

z

B

$W(z)$

z

Figure 9.3 Comparison of a spatial soliton pulse train (A) and a Gaussian pulse train (B). After B. E. A. Saleh and M. C. Teich, *Fundamentals of Photonics*. Reprinted by permission. © 1991 John Wiley & Sons.

replaced by x and its derivatives. One of the solutions of this nonlinear differential equation is

$$\Phi(x, z) = \Phi_0 \text{sech}\left(\frac{x}{W_0}\right)\exp\left(\frac{-iz}{4z_0}\right) \tag{9.10}$$

where W_0 is the original pulse width at $z = 0$, and $z_0 = \pi W_0^2/\lambda$ [15]. From Eq. (9.10), it is seen that the envelope is not a function of z. Other discussions apply as in the case of a temporal soliton.

Figure 9.3A shows what a spatial soliton would look like. The nonuniformity in the index of refraction of the fiber acts as a graded-index waveguide to compensate for the diffraction of light waves traveling in the z direction with unvarying width. For comparison, a Gaussian pulse would have varied in pulse width, as shown in Fig. 9.3B.

Spatial solitons have only recently been observed experimentally. At present, the demonstrated spatial solitons are continuous light beams that resist the wave property of diffraction or the tendency to broaden in space as a function of wavelength. In time, it might be possible to observe a soliton in both time and space. Such a light pulse may be called a light bullet.

Aitchison and Silberberg at Bell Communication Research demonstrated the spatial soliton in the following manner [16]. In a planar optical waveguide, a high-index dielectric layer is sandwiched between two low-index layers. Light waves entering

A

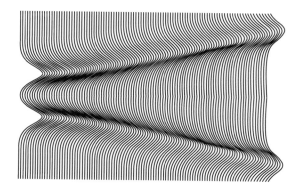

B

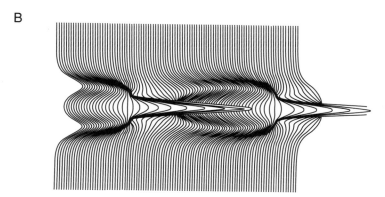

Figure 9.4   Spatial solitons. (A) Two solitons of opposite phases repel each other. (B) when in phase, they attract each other. After T. E. Bell, *Spectrum* **29** (1990) p. 57. © August, 1990 *IEEE*, with permission.

the waveguide suffer the usual optical diffraction which tends to spread the beam; however, by designing the nonlinear refractive index of the guide, which provides the self-focusing effect on the beam, and by adjusting the intensity of the light field, they balance these effects such that the beam remains parallel to the edges throughout travel.

If two solitons were placed within the guide, they would tend to interact; that is, their tails would overlap either constructively or destructively, depending on their relative phase relationships. Two such cases can be seen in Figs. 9.4A and B. In Fig. 9.4A, the local light fluxes are out of phase with each other and therefore "bend" themselves away or repel each other. In Fig. 9.4B, they are in phase and attract each other. These properties of spatial solitons might be useful as the building blocks for an integrated all-optical switching circuit.

Examples of Soliton Transmission Systems

The communication world expects that, with the advent of erbium-doped optical amplifiers, effective solitons can be developed for use in future undersea communication systems. Simulated soliton systems have been developed to probe the feasibility of the scheme. Actual lines have been built to test the properties of soliton transmission. We report on some of these experiments briefly.

Simulation of Soliton Transmission Systems

Mollenauer *et al.* reported a numerical simulation of soliton propagation that can traverse a great distance through a chain of lumped erbium fiber amplifiers connecting dispersion-shifted spans of fibers [17]. They showed that solitons are remarkably resilient to large variations in energy and dispersion as long as the characteristic length scale of those variations is considerably less than z_0 defined in Eq. (9.5). Lumped optical amplifiers can be used with solitons in ultralong-distance communications.

An Experiment

Mollenauer *et al.* [18] described an experiment in soliton transmission carried out in a recirculating loop such as that shown in Fig. 9.5 [18]. The recirculating loop consists of 75 km of optical fiber in three sections, each of which is followed by a high-gain erbium-doped optical amplifier that is pumped through a coupler with a solid-state laser at 1480 nm. Two of the erbium-doped optical amplifiers contain 2.4 m of fiber; the other is 3.1 m long. The extra gain is designed to compensate for additional

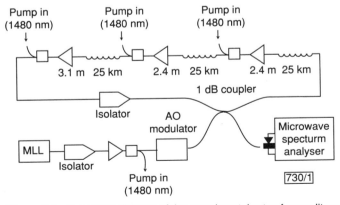

Figure 9.5 Schematic diagram of the experimental setup for a soliton experiment. A recirculating 75-km loop serves in soliton transmission experiments. Er-doped fiber amplifiers are pumped with 1480-nm diode pumps. Reprinted with permission from Mollenauer *et al.* [18].

losses due to the isolator and coupler. The system is excited by an external-cavity mode-locked diode laser that is amplified to a level appropriate for creating solitons in the loop. The output is coupled to the recirculating line through an acoustooptic modulator and a 1-dB coupler. The acoustooptic modulator gates out a pulse train large enough to fill the loop once. With the loop gain at unity, the pulse train is allowed to recirculate as many times as necessary to cover the required distance. The pulse shape at any moment can be checked with a microwave spectrum analyzer through another port on the 1-dB coupler. The results of the experiment show that at a distance of 5000 to 6000 km, the pulse shape is almost unchanged from the input except for a slight oscillation of the experimental points. Above 6000 km, the effective pulse width increases only slowly with total path traversed. This increase is expected as the jitter in pulse arrival time increase as a result of the Gordon–Haus effect, which serves as the upper limit on single-channel bit rate for a given distance. Gordon and Haus [19] suggested that through the nonlinear term, the amplified spontaneous emission (ASE) modulates the optical frequencies of the solitons at random to cause the jitter [19]. Limits on distance and bit rate of soliton transmission are set by the effect of noise. In the experimental system (see Fig. 9.5), the only important noise found is amplified spontaneous emission noise. The measured signal-to-noise ratio, or S/N, defined as the ratio of soliton pulse energy to path-average signal energy, is better than 200 : 1 at 9000 km. Stable transmission of 50-ps solitons over more than 10,000 km without noticeable dispersion has been observed. The transmission rate was 5×10^9 bits/s, which would be equivalent to about 100,000 digitized voice channels. Soliton transmission systems with erbium-doped amplifiers have a wide bandwidth to accommodate several channels simultaneously so that they are well suited to wavelength division multiplexing (WDM) applications. Up to five channels have been tried with very little cross-talk between them. The system is very cost effective and places practically no limit on the bit rate on an individual channel.

Time/Polarization Multiplexing with Solitons

Polarization Effects in Optical Fibers

Long-haul optical telecommunication system employing single-mode silica fibers is designed to work at a wavelength of about 1.55 μm, corresponding to the minimum fiber loss (about 0.25 dB/km or less). At this wavelength, the dispersion is about 10 to 12 ps/km-nm, too high for high-data-rate digital transmission. Dispersion-shifted fibers are used to counter this effect; however, even with the best designed single-mode fiber, the residual dispersion effect due to birefrigence can still limit the speed of transmission. Furthermore, the birefrigent effect is wavelength dependent. It can affect the polarization states of the fiber. A normal circularly polarized light beam may become elliptically polarized at the end. In coherent detection systems, this is unacceptable. Polarization-maintaining schemes should be used to correct this effect.

With solitons, not only does pulse shape remain unchanged after thousands of kilometers of traverse over the fiber, but the state of polarization also remains unaffected. Evangelides *et al.* showed that although the state of polarization continuously evolves in response to the residual birefringence, at any given point along the path, the same state of polarization always applies to the soliton as a whole [20].

Suggested Polarization Multiplexing Schemes

The finding that the state of polarization remains unaffected along the traverse path when soliton solutions are used suggests that a single wavelength signal containing orthogonally polarized waves, each with its own data modulation, can be multiplexed at the input, sent through the fiber as a soliton, and demultiplexed at the receiving end to reclaim two separate channels, thus doubling the capacity of a single fiber [20]. This requires simple and relatively inexpensive hardware at the transmitting and receiving ends and eliminates the expense of building another fiber line. Herman; discussion on solitary wave in nature is worthy reading [21].

Numerical Simulation

Before they experimented on this scheme, the authors made an exhaustive analysis of the system using numerical simulation. The simulations are based on solving the couple mode equations written for orthogonally polarized light beams of the same frequency, using practical parameters from their previous experimentation. They checked the pulse shape and measured the state of polarization at intervals by reading the angles of rotation with respect to the axis. To distances greater than 10,000 km, the deviation in polarization was less than 5° out of 180°. At intervals, they even introduced ASE noise as would occur in practice when Er-doped fiber amplifiers are used.

An Experiment

For the experiment, single-frequency source that could be separately polarized and modulated was needed. The pulse shape should look like that in Fig. 9.6. Two or-

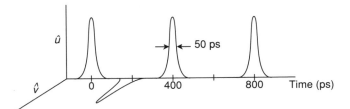

Figure 9.6 Pulse field used in the time/polarization multiplexing scheme. Two orthogonal polarized pulse trains are presented. After Evanglides *et al.* [20] with permission.

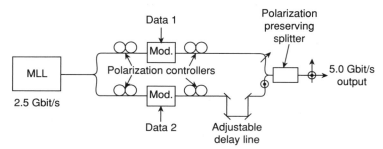

Figure 9.7 Transmission end of the time/polarization multiplex scheme, showing the separate modulation channels and the components. After Evanglides *et al.* [20] with permission.

thogonally polarized pulses are displayed on the time axis as shown. These pulses are displaced in time by a half-period and have a pulse width of 50 ps. One polarization is plotted along the vertical axis and the other on the horizontal axis. The time displacement ensures minimum coupling. The data rate for each is 2.5 Gbit/s which forms a beam with a 5Gbit/s stream. To generate this scheme, Fig. 9.7 was drawn. A single-frequency mode-locked laser (MLL) at 2.5 Gbit/s is fed into the Y-branch to divide the beam into two paths; one forms the upper and the other the lower branch. Each path passes through a polarization controller to ensure proper polarization before entering the respective modulator. The upper branch is modulated with data 1 and the lower with data 2. The two branches are recombined at the output Y-recombiner and sent to the fiber for transmission. An adjustable delay line (ADL) is used to adjust the proper time displacement. Extra polarization controllers are installed on each branch exit to control polarization state. The output waveform of the pulse should look like that shown in Fig. 9.6. The correctly adjusted multiplexer outputs are in orthogonal polarization, and the half-period time displacement achieves polarization division and also eliminates the potential at the receiving end for cross-phase modulation.

The multiplexed 5.0 Gbit/s signal is sent to a recirculating loop similar to that described in Fig. 9.5 for circulation as often as is required. At intervals, the state of polarization is checked and the pulse shape measured.

At the receiving end, the received signal in an arbitrary state of polarization must first be transformed so that the unwanted channel in each arm of the demultiplexer can be rejected by a polarization analyzer. A polarization controller is used to ensure that the overall gains for orthogonal polarizations of the two channels are substantially equal. The beam is then sent to a polarizing beam splitter to separate the channels. The polarization controller derives its input from one channel of the multiplexer. It is detected and filtered to control and adjust for maximum signal. A schematic of the demultiplexer is shown in Fig. 9.8.

In general, when there exists a difference in gain for orthogonal polarizations, the emerging polarization of the two channels will no longer be perfectly orthogonal,

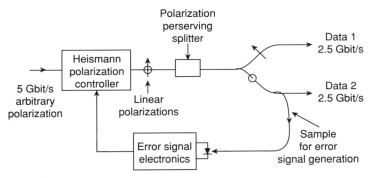

Figure 9.8 Receiving end of the time/polarization scheme with the necessary components. After Evanglides *et al.* [20] with permission.

and the receiving system must be modified slightly. In Fig. 9.9 the signal with arbitrary polarization is first split into two equal channels by a 3-dB Y-beam splitter. Each channel goes through a separate controller and a linear analyzer. Error detected in each channel is then fed back to its controller to correct the polarization.

There is another advantage of using orthogonally polarized solitons in the simulation study. In a single-polarization soliton multiplex system, to avoid soliton–soliton interaction over transoceanic distances, the solitons must be spaced at least five times the pulse width. With an orthogonal polarization multiplexing scheme, solitons can be spaced as little as 2.5 times the pulse width. Thus this scheme can achieve the highest possible channel bit rate.

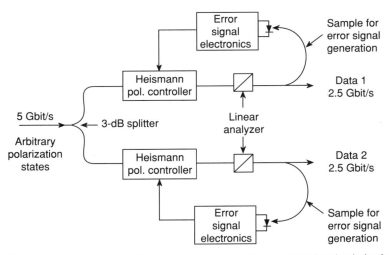

Figure 9.9 General demultiplexing scheme with arbitrary input signal polarization. After Evanglides *et al.* [20] with permission.

Recent Experiments on Soliton Transmission

Nippon Telegraph & Telephone Corporation of Japan (NTT) reported that it has completed a soliton data transmission experiment that has succeeded in sending soliton signals at 20 Gbit/s over a distance of 1020 km using erbium-doped fiber amplifiers (EDFAs). NTT is currently conducting experiments with 10-ps pulses. It is possible to increase the transmission speed to 100 Gbit/s by using 1- to 3-ps pulses, and to the terabit-per-second level by means of wavelength multiplexing [22].

Summary

We have discussed temporal and spatial solitons separately. Each has applications in its own field. From the statement that a soliton involves cancelation of terms in the propagation equation, one might get the impression that the soliton would be unstable and its operation critical. This is not true. Practically any minimum-bandwidth pulse of reasonable shape that has an approximately 2:1 ratio in peak light intensity will excite a soliton. Also, it was found that lumped erbium-doped amplifiers placed at intervals of less than 100 km will work equally well as distributed amplifiers.

Is it possible to have a pulse that is simultaneously a temporal soliton and a spatial soliton? If we can make such a soliton, it will be a light bullet that perhaps will revolutionize optical switching. Only time will tell.

The suggested time/polarization multiplexing scheme offered a simple and inexpensive way to double the carrying capacity of a single-mode fiber transmission system. It enhances the advantage of using solitons for transmission.

Solitary waves exist in nature as first described by Scott-Russell in water waves [1], but most of the subsequent research has concentrated on the theory of the formation on soliton waves. The technological innovation employing solitary waves in optical communication systems may be the first engineering application of this kind; however, back in 1967, Gardner *et al.* showed that solitary waves are readily formed in nature and only very special initial conditions can prevent the creation of solitary waves [23]. Many examples of solitary waves have since been found and investigated. Atmospheric solitary waves were observed to have been crossing the skies of Missouri, Illinois, Indiana, and Ohio in 1989, bringing increased atmospheric pressure, gusty winds, and rain. Ramamurthy *et al.* in 1990 reported on these effects [24]. The effects of solitary waves have been observed in the Andaman Sea near Thailand. For some time, photographs of this water from satellites and manned spacecraft have shown curious striations of water waves stretching for great distances and moving in parallel groups. It has also been speculated that solitary waves may exist in biological systems, although to date, the evidence remains inconclusive [25].

The extensive presence in nature of solitary waves, their stimulating characteristics for theoretical investigations, and the potential practical applications for technology may expand our interest in this subject in the near future.

Professor H. A. Haus of MIT recently published an article on soliton formation. He gave a simple explanation of the mechanism and presented a computer simulation showing that even with a square wave input pulse, a clean soliton can be carved out by the fiber's dispersion and Kerr nonlinearity [26, 27].

The Gordon-Haus effect due to amplified emission (ASE) noise which causes jetter in pulse that limits the operation range of a soliton wave can be minimized simply by inserting filters along the path of transmission. This makes the use of solitons for long-haul signal transmission even more attractive.

References

1. J. Scott-Russell, Report on waves. *Proc. R. Soc. Edinburgh,* pp. 319–320 (1844).
2. D. J. Korteweg and G. deVries, On the change of form of long waves advancing in a rectangular canal, and on a new type of long stationary waves. *Philos. Mag.,* pp. 422–443 (1895).
3. A. C. Scott, F. Y. F. Chu, and D. W. McLoughlin, The solitons: A new concept in applied science. *Proc. IEEE.* **61,** 1440–1483 (1973).
4. A Hasegawa and Y. Kodama, Signal transmission by optical solitons in monomode fibers. *Proc. IEEE* **69,** 1145–1150 (1981).
5. A Hasegawa, Amplification and reshaping of optical solitons in glass fiber. *Opt. Lett. B.* **8,** 650–652 (1983).
6. A. Hasegawa, Numerical study of soliton transmission amplified periodically by stimulated Raman process. *Appl. Opt.* **23,** 3302–3309 (1984).
7. L. F. Mollenauer, R. H. Stolen, and H. Islam, "Experimental demonstration of soliton propagation in long fibers: Loss compensated by Raman gain. *Opt. Lett.* **10,** 229–231 (1985).
8. L. F. Mollenauer, J. P. Gordon, and M. N. Islam. Soliton propagation in long fibers with periodically compensated loss. *IEEE J. Quantum Electron.* **QE-22,** 157–173 (1986).
9. D. Marcuse, In *Optical Fiber Telecommunications II* (Miller and Kaminow, Eds.), Chap. 3, Academic Press, San Diego, 1988.
10. R. H. Stolen and C. Lin, Self-phase modulation in silica optical fibers. *Phys. Res. A* **17,** 1448–1453 (1978).
11. R. A. Linke, Modulation inducted transient chirping in single frequency lasers. *J. Quantum Electron* **QE-21,** 593–597 (1985).
12. R. H. Stolen and E. P. Ippen, Raman gain in glass optical waveguides. *Appl. Phys. Lett.* **22,** 276–278 (1973).
13. C. R. Giles, E. Desurvire, J. R. Talman, J. R. Simpson, and P. C. Becker, 2 Gb/s signal amplification at $\lambda = 1.53$ μm in an erbium-doped single-mode fiber amplifier. *J. Lightwave Technol.* **7,** 651–656 (1989).
14. L. F. Mollenauer, J. P. Gordon, and S. G. Evangelides, Multigigabit soliton transmissions traverse ultralong distances. *Laser Focus World,* pp. 159–170 (Nov. 1991).
15. L. J. Andrews, T. Mukai, N. A. Olsson, and D. N. Payne (Eds.), Special issue on optical amplifiers: A joint IEEE/OSA publication. *J. Lightwave Technol.* **9,** No. 2 (1991).
16. J. S. Aitchison and Y. Silberberg, Observation of spatial optical solitons in a nonlinear glass waveguide. *Opt. Lett.* **15,** 471–473 (1990).
17. L. F. Mollenauer, S. G. Evangelides, Jr., and H. A. Haus, Long-distance soliton propagation using lumped amplifiers and dispersion shifted fiber. *J. Lightwave Technol.* **9,** 194–197 (1991).
18. L. F. Mollenauer, M. J. Neubelt, S. G. Evangelides, J. P. Gordon, J. R. Simpson and L. G. Cohen, Experimental study of soliton transmission over more than 10,000 km in dispersion shifted fiber. *Opt. Lett.* **15,** 1203–1205 (1990).
19. J. P. Gordon and H. A. Haus, Random walk of coherently amplified solitons in optical fiber transmission. *Opt. Lett.* **11,** 665 (1986).

20. S. G. Evangelides, Jr., L. F. Mollenauer, J. P. Gordon, and N. S. Bergano, Polarization multiplexing with solitons. *J. Lightwave Technol.* **10,** 28–35 (1992).

21. R. Herman, Solitary waves. *Am. Scientist* **80,** 350–361 (1992).

22. S. M. Dambrot, NTT advances soliton transmission. *Photonics Spectra,* p. 50 (Sept. 1992).

23. C. S. Gardner, J. M. Greene, M. D. Kruskal, and R. M. Miura, Method for solving the Korteweg–deVries equation. *Phys. Rev. Lett.* **19,** 1095–1097 (1967).

24. M. K. Ramamurthy, B. P. Collins, R. M. Rauber, and P. C. Kennedy, Evidence of very-large-amplitude solitary waves in the atmosphere. *Nature* **348,** 314–317 (1990).

25. H. C. Tuckwell, Solitons in a reaction-diffusion system. *Science* **205,** 495–498 (1979).

26. H. A. Haus, Molding light into solitons. *Spectrum,* pp. 48–53 (Mar. 1993).

27. H. A. Haus. Optical fiber solitons, their properties and uses. *Proc. IEEE,* **81,** (7) pp. 970–983 (1993).

CHAPTER 10

Phase Conjugators

Introduction

In this chapter we review another aspect of nonlinear optics, one that has opened up a new field of research into the physics of optical materials and has provided vast possibilities for device applications. It is optical phase conjugation.

The phenomenon of optical phase conjugation was first observed by Zel'dovich *et al.* in 1972 [1]. They directed an intense beam of red light from a pulsed ruby laser through a frosted glass plate which distorted the light wavefront badly. The beam was then directed into a long tube containing high-pressure methane gas. The "reflected" beam from the gas-filled tube passed through the frosted glass plate again. This output was not, however, distorted. In other words, the distortion introduced by the frosted glass was undone. They explained that the intense laser light interacted with the gas molecules in the tube by stimulated Brillouin scattering (SBS), and this SBS light was traveling backward. The nonlinear interaction produced a special mirror. The beam "reflected" by the gas faithfully carried all the distortions introduced by the frosted glass plate, but now in reverse, canceling the effects originally produced. The backward traveling wave can therefore be thought as the time-reversed replica of the incident wave [2]. Reflection from an ordinary mirror would have increased the distortion.

Reflections of a diverging wave from an ordinary mirror and a phase-conjugate mirror are compared in Figs. 10.1A and 10.1B, respectively. Light radiating from a point source at zero illuminates both mirrors. From the ordinary mirror, the reflected wave reflects the beam with a change in direction according to the law of reflection, as shown in Fig. 10.1A, and diverges further as it travels away from the mirror. The wave reflected from the phase-conjugate mirror redirects the diverging beam, forming a converging beam, and returns to the source at zero as shown in Fig. 10.1B. The beam is said to have retraced its path back to the origin in a time-reversed manner.

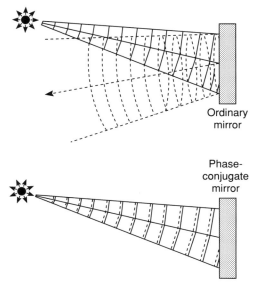

Figure 10.1 Comparison of the reflections from an ordinary plane mirror and a phase-conjugate mirror. Note that the angle of reflection from the former obeys the law of reflection. That from the latter does not. After D. M. Pepper *et al.* [16] with permission.

Many practical applications have been suggested. These include applications to radar for target identification and detection of moving targets, image processing, image restoration in optical fibers, phase-conjugate interferometers and image subtractors, and phase-conjugate fiberoptic gyros.

Phase Conjugation

The term *phase conjugation* is derived from the mathematical description of an electromagnetic radiation field in a dielectric medium. The electric field is written similarly to Eq. (2.1) in Chapter 2, with a change in notation: E is used instead of the wavefunction Φ, for a plain wave propagating in the z direction.

$$\mathbf{E} = \mathbf{A}(r) \exp i[\omega t - kz - \phi(r)] \tag{10.1}$$

Here, $A(r)$ is the amplitude, $\phi(r)$ is the phase, ω is the angular frequency, z is the direction of propagation, and k is the propagation constant defined as $d\omega/d\beta$. Equation (10.1) defines a wave traveling in the $+z$ direction with a phase angle ϕ. Defining a complex amplitude $\mathbf{A}_1(r) = A(r) \exp[-i\phi(r)]$, and assuming that $A(r)$ is a slow varying function, then Eq. (10.1) can be rewritten as

$$\mathbf{E} = \mathbf{A}_1(r) \exp i(\omega t - kz) \tag{10.2}$$

The electric field is now expressed in terms of a complex amplitude $A_1(r)$ which is a function of position and phase angle $\phi(r)$.

If the conjugate rule of complex algebra is applied, the conjugate of E, written as E*, becomes

$$E^* = A_1^*(r) \exp i(\omega t + kz) \tag{10.3}$$

Note that the phase-conjugate wave is found by taking the complex conjugate of only the spatial part of **E,** leaving the temporal part unchanged. Also note that Eq. (10.3) represents a wave traveling in the $-z$ direction. It is therefore more appropriate to think of the process as a reflection with a phase reversal. As the conjugate field expression can be obtained by leaving the spatial part of **E** unchanged and reversing the sign of the time t, we may refer to this process as *time reversal*.

Before we proceed to discuss the methods used to achieve phase conjugation, let us illustrate the four-wave mixing scheme. To produce phase conjugation in a nonlinear medium by four-wave mixing, three waves are mixed to yield a fourth wave. E_1 is a signal wave and E_2 an output wave, which we wish to make the phase conjugate of the signal wave E_1. The other two waves, E_3 and E_4, are a pair of pump waves. An interesting case occurs when we let all waves have the same angular frequency ω, or a degenerated special case if we let the pump waves be a pair of counterpropagating waves. Then, as all angular frequencies are the same, we may write

$$\omega_1 = \omega_2 = \omega_3 = \omega_4 \tag{10.4}$$

We wish to define the necessary condition to achieve phase conjugation. For this purpose, we need to bring the wavevectors into the picture.

The nonlinearity involved in phase conjugation is derived from the third-order polarization vector P_{nl} discussed in Eq. (2.9). We first rewrite P_{nl} for the third-order nonlinearity terms. For a case involving three waves, $E = E_1 + E_2 + E_3$, and

$$P_{nl} = \epsilon_0 \chi^3 E_j^3 = \epsilon_0 \chi^3 [E_1 + E_2 + E_3]^3 \tag{10.5}$$

where $E_j = A_j(r) \exp[i\omega t - ik_j \cdot r]$, $j = 1, 2$, and 3, and χ^3 is the third-order susceptibility. The polarization vector is made up of many frequencies, including $3\omega_1$, $3\omega_2, 3\omega_3, \ldots, \omega_4 = (\omega_1 - \omega_2 - \omega_3)$, etc. The amplitude $P_{nl}(\omega)$ of the component frequency ω (equals the permutations of different frequency components) involves six permutations. We assume that only one grating gives rise to strong interaction among the beams. The one that is particularly interesting in phase conjugation is the difference polarization which oscillates at

$$\omega_2 = \omega_4 + \omega_3 - \omega_1 \tag{10.6}$$

and

$$P_{nl}(\omega) = \epsilon_0 \chi^3 [E_3 E_4 E_1^*] \tag{10.7}$$

where E_1^* is the complex conjugate of E_1. Note that Eq. (10.4) satisfies Eq. (10.6), which is the frequency matching condition.

The expressions of the three beams are distinguished by their wavevectors $\mathbf{k}_j$. In the case of four-wave mixing for phase conjugation, the wavevectors come in two oppositely directed pairs.

$$\mathbf{k}_3 = -\mathbf{k}_4, \qquad \mathbf{k}_1 = -\mathbf{k}_2 \tag{10.8}$$

Thus the retroreflected wave is like a time reversal of the incident wave.

Equation (10.8) is the condition necessary to achieve phase conjugation in a four-wave mixing scheme in a nonlinear medium.

Figure 10.2 shows the operational schemes of a phase conjugator produced by nonlinear four-wave mixing. Two strong pump waves A_3 and A_4 counterpropagate into the nonlinear medium. The signal wave A_1 travels into the medium in the z

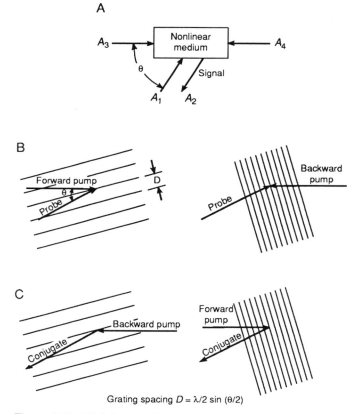

Grating spacing $D = \lambda/2 \sin(\theta/2)$

Figure 10.2 (A) Scheme of four-wave mixing through a nonlinear medium. (B) Formation of gratings as a result of interactions between the signal wave and the forward pump wave and the backward pump wave (the writing processes). (C) Formation of the conjugate wave when the reflections from the gratings are read out by the pump waves. After D. M. Pepper *et al.* [16] with permission.

direction. The generated fourth wave A_2, the phase-conjugated wave, travels in the $-z$ direction, thus retracing the signal in the opposite direction.

The ratio of the amplitudes of the reflected wave to those of the incident wave is

$$\left[\frac{\mathbf{E}_2}{\mathbf{E}_1^*} \right] = \epsilon_0 \, \chi^3 \, \mathbf{A}_3 \, \mathbf{A}_4 \tag{10.9}$$

which is proportional to the product of the amplitudes of the pump waves. As the pump waves are much stronger than the signal wave, amplification of the signal wave is possible even with very small χ^3. The power gain is derived from the pump supply. Equation (10.9) is also known as the phase-conjugate reflection coefficient. It is greater than unity for a conjugator.

A second-harmonic method to produce phase conjugation can be demonstrated in the same manner.

The function A_1 in Eq. (10.2) describes the spatial variation of the field as

$$A_1 = A(x,y) \, \exp \, i\{ -kz + \phi(x,y)\}$$

Multiply A_1 by $\cos 2\omega t$ and substitute in Eq. (10.2). The result can be written as

$$\mathbf{E} = \tfrac{1}{2}\{\mathbf{A}_1(x,y) \, \exp \, i[-kz + 3\omega t] + \mathbf{A}_1^* \, \exp \, i[kz - \omega t] \tag{10.10}$$

Note that $\mathbf{A}_1^*$ appears as in Eq. (10.3), which is the conjugate of $\mathbf{A}_1$.

Methods for Generating Phase Conjugation

There are several standard methods for producing phase-conjugate waves: stimulated Brillouin scattering (SBS), stimulated Raman scattering (SRS), the Kerr effect, and four-wave mixing in nonlinear optical media. Another possibility is the use of a non-linear transducer, which senses the signal and mixes it with the second-harmonic signal frequency electronically to produce phase conjugation. We review each method briefly as follows.

Stimulated Brillouin Scattering

Stimulated Brillouin scattering involves the generation of a coherent acoustic wave when an intense optical pump wave interacts with a weak signal wave within the nonlinear medium. The interaction is through the electrostriction effect and produces a sound wave. The input wave Bragg scatters from the sound wave to produce the scattered light wave. The intensity of interaction increases in proportion to the electric field intensity. These two couplings reinforce one another. Above a threshold condition, a large fraction of the pump light can be converted into the scattered light wave. The process in which the Brillouin gain is greatest is the one in which the scattered wave is the conjugate of the incident wave [1]. One can think of the process as creating, in the medium, a deformable mirror whose surface is just right to reverse the phase of the reflected wave from that of the incident wave. Thus, when the reflected wave retraces the incident path, the medium removes from it whatever phase errors were introduced in the first pass.

Stimulated Brillouin scattering produces phase conjugation very efficiently and has been observed over a wide range of wavelengths under both CW and pulsed conditions. But it requires a high optical intensity.

Stimulated Raman Scattering

Light waves can be stimulated and scattered by the Raman effect. In this case, the interaction is between the light wave and the longitudinal optical phonons in the solid. Raman scattering is also known as Stokes scattering. It causes the light wave to downshift by a frequency corresponding to the molecular vibrations of the molecules in the solid. The coupling between them is through the changes in polarizability with a molecular coordinate. This is different from Brillouin scattering, where the interaction is through the electrostrictive effect. The frequency shift of the phase conjugation via Raman scattering is nearly independent of scattering angle.

Kerr-Like Effect

A Kerr-like effect is characterized by the intensity-dependent index of refraction in the form

$$n = n_0 + n_2(E^2) \qquad (10.11)$$

where n_2 is the nonlinear index of refraction, which is related to χ^3, the third-order effect introduced in Chapter 2. An increase in n_2 with an increase in light intensity will cause a decrease in the speed of the light wave. This effect arises from a forced orientation shift of an elongated molecule along an applied electric field and changes the polarizability tensor of the medium.

The intensity-dependent index is also responsible for the self-focusing effect. The transverse variation in refractive index caused by the intensity variation of the beam propagating in the z direction may act as a lens formed in the material to concentrate the beam along the beam axis.

Four-Wave Mixing

Four-wave mixing is a convenient method to produce phase conjugation in nonlinear optical media. As described in the preceding sections, a nonlinear medium is pumped by a pair of counterpropagating light beams. When a signal is incident on the medium, a phase-conjugate wave is generated. This generated beam is traveling opposite to the signal beam and is the time-reversed replica of the signal beam, provided that the conditions of frequency and phase matching are met.

Phase conjugation in refractive media can be improvised in many ways. We describe these under Photorefractive Effect.

Second-Harmonic Mixing

In second-harmonic mixing a transducer is used to sense the signal. The signal collected from a transducer element is electronically mixed with a second-harmonic signal. The difference frequency signal so obtained has a component whose phase is the conjugate of that of the incident wavefront, as described in Eq. (10.10).

Analogous Holographic Process

The steps leading to production of the fourth beam in four-wave mixing are analogous to those of conventional holography. In holography the photogrphic emulsion is illuminated with light from an objective and a reference beam. The interaction between these beams produces an interference pattern in the medium or a hologram on the emulsion. The film is fixed and is ready for viewing. To view the hologram, it is exposed to the same reference beam again. The result is a three-dimensional reproduction of the object's image.

In a phase-conjugate medium, beams that overlap within the medium induce gratings in the volume where the overlapping occurs. These gratings can be spatial variations in the refractive index of the material. In four-wave mixing the nonlinear medium acts as the photographic emulsion. Two pumps are needed. The pump beam (say the forward pump) will interact with the signal (or probe) beam to yield a grating (on the emulsion), and the information is stored on it. An output beam is generated when the other pump wave (say the backward pump) reads the stored information by diffracting light from these gratings. If the direction of the readout beam is selected properly (counterpropagating), the diffracted beam will emerge as a wavefront-reversed replica of the input signal originally written on the grating. The output then yields a phase-conjugated or time-reversed beam. Now, the same signal wave will also interact with the backward pump wave to yield another set of diffraction gratings, which, when they interact with the forward pump wave, will also yield a conjugated wave of the same direction.

There is a difference between conventional holography and nonlinear optical phase conjugation. In conventional holography, the steps of recording, developing, fixing, and playing back occur as discrete steps; the equivalent processing steps in phase conjugation can occur simultaneously. Thus, the optical phase conjugator can respond to time-varying changes of an input beam instantly, and permits continuous tracking of the time-varying information.

Classification of Phase Conjugation in Refractive Media in Four-Wave Mixing

Phase conjugation can be classified as (1) externally pumped phase conjugation (EPPC), (2) self-pumped phase conjugation (SPPC), and (3) mutually pumped phase conjugation (MPPC) [3, 4].

Externally Pumped Phase Conjugation

The phase conjugation described in the preceding section involves the use of two external pumps that interact with the signal frequency to produce phase conjugation. It requires a pair of external counterpropagating beams for the pumps. The alignment of these pumps to achieve the phase-matching condition $\mathbf{k}_3 + \mathbf{k}_4 = 0$ is essential to obtain good-quality conjugation.

Equation (10.9) represents the reflection coefficient of the phase conjugation and can be used to design phase conjugators. Average reflectivities in excess of unity (or amplifying) are possible. The frequency of the conjugate wave generated is determined by the input beam and undergoes a well-defined overall phase shift on reflection. The conjugation is generally not threshold limited.

To implement externally pumped phase conjugation using a photorefractive medium, the scheme shown in Fig. 10.2 can be used. It requires two counterpropagating pumps to produce the gratings in the nonlinear crystal to interact with the signal wave to produce the conjugate wave. To produce good-quality conjugation, the oppositely propagating pumps must be aligned precisely; however, a mirror can be used to replace one of the pumps for the same purpose. The wave reflected from the mirror back into the nonlinear medium serves as the second pump. In this manner, alignment of the direction of propagation becomes less critical.

Self-Pumped Phase Conjugation

The inconvenience of using two external pumps can be eliminated if a self-pumping scheme is improvised. In this case, the signal wave, whose phase-conjugate replica is sought, acts to simulate the scattering interaction itself. In other words, the incident wave generates its own phase-conjugate wave. This has been demonstrated [5]. Generally, an input beam in excess of a given threshold intensity is required. The average reflectivity of this scheme is less than unity; the frequency of the conjugated wave may be shifted relative to the input beam and becomes material dependent. This is a result of the Stokes effect.

Various schemes for self-pumped phase conjugation are possible. In one case, a photorefractive resonator may be used between a pair of mirrors. Sustained oscillation between the two mirrors provides the counterpropagating beams needed for generating the phase-conjugate wave via four-wave mixing (Fig. 10.3A). It is known that a frequency shift (Stokes shift) is associated with SPPC. The Stokes shift may be positive or negative, representing an up- or downshift in frequency. For photorefractive crystals such as $BaTiO_3$ at an operating power on the order of 1 W/cm², the shift is about 1 Hz.

Other schemes for SPPC include the use of bidirectional oscillation in ring resonators (Fig. 10.3B) and oscillation between two parallel surfaces of a photorefractive crystal (Fig. 10.3C). In Fig. 10.3B, the oppositely directed oscillating beams in the ring resonator provide the pumping needed in four-wave mixing. Self-pumped conjugation can occur in a single cube of photorefractive crystals under conditions of parallel surface cuts.

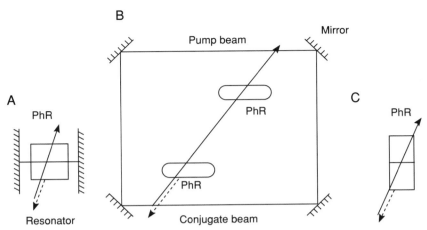

Figure 10.3 Schemes for producing self-pumped phase conjugation. (A) With a resonator; (B) with a ring resonator; and (C) with crystal boundaries.

Mutually Pumped Phase Conjugation

Mutually pumped phase conjugation requires two pump beams as does EPPC. But one of the pump beams is derived from the first pump of a self-pumped scheme. In this way, the two beams are conjugates of each other; they do not need to be aligned to be counterpropagating. The pump beam need not be plane waves. For example, stimulated Brillouin scattering self-pumping is employed to produce the second pump [6].

Polarization-preserving phase conjugation is another subject we wish to address. A phase conjugator employing photorefractive materials may respond to only one linearly polarized component of an incident electromagnetic wave, leaving the other components less affected. The resultant phase conjugation does not retroreflect all parts of an incoming electromagnetic wave, and is therefore not a true phase conjugator. In this case, the incoming wave should be separated according to the polarization orientation and separate phase conjugations performed.

Photorefractive Effect

Photorefractive materials play a major role in photonic phase conjugation. We wish to expand the discussion on this important subject.

The index of refraction of a medium can be changed by exposing the medium to an optical beam with high variable intensity [7, 8]. The presence of laser beams inside some electrooptic crystals leads to index inhomogeneity, which distorts the wavefront of the transmitted laser beam. Back in the 1960s, these changes were referred to as optical damage [9]. It was thought that the strong laser field had damaged the medium. But the effect is reversible and the damage does not seem to be permanent.

In response to an oscillating electric field of a monochromatic light wave, the bond charges of the isotropic dielectric medium will vibrate with the same frequency as the electric field. For low fields (less than 3×10^{10} V/cm), the induced polarization vector $\mathbf{P}_i$ of the dielectric medium is proportional to the electric field $\mathbf{E}_j$ and can be written as

$$\mathbf{P}_i = \epsilon_0 \sum_j \chi_{ij} \mathbf{E}_j \tag{10.12}$$

where ϵ_0 is the permittivity in free space, and χ_{ij} is a dimensionless linear susceptibility coefficient in tensor form. (We use tensor form in general equations. But for certain crystal symmetry and electric field orientation, the coefficients can be reduced to a simple manageable form for practical applications.) This notation is necessary to account for the nonuniformity of the coefficients in different directions.

If the electric field of the monochromatic light wave is sufficiently large, the dipoles of a dielectric medium will no longer be sufficient to describe the optical properties of the medium. To include the nonlinear effects, Eq. (10.11) is modified as

$$\mathbf{P}_i = \epsilon_0 \sum_j \chi_{ij} \mathbf{E}_j + \epsilon_0 \sum_{j,k} \chi_{ijk} \mathbf{E}_j \mathbf{E}_k + \epsilon_0 \sum_{j,k,l} \chi_{ijkl} \mathbf{E}_j \mathbf{E}_k \mathbf{E}_l + \ldots \tag{10.13}$$

Equation (10.12) is more general than Eq. (10.5) but is more complex. It includes higher-order nonlinear terms as well as the variation in the tensor coefficients in different orientations. In practical applications, however, we must take advantage of crystal symmetry to reduce the actual coefficient to a single number.

Note that the first term in Eq. (10.12) is a linear term similar to that in Eq. (10.11), which is linearly related to the electric field. The second term contains the power series of the second-order nonlinearities from which optical rectification, Pockel's effect, second-harmonic generation (SHG), sum and difference frequency mixing, and optical parametric amplification (OPA) and parametric optical oscillation (OPO) can be derived. The third term includes the power series dictating the effects of third-order harmonics generation (THG), the Kerr effect, third-order sum and difference frequency mixing, coherent anti-Stokes Raman scattering (CAR), general four-wave mixing, optical Kerr effect, stimulated Raman scattering, stimulated Brillouin scattering, phase conjugation, self-focusing, self-phase modulation (soliton), and so on.

Many crystals are found to process the photorefractive effect. These include $BaTiO_3$, $KNbO_3$ (KN), $LiNbO_3$, LiB_3O_5 (LBO), $\beta\text{-}BaB_2O_4$ (BBO), $Sr_{1-x}Ba_xMb_2O_4$ (SBN), $Ba_{2-x}Sr_xK_{1-y}Na_yNb_5O_{15}$ (BSKNN), $Bi_{12}SiO_{20}$ (SBN), $Bi_{12}GeO_{20}$ (GBO), GaAs, InP, and CdTe.

Physical Model of Photorefractive Medium

The crystals listed in the last section are known to have a photorefractive effect. The bandgap structure of the crystal can be represented, in general, by Fig. 10.4. Between the bottom of the conduction band edge E_c and the top of the valence band edge E_v,

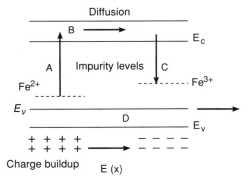

Figure 10.4 Energy-level diagram of a typical photorefractive material such as $LiNbO_3$. The processes that lead to the change in refractive index are shown in steps: (A) photoionization; (B) diffusion; (C) recombination; and (D) space-charge formation and internal field generation leading to refractive index changes.

there exist impurity bands. Usually both donor and acceptor types are present. When light exposure occurs, the following events take place in sequence:

1. Free charges are generated, either electrons or holes. As shown in Fig. 10.4, electrons are generated from the donor-level Fe^{2+} and raised to the conduction band, leaving behind fixed charges of opposite sign.

2. The elevated electrons then diffuse outward along the conduction band from the position of high density, where they were generated, until they are trapped by an ionized level such as the Fe^{3+} level, depositing their charges as they recombine and remain in place for a period after the light is removed.

3. The deposited charges then change the internal distribution of the field, which in turn modulates the index of refraction n.

The change in index of refraction can be expressed as

$$n(x) = -\tfrac{1}{2}n^3 r \, E(x) \tag{10.14}$$

where $E(x)$ is the electric field developed by the charges. This is the well-known Pockel effect.

Wave Mixing in Photorefractive Media

When two beams of coherent electromagnetic radiation interact inside a photorefractive medium, the periodic variation of the intensity caused by interference will induce a volume index grating [10]. Beam 1 is scattered by the index grating and the diffracted beam is propagating along the direction of beam 2. Similarly, beam 2 is scattered by the same grating and the diffracted beam propagates along the direction of beam 1. This leads to energy coupling; the scattering is known as self-diffraction.

Analysis of coupled systems in two-wave and four-wave mixing have been treated thoroughly by Yeh [4]. Besides reconfirming the conditions necessary for phase conjugation, Yeh discusses many schemes and a variety of phase conjugators for different applications are mentioned.

In fact, four-wave mixing is only one of the possibilities. Other schemes include self-pumped phase conjugation, in which the incident wave generates its own phase conjugation, and mutually pumped phase conjugation.

Brillouin-Enhanced Four-Wave Mixing

Under appropriate conditions, the backward-scattered light in stimulated Brillouin scattering may also occur in photorefractive media. This type of SPPC has received considerable attention because of its ability to achieve phase conjugation using pulsed lasers with extremely high reflectivities.

The first demonstration of phase conjugation by Zel'dovich *et al.* [1] was obtained by stimulated Brillouin scattering (SBS) [1]. Then it was found that using SBS to produce phase conjugation often requires a high input pump power, greater than the threshold. Other methods requiring less pump power were preferred. Research has shown that Brillouin-enhanced four-wave mixing can achieve high reflectivity in the region of 10^6. It is thus suitable for conjugation of very weak input signals [12] and is particularly useful in pulsed work. SBS-enhanced four-wave mixing can also efficiently transfer energy from a pump beam into the conjugated beam [12]. Research into the use of this technique has been revived in recent years.

The technique involved in this demonstration is near-degenerate four-wave mixing coupled with the backward-scattered light of stimulated Brillouin scattering. The incoming signal beam has a Brillouin frequency shift with respect to one of the pump beams. The interference produces an acoustic wave. This scatters the second pump beam to form the conjugate beam. In this process, the acoustic wave is amplified and its efficiency in scattering the pump beam increases until a significant fraction of the available power is extracted from the pump beam.

In a normal four-wave mixing scheme, large reflectivities greater than 100 are difficult to achieve. This is because to achieve conjugation extremely precise control of the conditions that lead to high reflectivity is required. For instance, in practice, instead of waves having the same frequency, the frequencies of the signal, the two pumps, and the conjugate are all slightly different, with shifts of typically one part in 10^5. The diffracted wave will not necessarily be conjugated. With Brillouin-stimulated scattering-enhanced four-wave mixing, the signal can be anti-Stokes shifted (i.e., upshifted) with respect to the pumps. When the pump intensity exceeds some critical intensity, the conjugate power will grow approximately exponentially. The maximum conjugate power is obtained if a Stokes-shifted signal is approached. If precautions, such as avoiding oscillations between the four-wave mixing cells and other spurious reflections are taken, very good quality phase conjugation can be achieved [13]. Thus, the Brillouin-enhanced four-wave mixing scheme (BEFWM) for generating phase conjugation is in favor again.

Applications of Optical Phase Conjugation

Aberration Compensation Using Phase Conjugation

Phase conjugators are used for aberration correction in many applications. A simple example is shown in Fig. 10.5, which compares the spatial properties of a phase-conjugate mirror with those of a conventional plane mirror. The experimental setup

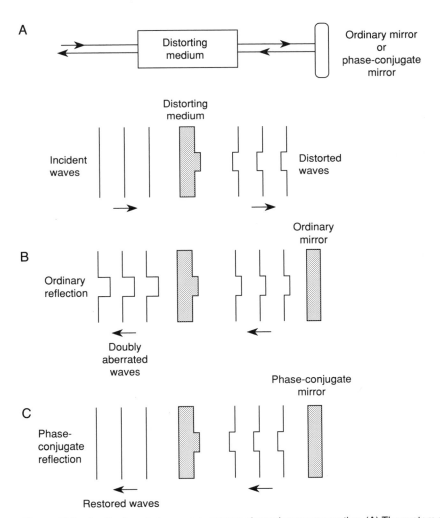

Figure 10.5 Application of phase conjugator to aberration compensation. (A) The system setup. (B) Reflected output waves from an ordinary mirror are doubly aberrated. (C) Reflected output waves from phase conjugate mirror have the aberrations removed. After D. M. Pepper *et al.* [16] with permission.

consists of a laser source that aims the beam through an aberrator, represented by a glass slab, where the wavefront distortions are introduced, as is shown in Fig. 10.5A. In Fig. 10.5B, the beam is shown to be reflected from a conventional mirror and back through the same aberrator again. The wavefront shown is doubly aberrated. In Fig. 10.5C, a phase-conjugate mirror replaces the conventional mirror and the wave reflected through the same aberrator shows no aberration. The aberration has been compensated. The wave reflected from the phase-conjugate mirror retraces its path and compensates for the aberration by advancing the same amount that the glass slab retarded the incident wave, leaving the emerging wavefront distortionless.

Many applications can be derived from this scheme. A modified system using this method has been used to measure atmospheric disturbances for weather prediction and forecasting [14].

Temporal Reversal of a Phase Sequence

A phase conjugator can be used to reverse a pulse sequence in time so that the last part of the sequence to enter the device is the first to leave (Fig. 10.6). In Fig. 10.6A, a pulse sequence is introduced onto an ordinary mirror. The reflected wave shows that the sequence of the pulse is preserved. In Fig. 10.3B, a phase-conjugate mirror is used. After the entire pulse has entered the crystal, the mirror is activated by turning on the two counterpropagating pumps. The emerging pulse has its sequence reversed. By passing through the device again, the emerging beam compensates for the distortion within the system.

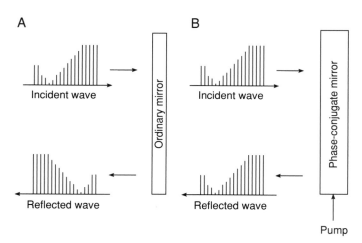

Figure 10.6 Temporal reversal of a pulse sequence. (A) Reflection from an ordinary plane mirror. (B) Reflection from a phase-conjugate mirror. Note that in (B) the sequence of pulses is reversed.

Transmission of High Power

Electrical power engineers are talking about building a solar power station in space, and then beaming the power via a microwave link or optical radiation toward a terrestrial site. The problem they face is obvious. Multimegawatt power transmission requires absolute accuracy and reliability in pointing the beam from the space to the earth station precisely. It seems that only the phase conjugator scheme could solve this problem. It is suggested that a space-based power-generating station would consist of a solar power generator, a high-energy laser oscillator, and a four-wave photorefractive mixing medium. The solar power that may be generated by solar cells acts as one pump source in the four-wave phase conjugation system. The high-power laser provides the other counterpropagating pump. The nonlinear medium could be contained in the laser cavity. An input signal from the earth station initiates a command to release the power to the earth station when the demand arises. The phase-conjugate process provides the coupling from the laser resonator, through the nonlinear medium, and returns the earth-directed output beam along the same path dictated by the original signal to the earth station. Power is sent only to the station that sent out the initial signal. Moreover, the energy circulating inside the spaceborne laser will be coupled out only when the pilot beam is present [15].

Fusion Targets

Use of laser power in the creation of fusion material has been proposed. Powerful multilaser beams are directed at a target pellet for a moment to raise the pellet temperature to millions of degrees, to fuse the elements together and derive energy from the fusion process. One suggested scheme is shown in Fig. 10.7. The system consists of a target, a laser amplifier, a phase-conjugate mirror and the necessary optics. The target pellet is illuminated by a weak optical source. The distorted beam is then sent through the optics to a laser amplifier. The amplified (but distorted) signal is sent through a phase conjugation mirror. The beam goes back through the laser amplifier again to increase its intensity and is refocused back on the pellet. These events are illustrated in Fig. 10.7 [16].

Phase conjugation can also be arranged to synchronize many pulses from a chain of laser amplifiers arranged in a parallel configuration simultaneously, thus increasing the power capability of the system.

Fiberoptic Transmission

In optical fiber transmission, as a result of fiber dispersion, pulse broadening occurs after a long-distance transmission. Pulse broadening occurs because the various frequency components that constitute a pulse propagate at slightly different speeds through the fiber. Pulse broadening usually limits the maximum useful distance that can be covered by the fiber if fiber loss has not already done so.

An attractive scheme has been suggested to alleviate the pulse broadening effect

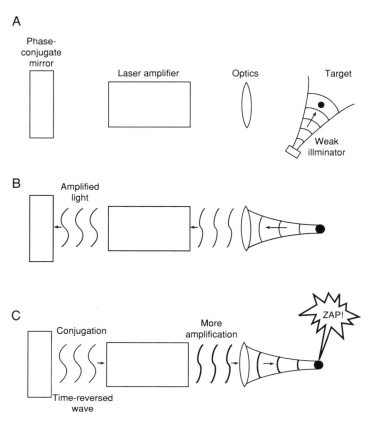

Figure 10.7 Application of phase conjugator to laser fusing experiment. (A) A weak light source is used to illuminate the target. (B) The reflection is led through a laser amplifier onto a phase-conjugate mirror. (C) The conjugated signal is led through the laser amplifier again and aimed at the pellet with amplified intensity. After D. M. Pepper *et al.* [16] with permission.

(Fig. 10.8). A four-wave mixing phase-conjugate mirror, pumped at the central frequency of the pulse, is placed midway along a long fiberoptic cable system. If it is assumed that both halves of the fiber links bear similar dispersion characteristics, as shown in the diagram, the phase-conjugated signal through the second half of the fiber will compensate for the original dispersion perfectly, thus eliminating the pulse broadening effect [17].

High-Power Phase Conjugate Diode Master Oscillator Power Amplifier

We have discussed the possibility of using semiconductor diode lasers to pump solid-state lasers in Chapter 5. Recent research further promotes the use of phase conjuga-

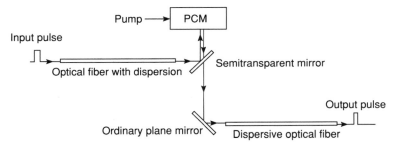

Figure 10.8 Fiberoptic transmission experiments with phase-conjugate compensation. Illustration of the pulse broadening effect resulting from fiber dispersion. PCM, phase conjugate mirror.

tion to increase the power output of the laser through the master oscillator power amplifier (MOPA) scheme.

The preceding sections have also shown that phase conjugation can be used to compensate for aberrations in solid-state media. As the phase distortions arising from optical path length differences among an array of gain media can also be expected to be eliminated, coherent coupling is possible. The coherent coupling is referred to as *beam combining*. It allows laser energy scaling beyond traditional limits imposed by the maximum available volume of a single active medium [17].

Stephens *et al.* investigated the possibility of using coherent coupling to build a high-power phase-conjugate diode MOPA according to the scheme shown schematically in Fig. 10.9 [18]. A semiconductor laser diode operating in a single longitudinal and transverse mode is used as a master oscillator. This is amplified by a ten-stripe, gain-guided, phase array amplifier. The output is directed into a phase-conjugate mirror (a BaTiO$_3$ crystal) arranged in a self-pumped ring configuration. With about 3 mW coupled into the amplifier, output powers as high as 52 mW are observed when the output is reflected from the conjugator and double amplified through the amplifier. The output beam quality is excellent: the primary far-field lobe has a divergence of 0.58°, slightly higher than the diffraction limit of 0.56°.

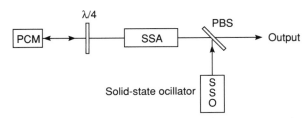

Figure 10.9 A master oscillator power amplifier (MOPA) scheme using phase conjugation. PCM, phase-conjugate mirror; PBS, polarization beam splitter; SSA, stripe slab amplifier; SSO, solid-state oscillator.

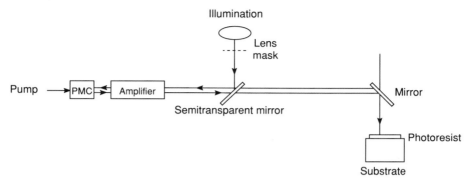

Figure 10.10   Transfer of a two-dimensional pattern in integrated microelectronic circuits by the photolithography method without physical contact with the substrate.

Photolithography

Photolithography is a technique in demand in modern integrated microelectronics. Projection of complex patterns onto photoresistive layers is of great technological importance in the integrated microelectronics fabrication industry. Projection systems using a conventional optical technique are very complex, because of the need for near-diffraction-limited, low-f-number performance. Non-contact-type photolithography is also highly desirable. Figure 10.10 suggests a system for this purpose. The process involves the use of a phase-conjugate mirror. The input beam from a laser passes through the mask, a semitransparent mirror, and then a laser amplifier. The laser amplifier may increase the power of the beam, but usually distorts its quality. After the beam is reflected from the phase-conjugate mirror, it travels back through the amplifier, undoing the distortions and emerging with the original beam information restored. When this beam is reflected from a semitransparent mirror, the intensified image of the mask pattern can be used to expose the photoresistor on the substrate. This system compensates for optical aberrations, has high resolution over a large field of view, eliminates laser speckle, minimizes beam spreading, and avoids physical contact with the substrate [19].

Note that in Fig. 10.10, only mirrors and perhaps a beam splitter or two are used. No optical lenses of any kind are involved. This imaging system is therefore lensless. The advantage of an imaging system without lenses eliminates the difficulty encountered in controlling the alignment of a large number of optical lenses otherwise deemed necessary. Many applications can be derived from lensless imaging.

Optical Resonator

If one or both of the reflecting mirrors of a laser cavity were replaced by a phase-conjugate reflector, we would have an optical resonator that behaves differently than the usual laser resonator. A unique property of phase-conjugate resonators is that such resonators do not possess a longitudinal mode that depends on cavity length. In an ordinary resonator, the round trip of the wave inside the resonator that has accumulated a total phase shift of two pi radians ($2n\pi$) can add constructively to sustain the oscillation. In phase-conjugate resonators, the total phase of the round trip is always zero. Consequently, a phase-conjugate resonator of length L can support any wavelength, as long as the bandwidth of the gain medium and the conjugation conditions are satisfied, independent of cavity length.

Another property of a phase-conjugate resonator is that as the intracavity distortion within the body of the resonator is compensated, the resonator becomes attractive for application to high-power oscillators [20].

In a four-wave phase conjugator using two counterpropagating pumps, the intensity of the output conjugate beam can exceed that of the input beam; optical amplification is thus possible. Moreover, phase-conjugate resonators are more stable than ordinary ones. The ability of a conventional resonator to store energy depends on the relation between radiation, cavity length, and mirror curvatures. A phase-conjugate resonator is free of such constraints.

Summary

We have described briefly the theory of phase conjugation in general and more specifically four-wave mixing and the photorefractive aspects of nonlinear optics. Note that we have touched on only a very small portion of a vast area of nonlinear optics with respect to both research and applications. For example, in using the pump scheme to generate another wave of different frequency, we leave a lot of stones unturned. We can easily manipulate the pump wave to achieve modulation and other desirable properties besides being the spatial conjugate of the input wave.

Phase conjugation is a practical, realistic approach to achieving high-quality beams in high-average-power solid-state lasers with a minimum increase in system complexity. Besides its time-reversal traveling capability to cancel distortions, the approach of phase-conjugate MOPA offers coherent coupling to produce efficient, long-lifetime, high-average-power solid-state laser sources. The fundamental question remains: Can good conjugation fidelity be maintained as the laser power is increasingly scaled to higher levels in the future?

Also, material having nonlinear optical properties is plentiful. Methods to grow large crystals at lower cost are always welcome. Progress in this field is occurring rapidly.

References

1. B. Ya. Zel'dovich, V. I. Popovichev, V. V. Ragul'skii, and F. S. Faizullov, Connection between the wavefronts of the reflected and exciting light in SBS. *Zh. Eksp. Theor. Fiz.* **15**, 160 (1972).
2. V. V. Shkunov and B. Ya. Zel'dovich, Optical phase conjugation. *Sci. Am.,* Dec. 1985.
3. R. A. Fisher (Ed.), *Optical Phase Conjugation,* Academic Press, San Diego, 1983.

4. P. Yeh, Photorefractive phase conjugators. Proc. *IEEE* **80**, 436–450 (1992).
5. J. Feinberg, Self-pumped continuous-wave conjugator using internal reflection. *Opt. Lett.* **7**, 486–488 (1982).
6. P. Yeh, T. Y. Chang, and M. D. Ewbank, Model for mutually pumped phase conjugation. *J. Opt. Soc. Am.* **85**, 1743 (1988).
7. V. L. Vinetskii, N. V. Kukhtarev, S. G. Odulov, and M. S. Soskin, Dynamic self-diffraction of coherent light 0 beams. *Sov. Phys. Usp.* **22**, 742–756 (1979).
8. N. V. Kukhtarev, V. B. Markov, S. G. Odulov, M. S. Soskin, and V. L. Vinetskii, Holographic storage in electrooptic crystals, beam coupling and light amplification. *Ferroelectrics* **22**, 961–964 (1979).
9. A. Ashkin, G. D. Boyd, J. M. Dziedzic, R. G. Smith, A. A. Ballman, J. J. Livinstein, and K. Nassau, Optically induced refractive index inhomogeneities in LiNbO$_3$. *Appl. Phys. Lett.* **9**, 72–74 (1966).
10. N. F. Andreev, V. I. Bespalov, A. M. Kiselev, A. Z. Matveev, G. A. Pasmanik, and A. A. Shilov, Wavefront inversion of weak optical signals with a large reflective coefficient. *Zh. Eksp. Teor. Fiz.* **32**, 639–642 (1980); *Sov. Phys. JETP Lett.* **32**, 625–629 (1981).
11. N. F. Andreev, V. I. Bespalov, M. A. Dvoretskii, and G. A. Pasmanik, Four-wave hypersonic reversing mirrors in the saturation regime. *Kvant. Elektron.* **11**, 1476–1479 (1984); *Sov. J. Quantum Electron.* **11**, 999–1000 (1984).
12. A. M. Scott and K. D. Ridley, A review of Brillouin-enhanced four-wave mixing. *IEEE J. Quantum Electron.* **25**, 438–459 (1989).
13. B. Ya. Zel'dovich and A. A. Shkunov, "Characteristics of stimulated scattering in opposite pump beams. *Kvant. Elektron.* **9**, 393 (1982); *Sov. J. Quantum Electron.* **12**, 223–225 (1982).
14. D. M. Pepper (Guest Ed.), Special issue on nonlinear optical phase conjugation. *IEEE J. Quantum Electron.* **QE-25**, No. 3 (1989).
15. V. V. Shkunov and B. Ya. Zel'dovich, Optical phase conjugation. *Sci. Am.* **253**, 54–59 (1985).
16. D. M. Pepper, D. A. Rockwell, and G. J. Dunning, Nonlinear optical phase conjugation. *IEEE Circuit Device*, pp. 21–35 (Sept. 1991).
17. D. A. Rockwell, A review of phase conjugate solid-state lasers. *IEEE J. Quantum Electron.* **24**, 1124–40 (1988).
18. R. R. Stephens, R. C. Lind, and C. R. Giuliano, "Phase conjugate master oscillator-power amplifier using BaTiO$_3$ and AlGaSAs semiconductor diode lasers. *Appl. Phys. Lett.* **50**, 647–649 (1987).
19. D. M. Pepper, Applications of optical phase conjugation. *Sci. Am.* **254**, 74–83 (1986).
20. D. M. Pepper, J. Feinberg, and N. V. Kukhtarev, The photorefractive effect. *Sci. Am.* **263**, 62–74 (1990).

CHAPTER 11

Photonic Components

Introduction

As optical fiber systems continue to grow and expand, and their range of applications broadens, there is an increasing demand for a variety of photonic components to be used in the system to fully exploit the capability of fibers. Optical functions to be performed by these components include the coupling of fibers to a process that modulates, multiplexes, switches, and coupled to the light source for transmission. At the receiving end, a combination of similar components are used in reverse order in conjunction with the demultiplexer and detector. Many other auxiliary components may be required between the light source and detector if other passive and active functions such as polarization change and amplification are needed. Thus, a complete fiber system may need more components to process than to generate and to transmit the signal.

Optical components in use with optical fibers may be classified by the material used to make them. The following types are described: the optical components, made from bulk dielectric material such as lenses, prisms, and mirrors; the graded-index (GRIN) rod lenses; and the guided-wave type of components made from $Ti:LiNbO_3$ planar waveguides. For the last components, special effects such as the electrooptic effect and the magnetooptic effect can be added to enhance the function for various purposes.

Photonic Components

For optical communication systems using optical fibers, it is natural to use optical components as much as possible if conditions permit. Lens, prisms, mirrors, and so on are usually rather bulky. Thus, only microoptic components are considered.

Microoptic Lenses

The microoptic lens is a conventional dielectric (glass) lens of miniature size. The lens is used to focus or to collimate the beam into an input fiber and/or to focus the output beam onto the output fiber. The principle and the design of good lenses are well known [1]. Only engineering refinements of some particular components may require special treatment.

Conventional homogeneous thin lenses have little to offer fiber systems compared with the GRIN rod lenses to be discussed later. Spherical ball lenses, which are simply transparent spheres, have been used in wavelength multiplexers primarily for laser-to-fiber coupling. They are easy to fabricate. When made with high-index semiconductor material [2], they have high numerical aperture required to collect light from a laser. Numerical aperture (NA) is defined by the expression

$$NA = \{n_1^2 - n_2^2\}^{(1/2)}, \qquad n_1 > n_2 \qquad (11.1)$$

where n_1 and n_2 are the refractive indices of the surrounding material and of the lens, respectively. NA is also defined as the sine function of the acceptance angle.

Ball lenses made from the magnetooptic material yttrium iron garnet (YIG) can serve as polarization rotators for isolators [3]. Molded glass lenses have been used for fiber connectors. Molding is a well-established practical technique for volume production of moderate-quality lenses [4]. Plastic lenses in hemispheric or parabolic form are molded onto LEDs to shape the radiation pattern.

Mirrors

A mirror is used to reflect the light beam. The surface of a mirror may be planar, parabolic, elliptic or spherical, and it is so named. The law of reflection, $\theta_1 = \theta_3$ applies. θ_1 and θ_3 are the incident and reflected angles, respectively; both refer to the normal of the surface. In fiberoptic systems, a thin glass plate can be used to split or to combine a beam. It is often made by depositing a thin semitransparent metallic or dielectric film on a thin glass plate. A beam that strikes the mirror surface at 45° to the normal will have about half of its intensity reflected into the original medium. The other half is transmitted through the mirror in the original direction, assuming the plate is so thin that both the loss and the refraction of the beam are negligible. Figure 11.1A shows this configuration. If the incident beam is in a medium whose refractive index is n_1, and the glass plate has an index of n_2, and $n_{22} > n_{21}$, then the reflected wave suffers pi radians of phase change. This is known as external reflection. The magnitude of the reflected beam is governed by Fresnel's equation [1].

A parabolic or paraboloidal mirror is used to focus a beam to a single point on the axis. The distance along the axis from the surface to the focal point is called the focal length. Paraboloidal mirrors are also used to make a parallel beam of light from a point source placed at the focal point, the reverse of the process described earlier. Figure 11.1B shows this arrangement.

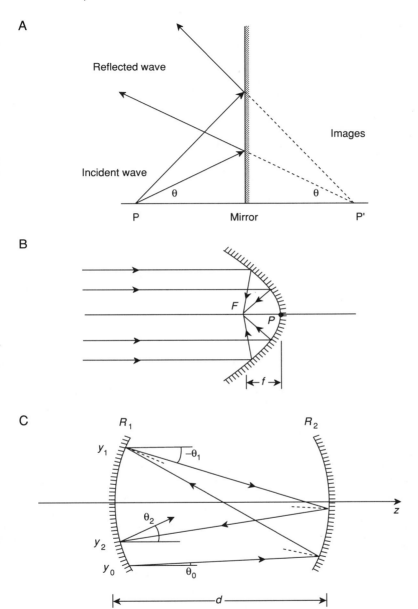

Figure 11.1 Reflections from optical mirrors. (A) Plane mirror. (B) Focusing of light by a para-boidal mirror. *f* is the focal length. (C) Two sections of a spherical mirror are separated by distance *d*.

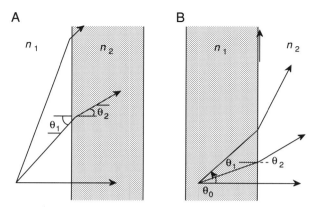

Figure 11.2   Relation between the angles of refraction and incidence in (A), external refraction, and (B), internal refraction.

Although a spherical mirror does not have the focusing property of a paraboloidal mirror, it is easier to fabricate. An optical resonator can be built using two concave sections cut out from a spherical mirror, placed facing each other and separated by a distance d, as shown in Figure 11.1C.

Prisms

A prism is used to refract a beam from its original direction. It is a flat dielectric body whose refractive index is different from that of the surrounding material. At an interface, a beam entering from one medium to the other suffers refraction, and the relationship between the angles of refraction and incidence obeys Snell's law: $n_1 \sin(\theta_1) = n_2 \sin(\theta_2)$. Both angles refer to the normal to the surface. Figures 11.2A and 11.2B show two cases, external and internal refraction, respectively. In external refraction, $n_1 < n_2$, a beam that is incident on the boundary from a medium of small to one of high refractive index will have the angle of refraction $\theta_2 < \theta_1$, and the refracted beam bends away from the boundary. In internal refraction, $n_1 > n_2$, a beam that is incident from a medium of higher to lower refractive index will cause $\theta_2 > \theta_1$; the refracted beam bends toward the boundary.

Polarizers

The electric field of the light beam is a vector quantity. Its direction in a coordinate system is used to define the polarization of the beam, for example, vertically or horizontally polarized beams. A device that modifies the state of polarization of the beam is called a polarizer.

The state of polarization of the beam in an isotropic dielectric material can be altered by selective absorption or selective reflection. The reflection of light from a

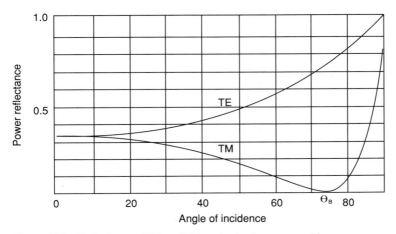

Figure 11.3   Reflectance of TE and TM modes in plane waveguide.

boundary between two isotropic materials is polarization dependent. The reflected power (reflectance) of TE and TM polarized plane waves at the boundary between air ($n = 1$) and GaAs ($n = 3.6$) as a function of the incidence angle θ is shown in Fig. 11.3. At the Brewster angle of incidence θ_B, light of the TM polarization is not reflected; only the TE mode is reflected. The Brewster angle is defined by the expression $\theta_B = \arctan\{n_2/n_1\}$.

In an anisotropic medium, selective refraction and selective absorption/reflection are also used to affect the state of polarization. Dichroic material is a known nonisotropic material whose molecular structure responds differently to the direction of electric field that is incident on it.

In fiberoptics, prisms made of anisotropic media can serve as polarizing beam splitters. One type, the Wollaston prism, is shown in Fig. 11.4. When unpolarized

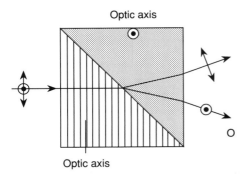

Figure 11.4   Wollaston polarizer, formed by cementing two pieces of anisotropic crystals in different orientations.

light refracts from the surface of an anisotropic crystal, out come two beams that are orthogonally polarized and refracted at different angles as shown. Wollaston prism [5] is made by cementing two anisotropic crystals in different orientations as shown in the diagram.

Diffraction Gratings

When a light beam passes through a thin transparent plane of periodically varying thickness, the incident wave is diffracted into multiple plane waves traveling in different directions. The angle of the emerging waves can be expressed as

$$\theta_P = \theta_i \pm p\left(\frac{\lambda}{\Lambda}\right) \qquad (11.2)$$

where $p = 0, \pm 1, \pm 2, \ldots$, is the diffraction order, θ_p is the angle of the pth wave, and θ_i is the incident angle, all with respect to the z axis. λ is the incident beam wavelength and Λ is the period of the thickness variation where $\lambda \ll \Lambda$. Such a medium is a diffraction grating. The diffracted waves are separated by an angle $\theta = \lambda/\Lambda$, as shown schematically in Fig. 11.5.

If two waves of different wavelengths, λ_1 and λ_2, are directed onto a diffraction grating transparency, the emerging beams are directed into two directions, making different angles with respect to the z axis. This property can serve as a spectrum analyzer. It can also be used as a filter; as θ_p is wavelength dependent, any desired frequency can be spatially selected.

A diffraction grating can be made of a transparent plate with periodically varying refractive index. Reflective diffraction gratings are fabricated by periodically ruling thin films of aluminum that have been evaporated onto a glass substrate. A diffraction grating can also be created on optical material using sound waves that diffracts light-waves in acoustooptic devices.

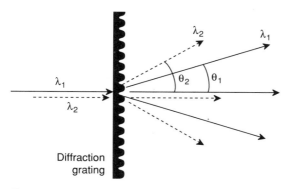

Figure 11.5   A diffraction grating directs two wavelengths λ_1 and λ_2 in two directions θ_1 and θ_2. It therefore serves as a spectrum analyzer or a spectrometer. After Saleh & Teich, *Fundamentals of Photonics,* with permission. © 1991 John Wiley, New York.

Optical Isolators

An optical isolator is a device that allows transmission of light in one direction only. It is used to prevent light in a system from reflecting back to the source, which could deteriorate its operation. The use of a Faraday rotator to change the beam polarization is one example (see Graded-Index Rod Lens Components).

Graded-Index Rod Lenses

The graded-index or gradient-index rod (GRIN rod) has a cylindrical structure. It has a refractive index profile that varies continuously in the radial direction given by $n(r)$. Although the GRIN rod lens was discovered as early as 1904, known as the Wood lens [6], its wide applications to imaging devices and optical fibers are only recent. It has a diameter of about 0.5 mm or larger. Lengthwise, it can be cut to any desired length for different applications.

The index profile of the GRIN rod has its maximum on the axis, and decreases as the square of the radius approximately as

$$n^2(r) = n_0^2\{1 - (\alpha r)^2 + b(\alpha r)^4 + c(\alpha r)^6 + \ldots\} \qquad (11.3)$$

where n_0 is the index at the fiber axis, α is a gradient constant, and b and c are high-order aberration coefficients that are usually small and can be neglected [7]. This approximation is known as the paraxial approximation. Equation (11.3) also applies to skew rays.

The properties of the GRIN lens can be summarized in Fig. 11.6, where ray paths for an object point on-axis and off-axis are shown in A and B, respectively. The

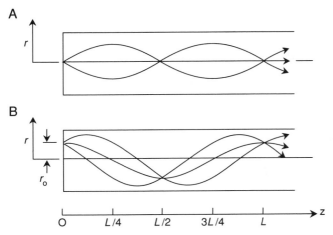

Figure 11.6 Ray paths of graded-index rod lenses. (A) On-axis rays. (B) Off-axis rays. After W. J. Tomlinson *et al.* [8] with permission.

following conclusions can be made: (1) the focusing property of the GRIN resembles that of a lens. Rays incident along the axis but at higher incident angles travel at a higher speed along paths further away from the axis, where the refractive indices are smaller. The increased speeds compensate for the longer paths taken than the axial ray and bring the rays to focus at $L/2$ and again at any integer multiples of $L/2$ along the axis. There is, however, no magnification of the image. Therefore it is a device of unity amplification. L is the period of an approximate sinusoidal path and $L = 2\pi/\alpha$, where α has been introduced in Eq. (11.3). The image formed at $L/2$ is inverted. (2) The image becomes erect at a length L. (3) For the off-axial rays, the focal point has been shifted to the opposite side of the axis, as shown in Fig. 11.6B; the image is inverted. (4) At intermediate points $L/4$ and $3L/4$, all rays emerging from a given object point are nearly parallel, as can be seen from Fig. 11.6B. This property can be used to collimate the rays. We discuss individual sections of GRIN lenses more thoroughly in the following sections.

GRIN rod lenses are made from a high-index base glass rod immersed in a molten salt bath, which allows it to go through a process of ion exchange. Through control of the ion species, the exchange time, and the temperature, any desired gradient profile can be produced. Because of the vast production capacity demanded by compact xerographic copiers, which use GRIN rod lenses, these lenses can be obtained commercially at a fairly reasonable price.

The Basic Building Block of Graded-Index Rod Lenses

The basic building block of GRIN rod lenses is a quarter-length unit ($L/4$), or a quarter-pitch lens. The end faces of a GRIN lens are planar. This makes it easier to mount with other components in the system. The focal length of a GRIN rod lens is about one to a few millimeters, and has a numerical aperture of about 0.1 to 0.5. A variety of microoptic components can be built using GRIN rod building blocks. For example, a GRIN rod can be used as a fiberoptic coupler. As an input coupler, it is capable of accommodating a larger beam width and focusing the beam to a smaller size, or vice versa. It is found that the aberrations of a GRIN lens are several times smaller than those of equivalent optimized conventional lenses [8].

If its length is doubled, making it a half-pitched GRIN lens, the GRIN lens can transfer a focused point at the entrance plane to an inverted one at the exit plane, thus making possible imaging of the original as in a copier.

Graded-Index Rod Lens Components

We describe some applications of GRIN rod lenses in optical fiber systems.

Connectors

Two sections of quarter-pitch GRIN rod lenses can be aligned and butted together into a sleeve to be used as a connector. A mechanical locking mechanism is provided

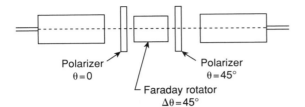

Figure 11.7 Schematic cross-sectional view of an isolator, showing two $L/4$ graded-index rod lenses, two polarizers with different polarization settings, and a Faraday rotator. After W. J. Tomlinson, *Optical Fiber Telecommunication, II.* (S. E. Miller & J. P. Kaminoa, Eds.) Reprinted by permission of Academic Press, 1988.

to prevent them from coming apart. As the focused input beam travels along the first $L/4$ section of the rod, the beam diverges at the end, enters the second section through an air gap, and finally is refocused at the output end. As the beam has diverged and is collimated at the junction of the two sections, the effects of misalignment and dust particles in the gap could become less important.

Isolators

To prevent light from reflecting back to the source to affect its performance, isolators are usually placed in the entrance path of the source. A scheme for an isolator is shown in Fig. 11.7. Here, two quarter-pitch GRIN rod lenses are aligned with a Faraday rotator in the middle. A Faraday rotator is a device that can rotate the polarization angle of a light wave traveling through it by applying a suitable magnetic field. A polarizer is attached to each side of the rotator; one is a fixed vertical polarizer and the other has a fixed 45° of polarization built in. Light traveling from left to right will have its polarization rotated 45° through the Faraday rotator and will pass through the polarizer to the output. The reflected light, if any, will pass the 45° polarizer and the rotator again, for a total 90° of rotation, and will be stopped by the first polarizer which passes only the zero-angle beam. Thus, the source is isolated.

Attenuators and Filters

Similarly, if the polarizers and the Faraday rotator are replaced by a number of disks made of plate glass, each coated with a thin metal film of varying thickness, a set of attenuators of different attenuations can be built. The orientation of the glass plates is purposely set at an angle to the beam axis to prevent reflection back to the input fiber. The glass plates are interchangeable so that a range of attenuations are available. As the beam is located in the collimated region, the location of the plates is less critical.

A variable attenuator can be built by replacing the plate with an attenuator plate equipped with spatially varying attenuation of the desired optical intensity.

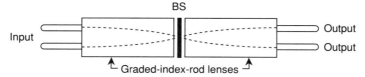

Figure 11.8   Directional coupler using two graded-index rod lenses. A beam splitter (BS) is inserted at the junction of the lenses to split the wave into different ports.

If the glass plate used between two quarter-pitch GRIN rod lenses is replaced by a thin-film filter that reflects all wavelengths except one wavelength channel, the combination becomes a filter.

Directional Couplers

A simple four-port directional coupler can be built using two sections of $L/4$ GRIN rod lenses with a beam splitter film between them. The input and output fibers are placed off-axis on both sides. The beam splitter film is located where the beam is focused. Near-normal incidence to the film is secured. This allows the device to be nearly insensitive to polarization of the input signal. Figure 11.8 shows such a scheme.

A variety of split ratios can be obtained by using different beam splitters. The excess loss is below 1.0 dB, and the crosstalk is typically less than -30 dB. The device is about 5 to 9 cm long and 0.5 cm in diameter.

Wavelength Division Multiplexers

The wavelength division multiplexing (WDM) scheme is useful in effecting multiple frequencies of carrier signal; hence the signal-carrying capacity of an existing optical fiber link is enhanced without having to install a new fiber. But this requires efficient couplers, a multiplexer at the input, and a demultiplexer at the output. With GRIN rod lenses, many schemes can be designed.

A simple multiplexer and demultiplexer pair using GRIN rod lenses and interference filters is shown in Fig. 11.9. It is essentially similar to that shown in Fig. 11.8, except the beam splitter film is replaced by an interference filter. An interference filter is a coated thin glass plate on which the phases of different waves can interact constructively or destructively to obtain the desired effect. In Fig. 11.9A is shown a multiplexer. Input waves at ports 1 and 3 are interfered at the interference plate to combine and output at port 4. In Fig. 11.9B, two wavelengths are input at port 1. The interference plate acts to divert the first wave to port 2 and the other wave to port 4. A sharp cutoff can be achieved with interference filters. With the recent developments in filter fabrication technology, a wavelength separation as little as 20 nm with an interchannel isolation of 30 dB is achievable [9].

An alternative scheme for a wavelength multiplexer uses only one GRIN rod lens

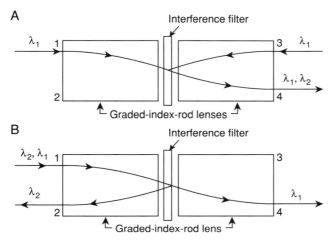

Figure 11.9 A pair of graded-index rod lenses with an interference filter as used in (A) a multi-plexer (λ_1 and λ_2 output together) and (B) a demultiplexer (λ_1 and λ_2 input together at port 1). From K. Kobayashi *et al.* [9] with permission.

as shown in Fig. 11.10. A coated filter and a mirror are placed at one end of the lens, leaving the other end free to attach many fibers. By means of diffraction gratings on the filter and by adjustment of the tilt angle of the mirror, two input waves can be combined into a single fiber output. This single-ended device uses one GRIN rod lens and thus simplifies the packaging.

Wavelength division multiplexers using GRIN rod lenses and filters have been built for more than two channels. Two or more cascaded-lens multichannel WDM couplers can be connected for either multiplexing (in series) or demultiplexing (in series and parallel) as shown in Figs. 11.11A and 11.11B, respectively. In Fig. 11.11A, four channels are entering the input terminals of different GRIN lens to be combined and output at a single multiplexed output terminal. Similarly, in Fig. 11.11B, mixed input consisting of multiplexed channels at one terminal is de-

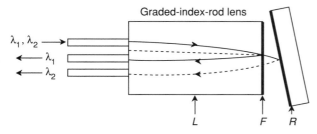

Figure 11.10 Single-ended wavelength multiplexer using a single graded-index rod lens. Adapted from Tanaka, Serizawa, and Tsujimoto, *Electron Lett.* **16**, p. 869. © 1980 *IEEE*.

A

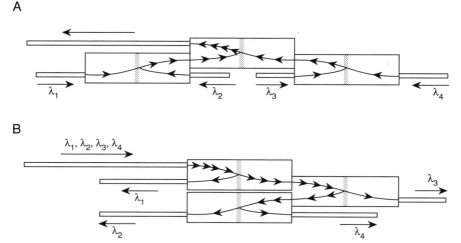

B

Figure 11.11   Multiple use of graded-index lenses for multiplexing and demultiplexing. (A) Four wavelengths output at one port; (B) one wavelength output at four ports. After K. Kobayashi *et al.*, *Microwave Communication Fiber and Integrated Optics* **2**, p. i (1979).

multiplexed into four different outputs from different GRIN lenses. With this type of structure for single-mode or multimode fiber devices, the insertion loss and crosstalk can be 0.5 and -35 dB, respectively.

Ishikawa *et al.* tried a seven-port coupler using a polished seven-port polygonal prism as a centerpiece and building a GRIN rod lens with spacer and filter on each port to form a coupler [10]. The polygonal surfaces are cut so that the reflections from each surface are directed to the next surface in succession. Six different wavelengths are combined into the coupler through six different ports. They combine to form a beam adding all wavelengths together and exit from the seventh port. The coupler can also be used as a demultiplexer by reversing the direction of transmission.

Other workers have tried with four ports [11]. With well-designed interference filters, high reflectivity at the desired wavelengths can be achieved with very little insertion loss. The implementation of this scheme is not without problems. The individual optical pieces must be individually manufactured, aligned, and cemented in place, making this design costly. The assembled coupler is also very bulky.

Waveguide Couplers

Optical components described under Photonic Components and Graded-Index Rod Lenses are all made of glass of special quality and shapes. A special graded-refractive-index profile is required to design GRIN rod lenses. Although they can be made as microoptical components, they are not suitable to build into integrated devices. The advantages of integrating optical components into systems have long been recognized: reduction in the overall size, shortening of the interconnections, and reduction of costs.

Planar Dielectric Waveguides

The layout of integrated electronic devices is usually planar because it is easier to process. To guide optical waves, we need to use dielectric materials. A medium of higher refractive index embedded in a medium of lower refractive index can trap the light inside the higher-index conduit, as in the case of optical fibers. The mechanism of light confinement is achieved by multiple total internal reflections at the dielectric boundary. Thus to implement an optical component suitable for integration we have to look into the planar dielectric waveguides.

Before we describe light guides suitable for integration (the rectangular or strip type of waveguides), we review some characteristics of a planar waveguide, typically the slab type. The geometry of a slab lightguide is shown in Fig. 11.12. The center core is a dielectric slab of thickness $2a$ with a refractive index n_1. It is sandwiched between two slabs (the claddings) of refractive index n_2, $n_1 > n_2$. The guide is oriented so that the direction of wave propagation is along the z axis, whereas the core and cladding thickness direction is along the x axis. Both y and z directions are extended toward infinity. Solution of Maxwell's wave equation reveals the existence of two independent modes, TE and TM, whose electric and magnetic field densities E and H are listed in Table 11.1, where u and v are a new set of parameters frequently used in wave solutions relating fiber parameters and transmission characteristics (see Ref. [12, Chap. 3]).

In Table 11.1, we use a set of parameters to simplify the mathematical expressions of the fields:

$$u = a[k_0^2 n_1^2 - \beta^2]^{1/2}, \; w = a[\beta^2 - k_0^2 n_2^2]^{1/2}$$

where k_0 and β_0 are the free-space propagation constant and phase constant, respectively (a is half-slab thickness). The field distribution for TE mode of a planar di-

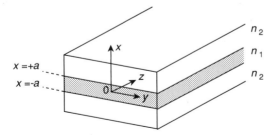

Figure 11.12 Slab geometry. Example of a dielectric slab waveguide consisting of a sheet of material with refractive index n_1 and thickness $2a$ between two layers of material with n_2, and $n_2 < n_1$.

TABLE 11.1
Wave Solutions of TE and TM Modes in a Plane Plate Waveguide[a]

	TE mode		TM mode	
	Even	Odd	Even	Odd
Inside the guide region, $n = n_1$	$E_y = A_e \cos \dfrac{u_x}{a}$	$E_y = A_0 \sin \dfrac{u_x}{a}$	$H_y = B_e \cos \dfrac{u_x}{a}$	$H_y = B_0 \sin \dfrac{u_x}{a}$
	$H_x = -\dfrac{\beta}{\omega\mu_0} A_e \cos \dfrac{u_x}{a}$	$H_x = -\dfrac{\beta}{\omega\mu_0} A_0 \sin \dfrac{u_x}{a}$	$E_x = \dfrac{\beta}{\epsilon_0 n_1^2 \omega} B_e \cos \dfrac{u_x}{a}$	$E_x = \dfrac{\beta}{\epsilon_0 n_1^2 \omega} B_0 \sin \dfrac{u_x}{a}$
	$H_z = \dfrac{1}{i\omega\mu_0}\dfrac{u}{a} A_e \sin \dfrac{u_x}{a}$	$H_z = -\dfrac{A_0}{i\omega\mu_0}\dfrac{u}{a} \cos \dfrac{u_x}{a}$	$E_z = \dfrac{iuB_e}{a\epsilon_0 n_1^2 \omega} \sin \dfrac{u_x}{a}$	$E_z = \dfrac{B_0}{i\epsilon_0 n_1^2 \omega}\dfrac{u}{a} \cos \dfrac{u_x}{a}$
In the cladding region, $n = n_2$	$E_y = C_1 e^{-w_x/a}$		$H_y = C_2 e^{-w_x/a}$	
	$H_x = \dfrac{\beta}{\omega\mu_0} C_1 e^{-w_x/a}$		$E_x = \dfrac{\beta C_2}{\epsilon_0 n_2^2 \omega} e^{-w_x/a}$	
	$H_y = \dfrac{1}{i\omega\mu_0} C_1 e^{-w_x/a}$		$E_z = \dfrac{i\omega}{a}\dfrac{C_2}{\epsilon_0 n_2^2 \omega} e^{-w_x/a}$	

[a] After C. Yeh, *Handbook of Fiber Optics* (1990). With permission of Academic Press.

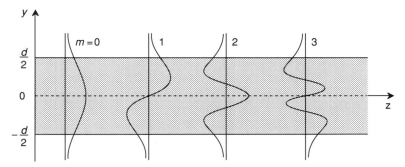

Figure 11.13 Field distribution for TE-guided modes in a planar dielectric waveguide for $m = 0, 1, 2,$ and 3.

electric waveguide is shown in Fig. 11.13 for $m = 0, 1, 2,$ and 3 modes. Note that the field distributions do not stop abruptly at the boundaries, but continue to exist into the second medium for a small distance. These residual fields, known as evanescent fields, play important roles in many devices to be described later.

From the field components listed in Table 11.1, we can find the eigenvalue equation of a particular mode of a planar lightguide by matching the field components at the boundary, from which most propagation properties such as phase constant, field distribution, and power relations of the mode can be derived [12].

Channel Waveguides

To build devices suitable for integrating lightguides into optical components, channel waveguides, such as the strip and ridge waveguides, are developed. Figure 11.14 shows some typical designs. The simplest type, the strip guide (Fig. 11.14A), consists of an elevated strip of dielectric material of higher index deposited on the substrate using an integrated technique. This thin and narrow strip of higher-index material confines and guides the optical wave along its length by the action of total internal reflection. In Fig. 11.14B, an embedded guide region of higher index n_g is diffused

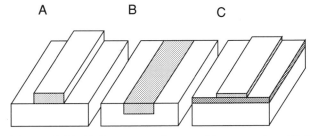

Figure 11.14 Various types of waveguide geometries: (A) Strip; (B) embedded strip; and (C) ridge-loaded strip.

into a substrate of index n_s. In Fig. 11.14C, a ridge-loaded guide is shown. Optical waves are confined to the region where $n_g > n_s$, assuming n_0, the index of the cover layer, is always less than n_g.

The substrate material can be either glass, semiconductor, or special light-sensitive material such as LiNbO$_3$. In glass substrates, the embedded region is formed by doping the substrate with the impurity diffusion or ion-exchange process. In semiconductor substrates, an epitaxial crystal growth process with a composition different from the substrate is used. The most advanced technique for fabricating waveguides employs Ti:LiNbO$_3$ which is made by diffusing titanium into a lithium niobate substrate.

The ridge guide, illustrated in Fig. 11.14C, consists of a substrate of index n_s, a guide layer with a higher index n_g, and a layer of index n_0, where $n_0 < n_g$, in the form of a pattern of ridges. The pattern layer defines the waveguide widths. The guide layer is used to accommodate other functional devices to be integrated, such as those needed for optical gain or electrooptic effect. Thus, a ridge-loaded waveguide cannot be used as a narrow-channel device. It is intended more for integrated optical devices made on semiconducting substrates.

Analysis of channel waveguides is complicated by the geometries of the guide. Readers may refer to specialized books for further details on this subject. To date, the design of optical channel waveguides is guided by approximations and experiment results.

Waveguide Bends

To use waveguides as circuit elements in optical systems, different configurations must be provided for different applications. Usually, a pair of channel waveguides can be designed for coupling and other uses. Although the separation between a pair of waveguides may be made to the order of 10 μm to make use of evanescent fields between waveguides for interaction, for the guides to be able to attach to input or output fibers, a separation of at least 125 μm is advised. At this distance, the inter-action field between the guides will be very weak. Thus, the combined requirements of keeping the guide separation at 10 μm and satisfying the interconnection require-ment at 125 μm require that a curved section(s) be placed inbetween. The minimum bend radius, however, is limited by the radiation loss. For Ti:LiNbO$_3$ single-mode waveguides, where the index change is ~0.5% and the mode size is about 7 μm at a 1.3-μm wavelength, radiation loss can be kept negligible for a bend radius of 20 mm. For semiconductor waveguides, bend radii of 1 mm or smaller are possible. This is because the index change is larger (up to 3%), and the mode sizes are correspond-ingly smaller.

A Y-branch waveguide is useful as a power divider. For a perfectly symmetric branch, the two output waveguides will divide the input power equally. Radiation loss caused by reradiation of the optical field can be kept low by using a branch angle much smaller than the ratio of the wavelength to the mode size. The branch angle is the angle sustained between the two arms of the Y-branches. For Ti:LiNbO$_3$ waveguides at 1.3 μm, the branching angle should be less than 1 radian.

Waveguide Directional Couplers

A single-mode optical waveguide coupler can be manufactured by depositing two similar waveguides in a planar, parallel configuration with a separation d between on a dielectric substrate. The substrate may be either glass or semiconductor such as GaAs. Recently, Ti:LiNbO$_3$ has become the most popular substrate with which to fabricate optical waveguides. This is because large, optical-quality crystals are commercially available, and the crystal has a reasonably large electrooptical effect.

Several parameters are involved in the design of optical waveguides: (1) the shape and size of the individual guide, (2) the thickness of the individual guide t, and (3) the refractive indices of the guides and of the substrate.

For shape, we prefer a simple rectangular configuration as it fits the planar integration process. The size is determined by its width w. Assuming that the guide material and the substrate have indices n_1 and n_2, respectively, and the top layer of the structure is air, then to support a mode of order m, the depth t of the guide layer must satisfy the expression

$$t = \left(\frac{\lambda}{2}\right)\left\{m + \frac{1}{2}\right\}(n_1^2 - n_2^2)^{-1/2} \tag{11.4}$$

where m is the mode number. If t is thinner than that limited by Eq. (11.4), it becomes impossible to support any mode at all [13]. If $n_1 = 1.5$ and $n_2 = 1.49$, then to support a single mode, the thickness of the guide must be $1.4\lambda < t < 4.3\lambda$. Practical lightguides are approximately 2λ thick.

Although the separation between guides does not appear directly in the design formula, it determines the coupling coefficient between the guides. The closer the guides are to each other, the larger will be the coupling, because of the larger evanescent field. But the fabrication process may limit the practical separation. Furthermore, the coupling between guides depends more on the length of the coupled guides.

The electric field distribution of a coupled light guide can be shown schematically as in Fig. 11.15. If $A_1(z) = \cos kz \exp(i\beta z)$ and $A_2(z) = i \sin kz \exp(i\beta z)$ are fields in lightguides 1 and 2, respectively, where k is the coupling coefficient (assume $k = k_{12} = k_{21}$) and β is the propagation constant for the lossless guides, then the power flow expressions are

$$P_1(z) = A_1(z)A_1^*(z) = \cos^2 kz \tag{11.5}$$
$$P_2(z) = A_2(z)A_2^*(z) = \sin^2 kz \tag{11.6}$$

Equations (11.5) and (11.6) indicate the oscillatory nature of the power flow. Complete power transfer from guide 1 to guide 2, as in a directional coupler, is possible if $kL = \pi/2$ or $L = \pi/2k$. It should be noted that for a longer guide length, power

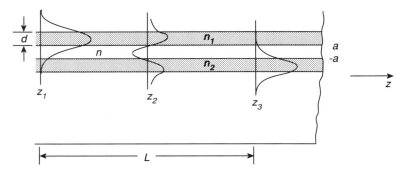

Figure 11.15   Electric field distribution of a coupled light guide. L is the coupling length.

begins to transfer back to guide 1 again periodically. The property of transferring power completely from guide 1 to guide 2 is a switching action. This happens when $L = \pi/2k$.

Coupling length L is therefore an important parameter in a coupler. If the coupling coefficient is known, L can be calculated. Otherwise, if the coupling length can be measured experimentally, the coupling coefficient can be computed. Theoretical prediction of the coupling coefficient is complicated by the geometry and the differences of the indices. For example, if L is 200 mm, then $k = 0.79$ l/mm.

A 3-dB directional coupler is a device that divides power equally between two lightguides. For a 3×3-μm guide with a separation of 3 μm, built on GaAs substrate with carrier density $n = 2 \times 10^{18}$/cm^3, and $\Delta n = 0.0005$, a 3-dB coupler or a power divider is obtained. The coupling length is now about 100 mm.

For other couplers, the power transfer varies as $\sin^2 \pi l/4L$, where l is the coupling length and L is that required for equal power division.

The power transfer ratio can also be affected by phase matching. Maximum power transfer occurs when $\beta_1 = \beta_2$, where β_1 and β_2 are the phase constants of A_1 and A_2 in the defining equations, respectively.

Other factors that influence the coupling property are the index difference δn, the bend radii, and the edge roughness of the guides. The refractive index difference must be less than 0.02. To limit the radiation loss to within 1 dB or less, the bending radii must be larger than 0.5 mm. The edge roughness of the guides must be held to within ± 500 Å for a 3-μm-wide guide.

Waveguide Electrooptic Devices

The usefulness of waveguides in fiber telecommunication services can be greatly extended if other effects, such as the electrooptic effect, the acoustooptic effect, and the magnetooptic effect, are included. In this section, we describe the electrooptic effect.

Pocket's Effect Devices

Many crystals, noticeably lithium niobate ($LiNbO_3$) and potassium dihydrogen phosphate (KDP), will respond to an applied electric field in certain crystal orientations to produce changes in the refractive index of the crystal [14].

Kaminow derived an expression that describes the electrooptic effect as

$$\Delta\left(\frac{1}{n^2}\right) = aE + bE^2 \qquad (11.7)$$

where a and b are coefficients of the linear and quadratic field effects, respectively. These coefficients are normally in tensor form. But for a certain crystal orientation, they can be expressed by a single term of the tensor.

The linear field term, aE, is the Pockel effect. The quadratic term, bE^2, is the Kerr effect. Both effects can be used to make devices for optical fiber applications. We describe only Pockel's effect in this section.

Let us choose a crystal orientation such that the electric field direction is in line with the crystal's z axis. This orientation reduces all tensor components to zero except $a (= r_{33})$. The linear field effect can be written as

$$n - n_0 = \tfrac{1}{2} an_0^3 E_z \qquad (11.8)$$

where n_0 is the refractive index before the electric field is applied. The components of n in the x and y directions are

$$n(x) = n_0 + \tfrac{1}{2}r_{33}n_0^3 E_z \qquad (11.9)$$
$$n(y) = n_0 - \tfrac{1}{2}r_{33}n_0^3 E_z \qquad (11.10)$$

Note that the refractive indices in the x and y directions are different. The velocity of propagation of a wave polarized along the x axis differs from that of a wave polarized along the y axis. After the waves travel through a length L of the crystal, there will be a phase difference between these components $\Delta\phi$ such that

$$\Delta\phi = \left(\frac{\pi}{\lambda}\right)r_{33}n_0^3 \frac{V}{L} \qquad (11.11)$$

where E has been replaced by V/L. Thus, the phase angle can be changed by a change in the applied voltage, a nice way to affect the phase angle of the wave, or phase modulation.

Table 11.2 lists the linear electrooptic coefficients of some crystals. By adding electrooptic effect to integrated waveguide structure, it seems as if we have added another degree of freedom to the device design. Many ingenious schemes can be developed to satisfy practically every application. We describe a few examples here.

Phase Modulators

The application of electric potentials to the crystal requires electrodes. Metallic strips can be deposited alongside the waveguides to serve this purpose. In high-speed

TABLE 11.2

Characteristics of Some Electrooptic Materials[a]

Material	λ_{min}	Linear electrooptic coefficient α^b (pm/V)		Retractive index		
		γ_{33} Ordinary	Extra-ordinary	n_0	n_0	ϵ_{12}
KDP (KH$_2$PO$_4$)	0.5	10.6		1.51	1.47	42
KD*P (KD$_2$PO$_4$)	0.5	26.4		1.51	1.47	50
LiNbO$_3$	0.5	30.8		2.29	2.20	18
LiTaO$_3$	0.5	30.3		2.175	2.180	43
GaAs	0.8	1.6		3.5	—	11.5

[a] After C. Yeh, *Handbook of Fiber Optics* (1990). With permission of Academic Press.

[b] Most materials have more than one linear electrooptic coefficient. We have quoted only the one used in Pockel's effect.

modulation systems, the addition of electrodes to the system can modify the modulation characteristics to a great extent. Care should be exercised in the design and placement of the electrodes.

If the electrodes are used to modulate the light wave on the guide with fast varying signals, the design is even more critical. If the electrical transit time along the interaction length is negligible compared with the period of the highest frequency of modulation, and if the electrical resistance of the electrodes is low, then the electrodes simply act as a lumped capacitance to load up the signal circuit. Otherwise, the electrodes must be considered as a part of the traveling-wave device whose structure, excitation, and termination must be considered carefully.

Under the lumped electrode criterion, the modulation bandwidth is limited by the *RC* time constant presented to the source. Typical electrode geometries result in a capacitance per unit length of about 3 pF/cm, thus limiting the bandwidth to about 2 GHz-cm for a 50-ohm load.

The ultimate modulation bandwidth can be improved if a traveling-wave electrode configuration can be adapted. Such an arrangement is shown in Fig. 11.16. In the center of the figure, a lightguide is embedded in a LiNbO$_3$ substrate. A pair of thin gold electrodes is deposited along the guide as shown [15]. One end of the electrodes is connected to the modulating signal through a coaxial table; the other end terminates with its characteristic impedance R. The electrical field of the traveling wave along the electrode then modulates the phase of the light wave in the guide. The phase change $\Delta\phi$ can be expressed as

$$\Delta\phi = \pi\left\{\frac{V}{V_\pi}\right\} \tag{11.12}$$

where V_π is the half-wave voltage, at which the phase shift changes by pi radians. $V_\pi = (d/L)(\lambda_0/r_{33}n_0^3)$ then becomes an important parameter of the modulator that

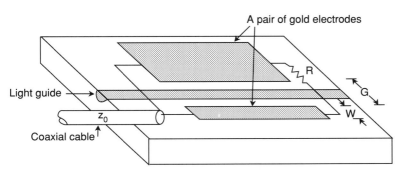

Figure 11.16 Traveling-wave phase modulator. After S. K. Korothy *et al. In Optical Fiber Telecommunications, II.* © 1988 Academic Press, with permission.

depends on the material property (n_0 and r_{33}) and the dimension ratio d/L for an optical wavelength λ_0. Equation (11.12) indicates that the phase change is linearly related to the applied voltage V. By varying V, phase modulation can be achieved. For an active length of 1 cm, a bandwidth of 5 GHz has been recorded.

Amplitude Modulators

In fiber telecommunications, amplitude modulation in the form of on/off switching can easily be implemented directly on the light source, an LED, or a laser; however, direct on and off switching of the source leads to wavelength broadening and frequency chirping. This is particularly troublesome at high data rates and when long repeater span is necessary. Even more serious is when the interchannel spacing in multichannel mixing is reduced. To alleviate the problem of spectral broadening, an external modulation is used. To implement external modulation, waveguide electrooptics is a good choice. In general, there are two ways to implement the intensity modulation. The first is to use the waveguides as the arms of an interferometer with deposited electrodes alongside or under the waveguide. The waveguides are connected at both the input and output ends by Y-branches. Input light waves to the input Y-branch divide the power equally among the two waveguides. They recombine at the output Y-branch. With no voltage applied to the electrodes, the input and output light-waves are in phase to present an intense output. By applying enough voltage to change the phase difference between the branches to π radians, the output light intensity becomes zero. Any voltage inbetween will change the intensity accordingly, thus achieving intensity modulation (Fig. 11.17). More generally, the transmission function of the interferometer is proportional to the square cosine function of the differential phase shift in the two arms.

The second method is to use a directional coupler to affect the intensity. Two identical waveguides are coupled together by depositing them in close proximity, say a few micrometers apart, again with electrodes deposited along the interaction region. The coupling length is initially chosen such that input light into the first waveguide

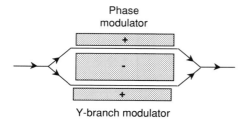

Phase
modulator

Y-branch modulator

Figure 11.17 An interferometer-type intensity modulator. After S. K. Korothy, **QE-22**, p. 252.
© 1986 *IEEE*, with permission.

will cross over to the second guide completely. Through the application of voltage to
the electrodes, a quantity of light power proportional to the applied voltage is shifted
to the first guide. Thus the intensity of the output can be varied. Similar schemes can
be used to switch signals as will be described in later chapters.

Polarization Control

The electrooptic effect can be described as a change in the refractive index of a dielectric material caused by the application of voltage to the electrode. In fact, it is the
change in its dielectric coefficient that counts. The dielectric coefficient is a complex
tensor for inhomogeneous material known as the dielectric tensor. In practical applications, we try to orient the crystal so that only a single element of the tensor appears.
Then we can derive the change in n as

$$\Delta n = -\left(\frac{V}{2d}\right)r_{33}n^3 \tag{11.13}$$

where r_{33} is the electrooptic coefficient of the crystal. For example, the largest r_{33}
coefficient in $LiNbO_3$ has a value of 31×10^{-12} m/V. For III–V compounds, it
becomes r_{41}, which has a value of 1.3×10^{-12} m/V. From a comparison of these
numbers, it might seem obvious that using $LiNbO_3$ has an advantage over use of the
III–V semiconductor. The advantage soon disappears when the devices are made.
Waveguides made of semiconductors are smaller in size and require less voltage to
operate.

To change polarization, we need to introduce changes in the spatial orientation of
the axes that define the polarization. The electrooptic effect does this task.

More about Polarization

Standard single-mode fibers do not preserve the state of the polarization. In conventional communication systems where direct detection is involved, the state of polarization of the wave is of no concern. But in sophisticated systems, where the process-

ing of optical signals may involve the use of birefringent crystals such as $LiNbO_3$, which is polarization sensitive, care must be taken that the proper polarization states are used. Also, in coherent detection schemes, only waves of identical polarization state can produce the best beat signal. In other devices, the orthogonality of polarization should be recognized or separated. Thus we need to control and to maintain the state of polarization of light waves at all times.

The state of polarization of light waves can be written in terms of the Jones vector (see, for example, [15])

$$\begin{bmatrix} E_x \\ E_y \end{bmatrix} = \begin{bmatrix} \cos \theta \\ \sin \theta \exp (i\Phi) \end{bmatrix} \qquad (11.14)$$

where E_x and E_y are the field components of the TE and TM waves, θ is the angle of the plane of polarization with respect to the x axis, and Φ is the phase between the TE and TM components. For $\theta = 0°$ and $\Phi = 0°$, it is a horizontally polarized light. If $\theta = 45°$ and $\Phi = \pm 90°$, it represents a circularly polarized light wave. The plus and minus signs indicate right and left circularly polarized light, respectively.

Polarization control is to adjust both θ and Φ to achieve the desired polarization. This can be done by adjusting V on the electrodes as shall be described in the following section.

TE and TM Mode Converters

The basic element for polarization control is the electrooptic TE to TM (or TM to TE) mode converter. It makes use of the different indices of birefringent crystals in response to the direction of propagation of TE or TM light waves to affect the state of the polarization. Specially designed electrode configurations. such as a periodic interaction structure to affect the phase-matched coupling between these asynchronous modes, are used. Figure 11.18 shows one example [16]. A Ti-diffused light-

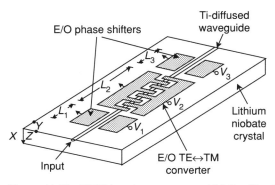

Figure 11.18 TE/TM mode converter on $LiNbO_3$. After R. C. Alferness and L. L. Buhl, *Applied Phys. Lett.* **38.** © 1981 APS.

guide is fabricated on a LiNbO$_3$ substrate shown in the center of the graph. Three sets of electrodes are deposited on top of the guide. Two of these (one on each side of the center structure) are of simple structure, used for shifting the relative phase between the TE and TM components to be $+$ pi radians when proper voltages V_1 and V_3 are applied to these plates. The central interdigital structure is the mode converter. It provides a length for interaction between the light wave and the periodic structure to achieve phase-matched coupling. Phase-matched coupling can be achieved only at the single wavelength for which the structure is designed. The condition for phase-matched coupling is

$$\Lambda[n_{\mathrm{TM}} - n_{\mathrm{TE}}] = \lambda_0 \tag{11.15}$$

where Λ is the period of the structure and n_{TM} and n_{TE} are the effective refractive indices of TM and TE polarization, respectively.

The first section of the phase shifter is provided to allow arbitrary transformation of any input polarization. The last phase shifter may be omitted if only linear output polarization in either the TE or TM mode is needed.

Polarization-Selective Coupler

The coupled waveguide described earlier can be modified to act as a polarization-selective coupler as shown in Fig. 11.19. Two Ti-diffused waveguides are coupled with a coupling length L. The electrodes are deposited along the coupling length as shown. A mixed TE/TM wave input at one end (port 2) will be separated into TE and TM at the outputs (ports 3 and 4) by choosing the diffusion parameters to change the interwaveguide coupling coefficient that affects the polarization. The appropriate fabrication parameters and interaction length L are $k_{\mathrm{TE}}L = \pi$ and $k_{\mathrm{TM}}L = \pi/2$. Polarization selection can also be achieved by varying phase mismatch, measured by ΔB. The electrooptically induced $\Delta\beta$ in lithium niobate enhances the selection.

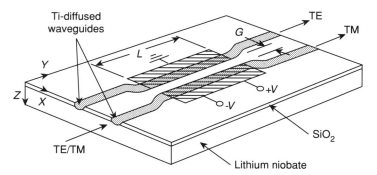

Figure 11.19 Polarization-selective coupler. After Schlak *et al., Electron. Lett.* **22**, 883. © 1986 *IEEE*, with permission.

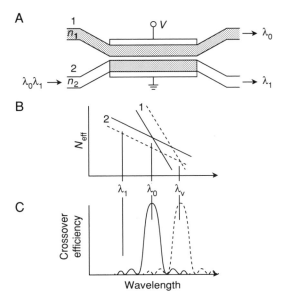

Figure 11.20 Directional coupler filter. (A) Waveguide directional coupler; (B) index profile; and (C) filter characteristics. After H. F. Taylor, *Opt. Communication* **8**, 428. © 1973 *IEEE,* with permission.

Directional Coupler Filter

A directional coupler with different index properties built on a semiconductor substrate can be used as a wavelength division multiplexer, as shown in Fig. 11.20A [17]. The top waveguide is built with index n_1; the bottom one with n_2. The coupling length is L. The design criterion is to achieve a finite power transfer at the designed wavelength to result in effective phase matching at that wavelength. The effective index profiles are shown in Fig. 11.20B. Only at the phase-matching wavelength is coupling between the two waveguides possible. At wavelengths different from λ_0 the coupling efficiency decreases. The filtering property is shown in Fig. 11.20C. A narrow bandwidth can be obtained. At 1.5-μm wavelength, on an InGaAsP/InP substrate, a 170-Å bandwidth has been reported [18]. The bandwidth can also be tuned by changing the applied voltage.

Acoustooptic Devices

Under Diffraction gratings, we described a mechanical diffraction grating device fabricated by inscribing lines on aluminum film deposited on a thin glass plate to create a grating transparency of periodically varying density. An elegant way to generate gratings is through acoustic means. The Bragg cell is the fundamental unit of this creation. There are several advantages to the Bragg cell: (1) Mechanical engraving is not used. (2) Gratings can be fine and close to each other. (3) The period of the grating can be changed by varying the acoustic frequency. (4) Many sound frequencies can be combined to generate gratings of different periods to produce multiple diffraction effects.

We describe the Bragg cell and the narrow film gratings in this section. Other configurations are described under different topics to avoid duplication.

The Bragg Cell

The interaction of a light wave (photons) and a sound wave (phonons) is known as the Bragg diffraction effect [19, 20]. The existence of such interactions was predicted by Brillouin in 1922 [21]. Experimental verification by Debye and Sears followed [22]. Generalized multiple scattering was furthered by Raman and Nath [23].

A Bragg cell is an acoustooptic diffraction grating used to modulate an optical wave with an acoustic wave. The optical properties of $LiNbO_3$ are locally varied with a strong acoustic wave to induce spatial perturbations with a periodicity equal to the acoustic wavelength. As optical frequency is many orders of magnitude higher than acoustic frequency, the acoustic wavefront seems stationary to optical waves. Local acoustic perturbation acts like a moving diffraction grating to the incident optical wave. Optical waves incident at the Bragg angle are reflected by the stratified planes created by the acoustic wave so that the Bragg angle θ_B satisfies the relation $\sin \theta_B = \lambda/2\Lambda$, where λ is the wavelength of light in air, and Λ is the acoustic wavelength in the medium. Constructive or destructive interference occurs when the total phase shift reaches 2π radians and introduces new frequencies to the diffracted wave. The frequency can be upshifted as $\omega_+ = \omega_0 + \Omega_a$ or downshifted as $\omega_- = \omega_0 - \Omega_a$, where ω_+ and ω_- are the new frequencies, ω_0 is the original light frequency, and Ω_a is the acoustic frequency.

These properties can be used to build filter, modulators, switches, and many other devices.

Thin-Strip Acoustic Transducer

As discussed in Chapter 2, if the width of the acoustic transducer L is reduced to become a narrow strip configuration, the spectrum of the sound wave in the medium broadens to one containing many plane waves of m order, $m = 0, \pm 1, \pm 2, \ldots$. The Bragg condition is also modified to become

$$\sin\left(\frac{\theta}{2}\right) = \frac{\lambda}{2\Lambda} \tag{11.16}$$

(Note that the Bragg angle is now $\theta_B = \theta/2$.) Each sound plane wave will interact with the incoming optical beam separately, and the resulting reflected beams are separated by λ/Λ. The angle of deflection is controlled by the acoustic frequency. As θ is small, $\sin \theta \approx \theta$, or $\theta = m(\lambda/v_s)f$, where v_s $(=\Lambda f)$ is the speed of sound. For each f, there is an identifiable point.

If, now, additional acoustic frequencies are added to the system, the number of resolvable spots can be increased. This property can be used to build an acoustooptic switch and an acoustic spectrum analyzer.

Tunable Acoustooptic Filter

The fact that Bragg condition relates the angle of the deflected light wave, the acoustic wavelength Λ, and the optical wavelength λ suggests that if θ and Λ are specified, reflection can occur only for a single optical wavelength. Thus for an input optical wave composed of a broad spectrum of wavelengths, only one wavelength will be selected on Bragg reflection. This wavelength-selection property acts as a filter. As angle θ can be varied by changing the sound wavelength, this becomes a tunable filter by tuning the sound frequency. With the improvements in crystal technology, reliable, acoustooptic filters are beginning to replace conventional grating technology in a wide range of applications.

A Bragg cell can also be used as an isolator, a frequency shifter, and many other devices.

Acoustooptic Modulator

For a weak sound, the intensity of the reflected light in a Bragg cell is proportional to the intensity of the acoustic wave. By varying the sound intensity, the optical wave can be modulated accordingly, producing a linear analog modulation of light. Yeh *et al.* used a Bragg cell as a single-sideband modulator [24].

The bandwidth of the modulation is governed by the acoustic frequency as described in Chapter 2. The maximum bandwidth depends on the highest acoustic frequency that can efficiently modulate the optic wave for a given crystal. When the amplitude of an acoustic wave is varied, the resultant amplitude modulation of the optical wave will have a frequency band covering $f_0 \pm B$, where B is the band of frequencies that modulate the acoustic wave of frequency f_0. Under the restriction of the Bragg condition, however, only one frequency can satisfy. Thus no modulation can be achieved theoretically. Fortunately, in practice light waves have a beamwidth b or an angular divergence $\Delta\theta$ so that $\Delta\theta = \lambda/b$. To match the Bragg condition, one only has to make the bandwidth B satisfy

$$\Delta\theta = \frac{\left(\dfrac{2\pi}{v_s}\right)B}{\left(\dfrac{2\pi}{\lambda}\right)} = \frac{\lambda B}{v_s} \tag{11.17}$$

Thus

$$B = \frac{v_s \, \Delta\theta}{\lambda} = \frac{v_s}{b} \qquad (11.18)$$

If b/v_s is defined as T, the transit time of the sound wave across the light beam, then

$$B = \frac{1}{T} \qquad (11.19)$$

The maximum bandwidth is thus limited by the transit time.

For example, for a fused quartz transducer with $n = 1.46$, the sound speed is 6000 m/s. If f_0 is set at 50 MHz and a He–Ne laser is used as the light source, $\lambda = 633$ nm, the angular divergence is about 1 milliradian. Then the maximum achievable modulation bandwidth is approximately 3×10^7 Hz.

At high sound power level, however, saturation occurs. Modulation of signals ceases; however, operation between zero and the saturation level then resembles that of an on–off switch and can be so used [25].

Magnetooptic Effect

The Faraday effect, the rotation of polarization of the electric field in certain materials in the presence of a magnetic field, can be used to build magnetooptic devices. The angle of rotation is proportional to the length of the interaction region, and the rotational power P (angle per unit length) is proportional to the component B of the magnetic flux density in the direction of wave propagation, or $P = VB$, where V is the Verdet constant. The sense of rotation for $V > 0$, is the right-hand-screw sense in the direction of propagation. If, however, the direction of propagation were reversed, the sense of rotation would not reverse with it. Thus, the total rotation can be doubled if the beam passes the interaction region twice by a reflecting mirror. This is known as the nonreciprocal property of the Faraday effect.

Magnetooptical switches are described in Chapter 12.

Polarization-Independent Fiberoptic Components

The problem of polarization dependence of the insertion loss of connectors, couplers, switches, and attenuators becomes increasingly important for fibers operating at high bit rates. In these systems, even a few tenths of a decibel in loss uncertainty in components is critical to the design and operation of the system. For example, erbium fiber amplifiers require highly polarization-independent components to maintain care-free operation. As it is usually impractical to install fiber links with polarization-maintaining fiber, considerable effort is focused on polarization-independent fiberoptic components.

Sources of Polarization Dependence

Both material design and the processes for manufacturing the fiber and components may contribute to the dependence of polarization on components. Semiconductors or nonlinear crystals such as lithium niobate and lithium tantalate are limited in operation to one polarization state. Mechanical stresses, such as microbending and poor fusion splices on fiber elements, can introduce polarization dependence. The refraction and reflection of light at interfaces are usually polarization dependent. The presence of multilayer coatings and absorption materials will all cause polarization dependence of the components.

Effect of State of Polarization on Optical Fiber Systems

The effect of state of polarization (SOP) on optical fiber systems is different for different components. For assemblies such as attenuators, isolators, and optical connectors, polarization dependence appears as a change in insertion loss as a function of SOP and, hence, adds uncertainty to the specifications of the components. For components such as couplers and wavelength division multiplexers, polarization dependence can result in a variation in splitting ratio or in isolation as the SOP varies. All these effects cause the SOP to change randomly in time with temperature and other operating conditions and, thus, are serious in practice.

Design Guidelines

1. Components composed of nonlinear crystals that work only with one state of polarization have to provide the right polarization by adding a polarization control element as needed.
2. Design components with interfaces that maintain the angle of entrance of light waves as small as possible. For a tilt angle below 10°, for example, a fiber attenuator may sustain a loss from polarization change to within a hundredths of a decibel. Be careful even with the multilayer coatings to be applied on the device.
3. Avoid adding mechanical stress and strain to the device which may cause polarization variations with operation.
4. Always ask whether this device will cause polarization changes when designing or ordering components.

Summary

In this chapter, we reviewed the information about devices that have been built to serve the increasing demands for better and faster communication systems as well as other scientific, industrial, and medical applications. We started with the description of optical elements, microlenses, and diffraction gratings which have long been used in building optical components. Then the GRIN rod was introduced. Many interesting devices were described using its special property. The possibility of using integrated waveguides in devices was introduced. With the addition of optoelectric, acoustooptic, and magnetooptics effects, the number of devices possible increases manyfold.

Acoustooptic devices are now found in a wide range of applications in science and medicine. The ability of acoustooptic devices to perform modulation, diffraction, filtering, and analysis at high speeds has broadened the range of application of these devices. Ultrasonic imaging in medicine is only one example. Biological laboratories use these devices to test delicate samples nondestructively. A wide range of signal processing operations can be initiated using acoustooptic principles.

This study is not exhaustive. New devices, using entirely new principles, will undoubtedly be made as technology progresses. We should remain alert to these developments.

References

1. M. V. Klein and T. E. Furtak, *Optics,* 2nd ed., Wiley, New York, 1986.
2. J. Lipson, R. T. Ku, and R. E. Scotti, Opto-mechanical considerations for laser coupling and packaging. *Proc. SPIE* **554,** 308–312 (1985).
3. T. Sugie, and M. Saruwatari, An effective nonreciprocal circuit for semiconductor laser-to-optical-fibre using TIG sphere. *J. Lightwave Technol.* **LT-1,** 121–130 (1983).
4. S. Kubota, A lens designed for optical disk system. *Proc. SPIE* **554,** 282–289 (1985).
5. See, for example, A. Yariv and H. Winsor, Proposal for detecting magnetostrictive perturbation of optical filters. *Opt. Lett.* **5,** 87–89 (1980).
6. R. W. Wood, *Physical Optics,* Macmillan, New York, 1905.
7. S. E. Miller, Light propagation in generalized lens-like media. *Bell System Tech. J.* **44,** 2017–2064 (1965).
8. W. J. Tomlinson, Application of GRIN-rod lenses in optical fiber communication systems. *Appl. Opt.* **19,** 1127–1138 (1980).
9. K. Kobayashi, R. Ishikawa, K. Miniemura, and S. Sugimoto, Microscopic devices for fiber optic communications. *Fiber Integr. Opt.* **2,** 1 (1979).
10. S. Ishikawa, F. Matsumua, K. Takahashi, and K. Okuno, High stability wavelength division multi/demultiplexer with polygonal structure. *Proc. SPIE* **417,** 44 (1983).
11. J. Lipson, C. A. Young, P. D. Yeates, J. C. Masland, S. A. Wartonick, G. T. Harvey, and P. H. Read, A four-channel lightwave subsystem using wavelength multiplexing. *J. Lightwave Technol.* **LT-3,** 16–20 (1985).
12. C. Yeh, *Handbook of Fiber Optics, Theory and Applications,* Academic Press, San Diego, 1990.
13. L. Levi, Applied Optics, Vol. 2, Wiley, New York, 1980.
14. I. P. Kaminow, *An Introduction to Electrooptic Devices,* Academic Press, New York, 1974.
15. R. C. Alferness, Waveguide electrooptic modulators. *IEEE Trans. Microwave Theory Technique* **MTT-30,** 1121–1137 (1982).
16. A. Yariv and P. Yeh, *Optical Waves in Crystals,* Wiley, New York, 1984.
17. R. C. Alferness and L. L. Buhl, Electro-optic waveguide TE–TM mode converter with low drive voltage. *Opt. Lett.* **5,** 473–475 (1982).

18. S. Lindgren, B. Broberg, M. Oberg, and H. Jiang, Integrated optics wavelength filter in InGaAsP/InP. Technical Digest of the International Conference on Integrated Optics and Optical Fiber Communications, pp. 175–178, Venice, 1985.

19. C. S. Tsai, *Guided-Wave Acoustooptics,* Springer-Verlag, Berlin, 1990.

20. A. Korpel, Acousto-optic—A Review of Fundamentals. *Proc. IEEE* **69,** 48–53 (1981).

21. L. Brillouin, Diffusion de la lumimiere et des rayons X par un corps transparent homogene. *Ann. Phys. (Paris)* **17,** 88–122 (1922).

22. P. Debye and F. W. Sears, On the scattering of light by supersonic waves. *Proc. Natl. Acad. Sci. USA* **18,** 409–414 (1932).

23. C. V. Raman and N. S. N. Nath, The diffraction of light by high frequency sound waves. *Proc. Indian Acad. Sci.* **2,** 406–420 (1935); **3,** 75–84 (1936); **3,** 119–125 (1936); **3,** 459–465 (1936).

24. H. Commins, N. Knable, L. Gumpel, and Y. Yeh, Frequency shifts in light diffraction at ultrasonic wave in liquid medicine. *Appl. Phys. Lett.* **2,** 62–66 (1963).

25. N. J. Berg and J. N. Lee (Eds.), *Acousto-optic Signal Processing,* Marcel Dekker, New York, 1983.

CHAPTER 12

Photonic Switches

Introduction

In the information age of the 1990s, light waves are being adapted widely as transmission media for telecommunications and local information networks. To take full advantage of optical fibers and components for applications in these systems, the development of photonic switches and switching schemes becomes increasingly urgent. We look into these subjects in this chapter.

A photonic switch is a device that establishes and releases an optical path of lightwave transmission at the command of a prescribed control. In its simplest form, it turns the light beam on and off, just as a mechanical toggle switch turns an electrical light bulb on and off with the flip of a handle. The command that controls a photonic switch can be either mechanical, electrical, electronic, magnetic, or optical. The type of control depends on the size, power, and speed of the transmission system. Fast response, good reliability, and energy efficiency are common objectives in choosing the type of control.

A simple switch has the dimensions 1×1, that is, one input line and one output line. Larger communication systems require multidimensional switches of $N \times M$ dimensions. In this chapter, we explore ideas for making photonic switches and discuss the likehood of developing these ideas to accomplish multidimensional switching. For a good multidimensional switch, low crosstalk between lines is important. Also, the overall size of the switches should be considered.

Why photonic switching? For a modern telephone substation where many electronic devices are linked together, simply switching between lines is overwhelming. High-speed, efficient switching systems are in big demand. Modern electronic circuits may not be able to meet all the requirements. The interest in photonic switching systems derives from the limitations of electronic switching and the hope that optical switching will overcome most of these limitations.

Several advantages to the use of all-optical switching in optical communications systems can be visualized immediately. First, in electronic switching, optical signals must first be converted to electrical signals before the electronic switching technique can be applied. After processing, if further transmission is needed, the signals have to be reconverted to optical form. The need for these double optoelectronic interfaces at each network node can be avoided if optical switching is adopted. Second, all-optical switching technology is likely to provide very large communication bandwidths with all the added optical advantages such as low crosstalk and low distortion. And finally, as the network expands into multiplexing services, all-optical switching may be simpler to implement.

In the following sections, we point out the limitations inherent to electronic switching; characterize a modern photonic switching system; and describe the different types of switches and various control mechanisms. Optical switches using other than all-optical controls are included.

Limitations Inherent to Electronic Switches

The fundamental limitations of electronic switches can be summarized:

Minimum switching time. Electronic switches are used to switch electrical signals. The control is either by electrical, mechanical or electronic circuits. Even with electronic control, using semiconductor enabling logic circuits, the speed of control is limited by the speed of electrons moving along the circuit. The fastest device made with GaAs field effect transistors (FETs) has a minimum switching time of about 20 ps.

Minimum switching power. The minimum switching power required to operate a silicon-on-sapphire (SOS) or complementary-symmetry metal–oxide–semiconductor (CMOS) is about 1 μW.

Minimum energy per pulse. The minimum energy per operation of these electronic devices ranges from 10 to 20 fJ.

Time required for signal conversion. To switch optical signals by electronic switches, the optical signals are first converted into electrical signals using photodetectors, switched electronically, and converted back into light again using LEDs or lasers. These double-conversion processes introduce unnecessary power loss and time delay. Increases in the error rate can also be anticipated. System reliability could be impaired.

Comparatively, photonic switching has many advantages over electronic switching. For example, the minimum energy needed for photonic switching is 21 photons, although in principle one photon is enough. The number 21 (photons) is based on the probability that a randomly Poisson distributed n number of photons will guarantee

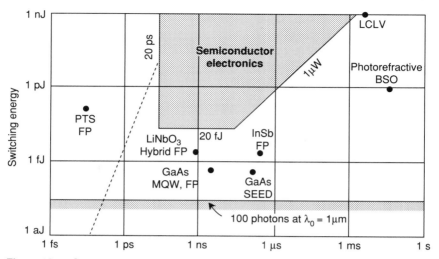

Figure 12.1 Switching energies and switching times of several switching devices. One hundred photons at $\lambda = 1$ μm serve as the limiting energy for optical switching devices. The region bounded by lines 20 ps, 20 fJ, and 1 μW describes the limits for electronic devices. The dashed line indicates the thermal transfer limit for heavy-duty-cycle devices. Some optical switches are represented on the graph as dots followed by the device type. LCLV, liquid crystal light valve; FP, Fabry–Perot; SEED, self-electrooptic effect device; MQW, multi-quantum-well; PTS, polymerized diacetylene, an organic material; BSO, bismuth silicon oxide. Adapted from Smith and Tomlinson [2].

delivery of at least one photon, with an average of one error in every 10^9 trials. The fundamental limit on the minimum switching time arises from energy–time uncertainty. By quantum-mechanical theory, the product of the minimum switching energy E and the minimum switching time T must be greater than $h/4\pi$ ($E * T > h/4\pi$), where h is Planck's constant. Switching energy can also, in principle, be much smaller than that in semiconductor electronics. This is demonstrated in Fig. 12.1 [1]. Here, grids of constant switching energies and constant switching times are arranged in a two-dimensional plot. Limits on the switching energy and time for all optical switches are bounded by a few lines. Switching energy must be above the 100-photon line; 100 photons is used as the practical limit instead of 21 photons to be safe. Data points of various devices are adapted from the literature [2]. For repetitive switching operation, points must lie to the right of a slant dashed line known as the thermal-transfer line. Within these boundary lines, one can find the operating range of some all-optical switches now available. On the same graph, limits for electronic switching operation are boxed in by the 1-μW, 20-fJ, and 20-ps lines. Relative advantages of photonic over electronic switching can easily be noted.

Lightwave transmission systems based on low loss, wide-bandwidth optical fiber, and high-speed optoelectronic repeaters are penetrating worldwide communication networks. The virtual elimination of electrical coaxial long-haul systems seems inevitable. Optical technologies are beginning to set their sights on switching within the network. At present, however, it is unlikely that optical processing can become practical enough to replace electronic switching completely. Rather, the present approach is to take advantage of the best characteristics of both optical and electronic technologies.

Present Status of Photonic Switching

Industrial and academic laboratories around the world are working on a variety of photonic switching architectures and devices. They are working along two paths: guided-wave photonics and free-space photonics. Guided-wave photonic switching is based on directional couplers. This technique is more highly developed. Both wavelength division multiplexing (WDM) and time division multiplexing (TDM) are being adapted to use the bandwidth fully. Because of size limitations, they are not suitable for large-scale integration. We describe some practical devices in the different sections.

Free-space photonic switching can support a large number of users, in parallel, on a large number of low-speed channels. Self-electrooptic effect devices make free-space photonic switching very attractive in building large-scale arrays of switches. The weakness of these devices is that their operation requires too much energy.

Switching Parameters

Before we proceed to describe different photonic switching devices in development, let us define some parameters that characterize the switch and use them to compare the various schemes.

Switching time: the time necessary to be reconfigured from one state to another of each individual switch including the on-time and off-time. In some devices, the on-time and off-time may be different.

Propagation delay time: time taken by the signal to cross the switch. In a complex switching network, the delay may be long.

Throughput: maximum rate at which data can flow through the switch when it is connected. Parallel operation of optical switching system may increase throughput dramatically.

Switching energy: energy needed to activate and deactivate the switch. Depending on the operating mechanism, switching energy may vary over a wide range.

Power dissipation: energy dissipated per second in the process of switching. Power dissipation may ultimately limit the size of the switching network.

Insertion loss: drop in signal power introduced by the connections. The insertion loss of individual switches should be kept as low as possible.

Crosstalk: undesirable power leakage to other lines. In multiplexing systems, crosstalk becomes one of the most important factors to consider. An isolation of -30 dB may be required against neighboring channels in crosstalk.

Size: number of input and output lines, and the direction of data transfer, either uni-
or bidirectional. Integrating several switches into a switching network is desirable
for a complex network.

Physical dimensions: actual physical size of the complete switching system.

Cost: The ultimate success of optical switches in replacing electronics installations
is dependent on the cost. Integration and mass production of optical components
may make the changeover very attractive.

Types of Switches

In describing the various technologies of photonic switching, we first make the distinction between passive-optical-path devices, waveguide devices, bistable devices,
and hybrid devices. Passive-optical-path devices exhibit optical attenuation and
crosstalk in the path and the transmission may be bidirectional. Waveguide devices
may involve directional couplers with or without nonlinear optical materials in the
path; electrooptic effect devices are included in this class. Bistable devices include
passive- and active-path devices. Hybrid devices combine the best properties of each
element in a usable switching device. Other devices involving active optical paths
may use clocked, level restoration and digital operations. The transmission mode of
these devices is unidirectional.

The Building Block

The building block of passive-path switches is a two-port, on–off switch, as in the
1×1 configuration shown in Fig. 12.2A. It involves two lines that can be engaged
or disconnected by a control. For large telecommunication switches with a large
number of input and output lines, a two-dimensional array of $N \times N$ crossbars can

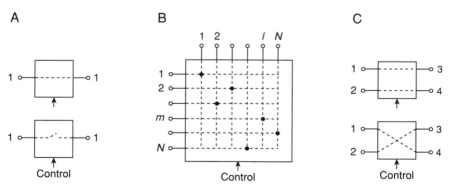

Figure 12.2 Building block of passive-path optical switches. (A) A 1×1 on–off switch. (B) Two-
dimensional $N \times N$ crossbar switch. (C) Four-port module of a 2×2 crossbar switch. Switch
positions can be in either the straight-through (bar) state or the crossover (cross) state.

be built as shown in Fig. 12.2B. Any input line can be connected to a free output line at the command of a control.

For telecommunication switching, a three- or four-port module, shown in Fig. 12.2C, is of more interest. Here, two inputs ports 1 and 2 can be connected directly to their respective output ports 3 and 4. This is called the straight-through state. Or they can be cross-connected to ports 4 and 3, known as the crossover state. The control is dictated by another mechanism to be described later. As the number of cross points has grown rapidly with size, more efficient pipeline matrices and parallel operation formats have been designed [3].

Mechanically Controlled Optical Switches

One family of mechanically controlled photonic switches involves an optical fiber that is attached to a rotating wheel and can be aligned with a number of optical fibers attached to a fixed wheel (Fig. 12.3A). The fibers are placed in V-grooves. An index-matching liquid is used in the movable space for better optical coupling.

Devices that deflect light in different directions through the use of rotating mirrors constitute another example of mechanically controlled switches as shown in Fig. 12.3B. The advantages of these devices are low insertion loss, low crosstalk, and insensitivity to wavelength or polarization. Both single- and multimode fibers can be accepted. The major limitation is that these devices are inevitably slow (order of milliseconds) and imprecisely timed in operation. They are also physically bulky.

Electrooptically Controlled Switches

A different class of electrically controlled switches can be made using the electro-optic effect on a pair of optical waveguides. Electrooptic materials, such as potassium dihydrogen phosphate (KDP), potassium deuterium phosphate (KD*P), and lithium niobate crystal (LiNbO$_3$), alter their refractive indices in the presence of an electric field [4]; this is known as Pockel's effect (see Chapter 11, Pockel's Effect Devices). The change in refractive index is usually expressed in terms of Pockel's coefficient r as $n(E) = n - \frac{1}{2}rn^3E$, where n is the index at zero field. Typical values of r lie in the range 10^{-12} to 10^{-10} m/V.

To build a waveguide switch using electrooptical control, a directional coupler is made by diffusing titanium into a lithium niobate substrate so that there is a region where the two bands run parallel and very close to each other. These form a waveguide whose evanescent fields interact. A pair of electrodes is deposited alongside the waveguide and a voltage source is applied across the pair. The combination now works as a phase modulator as follows. Let the interaction length of the coupler be L and the coupling coefficient k. Switching is accomplished by changing the relative refractive indices in the guides by changing the applied voltage, which affects the phase angle. Crossover occurs when $kL = \pi/2$. As shown in Fig. 12.4A, when no voltage is applied to the electrodes, input and output fibers are connected straight through. When voltage is applied, input and output cross over.

A

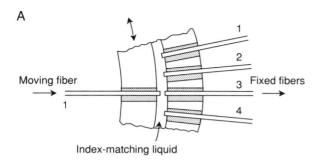

B

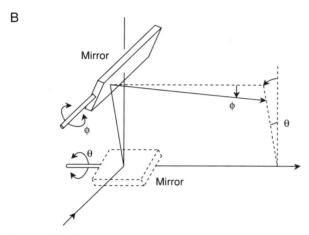

Figure 12.3 Mechanically controlled optical switch. (A) An optical fiber is mounted on a rotary wheel that can be aligned with a number of fixed optical fibers on a V-grooved disk. (B) Rotating mirrors are used to deflect the light beam in different directions for connection. After B. E. A. Saleh and M. C. Teich, *Fundamentals of Photonics*. Reprinted by permission © 1991. John Wiley & Sons.

In the system shown in Fig. 12.4A, it was found that very tight fabrication tolerances (a precise control of the coupling length) have to be observed in the manufacturing process because the cross state is not under electrical control. By addition of another pair of electrodes and connection of these in different ways (straight-through and crossover as shown), the control can be made easily (Fig. 12.4B). For the straight-through case, the two adjacent electrodes have the same polarity. For crossover, they bear opposite polarities. With this provision, the need for precision control of the coupling length is virtually eliminated. The crossover state and the straight-through state can be attained precisely by applying two appropriate voltages [5]. A longer coupling length is required for this arrangement.

A

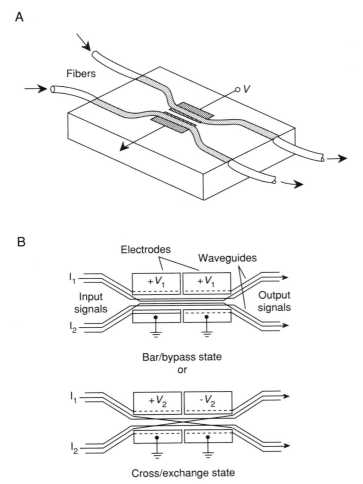

B

Figure 12.4 Electrooptic switches. (A) Waveguide directional coupler on top of a pair of electrodes deposited on LiNbO$_3$ substrate to form a four-port device. Electric potential is applied to the electrode to affect switching. (B) Adding one more pair of electrodes to the device and applying alternately positive and negative potentials to the pairs can alleviate the critical requirement of waveguide dimension adjustments. After R. C. Alferness *et al.* [4], with permission.

Although the insertion loss of the switch is low, because of the difference in the geometric structures of the guide and the fiber, coupling loss may become a problem. Polarization change along the guide length must also be considered in the design. The limit on the number of switches per unit area is governed by the relatively large physical dimensions of each directional coupler and the planar nature of the interconnections within the chip. The device is long, typically in the millimeter range. Fabrication of large arrays poses major problem. To date, up to 8 × 8 arrays on a single chip with impressive results have been reported [6]. Waveguides of different configurations (instead of parallel) may have to be tried.

A device of this structural configuration can switch very fast, well into the nanosecond regime. But it presents substantial capacitance to the driving circuit. Other disadvantages include polarization sensitivity, large physical size, and the need for another drive voltage. Low coupling to fibers and crosstalk can also be problems. The insertion loss is relatively low.

An optical cross-point switch using coupling between two polished fiber coupler blocks through a high-index interlay waveguide has been reported [7]. Because of the high refractive index of the interlay waveguide, the input light can be either coupled into the second fiber (cross-coupled) or recoupled into the first fiber (straight through) by modifying the propagation parameters of the waveguide. In one experiment, a liquid crystal interlayer was deposited on both coupler blocks and a 700-Hz square-wave voltage was applied to electrodes embedded in the blocks to rotate the direction of polarization of the input optical field to effect switching. Although the switch-on speed was several milliseconds, the switch-off time was a few seconds, too long for practical application. These experiments do, however, prove that interlay waveguide structure permits the use of materials of much higher index to effect switching.

Acoustooptically Controlled Switches

An acoustooptic switch makes use of sound waves to control light transmission. The refractive index of an optical medium is altered by the presence of sound, which modifies the effect of the medium on light (see also Chapter 2).

In Chapter 2, under Interaction of Light and Sound, the general principle of photon–phonon interaction was described. Interaction of the light wave (photons) and the sound wave (phonons) takes place in the crystal, which is transparent to the optical beam traveling in the z direction while a quasi-stationary acoustic wavefront is traveling across it in the x direction. This is known as Bragg diffraction effect [8]. As shown in Fig. 2.6A, a beam of light is directed at an acoustic medium containing a high-frequency sound wave. The quasi-standing plane wavefront of the sound wave becomes a dynamic graded index with a period Λ, the wavelength of the sound. Optical waves incident at the Bragg angle are reflected by the stratified parallel planes created by the acoustic wave, so that for an incident angle equal to the Bragg angle θ_B, constructive interference results and the intensity of the reflected wave is maximum. This relation is $\sin \theta_B = \lambda/2\Lambda$, where λ is the wavelength of light in the optical medium, and Λ is the acoustic wavelength. The new direction of the reflected wave makes an angle $2\theta_B$ with the original beam. Figure 2.6A shows two possible directions depending on the direction of acoustic wave propagation. We may also associate these cases with the incident angle of the optical wave. The downshifted (top) and the upshifted (bottom) interactions are for negative and positive entrance angles, respectively. The frequency of the downshifted wave becomes $\omega_- = \omega_0 - \Omega_a$, and that for the upshifted, $\omega_+ = \omega_0 + \Omega_a$, where Ω_a is the acoustic frequency. Bragg angle relationships hold true in acoustic media.

For a narrow Bragg cell, as discussed in Chapter 2, as the width of the acoustic transducer L is reduced to become a thin-strip configuration, the spectrum of the sound wave in the medium broadens to one containing many plane waves traveling in many directions. Each plane interacts with the optical beam and diffracts light in directions including $\pm\,\theta$. This modifies the Bragg condition to

$$\sin\left(\frac{\theta_B}{2}\right) = \frac{\lambda}{2\Lambda} \tag{12.1}$$

known as Raman–Nath diffraction. For clarity, in Chapter 2 we modified by introducing acoustic frequencies as shown in Fig. 12.5. Several acoustic frequencies are shown and the resulting diffracted beams are shown to be separated by λ/Λ on the diagram. For small angles, $\sin\theta \approx \theta$. The angle of deflection is controlled by the frequency of the sound, or $\theta = (\lambda/v_s)f$, where v_s is the speed of sound and $f = \Omega_a/2\pi$. Thus, different acoustic frequencies deflect the optical beam at different angles and appear as separated spots as shown. The number of spots can be increased by increasing the number of acoustic frequencies connected to the transducer. This property can be exploited to construct acoustooptic switches. Figure 12.5A shows such a switch. Each frequency component of the sound wave deflects the light in a different direction, thus forming an $1 \times N$ switch.

In acoustooptic switches, one is interested in knowing the number of resolvable or nonoverlapping spots that can be displayed at the output. As is evident in similar problems in acoustooptic modulation (see Chapter 11, Acoustooptical Modulator), the number of resolvable spots depends on the beam widths of both the optical and acoustic beams. By defining $N = \Delta\theta/\delta\theta$, where $\Delta\theta$ is the angular variation of the acoustic beam with a bandwidth of B ($\Delta\theta = \lambda B/v_s$), and $\delta\theta$ is the light beam divergence ($\delta\theta = \lambda/b$), we have, to satisfy the Bragg condition $\theta = \lambda/\Lambda$,

$$N = \left(\frac{b}{v_s}\right)B \tag{12.2}$$

or

$$N = TB \tag{12.3}$$

where N is the number of resolvable spots, $T = b/v_s$ is the transit time of sound through the light beam, and $B = \delta f$. This condition is shown in Fig. 12.5B. To have a large N, first, we need to have an acoustic source containing many frequencies or a large bandwidth B. Then the optical beam spread b must be small, or a material of high saturated velocity must be used, or both. Note that the requirement of larger T for a switch is the opposite of that required for an acoustic modulator, where the bandwidth $B = 1/T$. Acoustic cells with $N = 2000$ are available for switching.

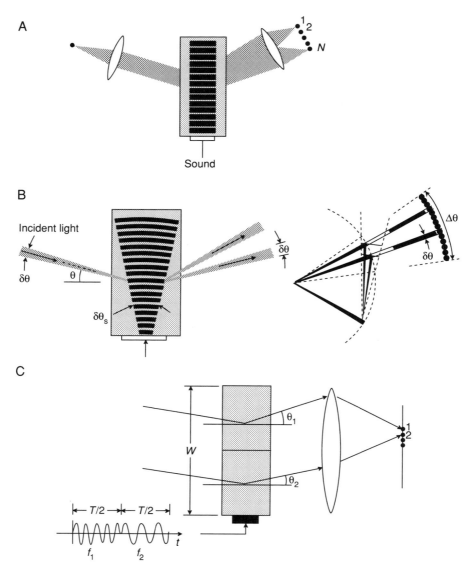

Figure 12.5 Acoustooptical switches. (A) If the acoustic medium is a narrow stripe, the spectrum of the sound wave broadens so that it contains many plane waves. Each wavefront interacts with the optical wave and is deflected in a separate direction. If N acoustic frequencies are present, N distinct points on the screen can be identified corresponding to each frequency. (B) The number of resolvable points can be determined from the ratio of acoustic wave bandwidth to optical beam divergence as defined. (C) Two or more time-divided acoustic waves can be injected on the same transducer to route different signals to different points. After B. E. A. Saleh and M. C. Teich, *Fundamentals of Photonics*. Reprinted by permission © 1991 John Wiley & Sons.

The intensity of the reflected light in the acoustic medium is roughly proportional to the intensity of the sound wave and becomes saturated at higher sound levels. To operate the device as a switch, it is desirable to operate at the saturation level, which will then turn the reflected light on and off securely. The rate of switching is limited to the transit time for the acoustic beam to sweep across the optical beam width. It should not be too long as to increase the switching time of the switch.

Acoustooptic routing can be achieved by subdividing the acoustic inputs into different time intervals such as $f_1(T/2)$ and $f_2(T/2)$ and impressing them together on the transducer. The diffracted optical beam will be switched to spot 1 (θ_1) for $T/2$ and then to spot 2 (θ_2) for $T/2$ for separate routing. This scheme is shown in Fig. 12.5C. Other schemes can also be instituted.

Magnetooptically Controlled Switches

The Faraday effect, rotation of polarization of the electric field vector in certain materials in the presence of a magnetic field, can be used to build magnetooptic switches (see Chapter 11). The angle of rotation is proportional to the length of the interaction region, and the rotatory power P (angle per unit length) is proportional to the component B of magnetic flex density in the direction of wave propagation, or $P = VB$, where V is the Verdet constant. The sense of rotation for $V > 0$ is in the right-hand-screw sense in the direction of propagation; however, if the direction of propagation is reversed, the sense of rotation does not reverse with it. Thus the total rotation can be doubled if the beam passes the interaction region twice by a reflecting mirror. This is known as the nonreciprocal property of the Faraday effect.

Magnetooptic polarization switches using passive 45° polarization rotation have been developed [9]. By the use of materials with large Verdet coefficients, the magnetization characteristic saturates easily. This gives the rotator maximum reproductivity when driven to saturation. The switch can be made bistable, and the drive power needs only to provide a change of state of the switch. Insertion losses of 0.2 dB have been reported.

A stripe-domain film light deflection element, as a crossbar switchboard for optical fibers, has been reported [7]. A thin film, about 25 μm thick, of bismuth-substituted lutetium iron garnet is grown expitaxially (by LPE) on a 600-μm-thick transparent nonmagnetic single-crystal garnet substrate (gadolinium gallium garnet). On this film, long, straight, regularly spaced magnetic domains can be formed spontaneouly as shown in Fig. 12.6A. Both the orientation and spacing of these domains can be changed by applying an external magnetic field. The magnetic field is supplied by small, rapidly responding stripline coils. The stripes align themselves parallel to the direction of the field, and their spacing varies inversely with the strength of the applied field. A stripe-domain film acts as a two-dimensional light deflector. When light is passed through the film, the state of polarization of the film rotates by an amount VD, where V is the Verdet constant and D is the thickness of the film. Figure 12.6A illustrates the deflection of light after passing the stripe with different applied magnetic fields. A 4 × 4 fiber-in, detector-out switchboard for video signals

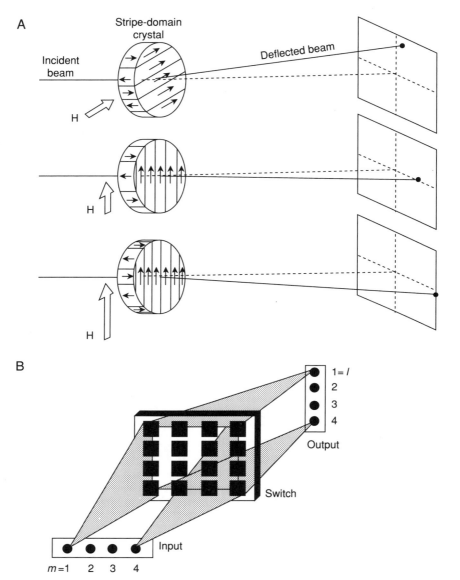

Figure 12.6   Magnetooptic switches. (A) A stripe-domain film acts as a two-dimensional light deflector in the presence of the magnetic field. The deflected light can be rotated with the applied field. (B) A 4 × 4 magnetooptic crossbar switch consisting of a thin film of magnetooptic material is deposited on a nonmagnetic substrate. A magnetic field is applied by using two intersecting conductors carrying electric current. Light beams passing through the thin film are rotated according to the applied field. The system operates in a binary mode. After B. E. A. Saleh and M. C. Teich, *Fundamentals of Photonics*. Reprinted by permission © 1991 John Wiley & Sons.

has been built for demonstration (Fig. 12.6B). This system can be used in an image system.

Switching speeds of 100 ns are possible.

Integrated Electrooptic Switches

In an electrooptic switch, electrically controlled nonlinear optical materials that alter their refractive indices are used to control the light. As large-bulk nonlinear crystals of that nature are difficult to make, the most promising technique is to use integrated electrooptic waveguides or integrated interferometers. For example, variations in refractive index with light intensity (Kerr effect) and dependence of the absorption coefficient on the applied light intensity can both be used to build switches. Figure 12.7A is an on/off switch in which one arm of the Mach–Zehnder interferometer contains a material exhibiting the optical Kerr effect. The refractive index of the

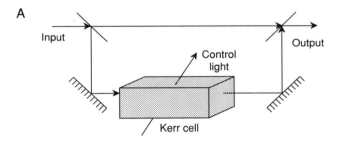

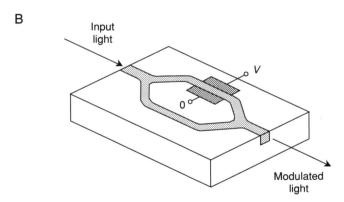

Figure 12.7 Optooptic switches. (A) An all-optical on–off switch using Mach–Zehnder interferometer with one arm containing material exhibiting the optical Kerr effect. (B) A Mach–Zehnder interferometer changes the modulated output into intensity-modulated light, which can be used as an optical switch. After B. E. A. Saleh and M. C. Teich, *Fundamentals of Photonics*. Reprinted by permission © 1991 John Wiley & Sons.

material is a function of E^2, that is, $n(E) = n - \frac{1}{2}sE^2$, where s is the Kerr coefficient. Typical values of s are 10^{-18} to 10^{-14} m²/V². The interferometer is a phase modulator. With no light on the cell, the output is balanced to register a zero output. In the presence of light, the transmittance of the interferometer is switched so that the output now registers 1.

Figure 12.7B illustrates an integrated version of a 1 × 1 switch (or modulator scheme) using a Mach–Zehnder interferometer with an electrooptic phase shifter. The light guides are made of Ti deposited on LiNbO₃ substrate. The phase control on one of the arms is activated by the voltage applied to the electrodes. The electric field is adjusted so that the output optical power can either be full (1) or zero. Large $N \times N$ switches can theoretically be built. But practically, the switch size is limited by the space required for each unit. Larger sizes may require a multilayer arrangement.

Optooptic Switches

Optooptic switches use light to control light. Figure 12.8 shows an all-optical switch. A single beam of light may control its own transmission. A light beam consisting of high- and low-power components is fed into a directional coupler at the input. If the waveguide material is nonlinear, its refractive index changes with the light intensity. The coupling length is selected so that when the input power at port 1 is low, light is channeled into the other waveguide at port 2. when the input power is high, by

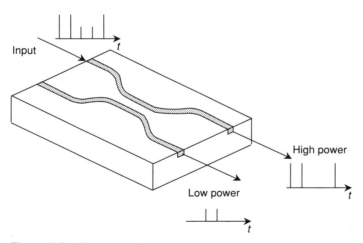

Figure 12.8 Waveguide directional coupler for separating low and high light powers. Directional coupler deposited on nonlinear optical material exhibiting the Kerr effect is used to separate beam input containing low- and high-power components. Input of low power entering the waveguide is channeled into the other waveguide. A beam of high power remains in the same waveguide. After B. E. A. Saleh and M. C. Teich, *Fundamentals of Photonics*. Reprinted by permission © 1991 John Wiley & Sons.

the optical Kerr effect, power remains in the same guide. Thus, signal light containing low- and high-power components can be segregated or switched as shown in Fig. 12.8.

The switch time of all-optical switches is very short, on the order of a few femtoseconds. Such speed cannot be attained by semiconductor electronic switches. Switching energy is another matter. Although some optical switches require very little energy, all optical switches using nonlinear optical medium usually require much more energy. A practical limit that all optical switches have in common is the problem of heat dissipation. It is a result of the weakness of nonlinear optical materials available at present time. The dashed line in Fig. 12.1 indicates this practical limit. The availability of better materials may change this situation.

Bistable Optical Switches

A bistable device has two states in output power for all input power levels: a "1" state for high input levels or an "0" state for small power levels. In other word, as the input power is increased from zero, the output power remains at zero until a critical input power is reached. At this point, a sudden jump in output to level 1 is observed. Output remains at this level with further increase in input power. The output remains at the high "1" state until the input power is reduced beyond the point where the original jump took place; then it suddenly jumps back to the original "0" state. The full excursion thus forms a hysteresis loop. In a sense, it is a switch that turns the output on and off by controlling the input. If a light source is used to control the input light, it is an optically controlled switch. Digital computers can use optical bistable devices to construct logic gates, memories, and switches. Communication systems can use these devices to route signals to their destinations.

Electronic devices perform these switching and routing services successfully in both the communication and digital computer fields. With the advent of optical transmission technology, however, electronic devices begin to reveal their limits, particularly in terms of speed and complexity. It is with this quest that we investigate optical bistable devices to seek a better way to build switches that can serve the new demands.

Operation of an optical bistable device requires two special features: nonlinearity and feedback. Nonlinearity alone is not sufficient to ensure bistability. It is the feedback that permits the transmission of light through nonlinear optical material to become multivalued. Fortunately, both features are available in optics [2].

A simple bistable optical device is shown in Fig. 12.9A. It consists of a Fabry–Perot resonator with a pair of partially transparent concave mirrors facing each other and aligned with the optical axis. A piece of nonlinear optical material is placed in the resonant cavity. Input light transmitted through the mirror bounces back and forth between the two mirrors and the nonlinear material and eventually leaks out of the resonator. Optical nonlinear material is characterized by the change in refractive in-

A

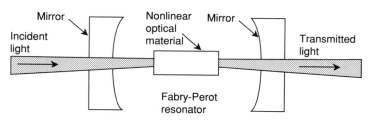

B

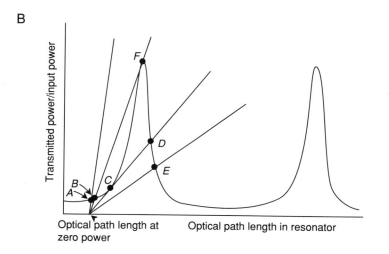

C

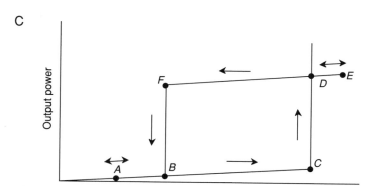

Figure 12.9 Bistable optical switches. (A) Fabry–Perot resonator containing a nonlinear optical material is used to exhibit optical bistable states. (B) With the resonator tuned to resonance, a resonance curve in output is observed. The procedure for constructing the bistable states is as follows: An operating point 0 is first chosen along the zero-output power line. At 0, a set of radial straight lines with slopes inversely proportional to input powers are drawn. These lines intersect with the resonance curve to define the operating points of the device. (C) The constructed bistable states, or the hysteresis loop, are indicated. After P. W. Smith and J. W. Tomlinson, [2], with permission.

dex on exposure to intense light (Kerr effect). The optical path length of the device, which is the product of n, the refractive index, and L, the physical length of the cavity, thereby varies with light power. If successive paths of the light beams in the resonator fall in phase, they interfere constructively and register a resonance where the intensity peaks. Resonance occurs when the path length of the resonator corresponds to any multiple of a half wavelength. The change in intensity, in turn, affects the refractive index of the nonlinear material, which provides the feedback. Conditions favoring bistable operation are therefore established.

A plot of transmitted light power versus input power (or optical path length) is shown in Fig. 12.9B. To recognize the abrupt change in the output–input power relationships, let us tune the cavity so that it is slightly off-resonance. The resonance curve is shown on the graph. An operating point A is chosen, slightly to the left of the resonant setting on the zero-output power line. From a point 0 on the axis, a set of radial straight lines whose slopes are inversely proportional to different input powers are drawn. Increasing input power is the means to operate the device on different lines with decreasing slopes. These lines intersect the resonance curve at several points as the input power is increased in steps. At low input power corresponding to point B, the transmission is low. With increasing input power, the power in the resonator begins to build up, but the transmission remains relatively low, until point C is reached. As light intensity reaches its peak, the nonlinear material becomes much more transparent, and the output power increases abruptly to reach point D along the load line. The device has been suddenly switched to a high transmission state corresponding to point D. A further increase in input power stops the transmission from increasingly continuously but remains at the high level D.

As the input power is reduced, the device remains in the high state until point F is reached and suddenly returns to the lower state to rejoin the original curve at B. The path is plotted in Fig. 12.9C. This is the bistable hysteresis loop. The size of the hysteresis loop depends on the operating point 0. If the resonator is tuned closer to resonance, moving point A toward the peak of the resonance curve along the zero-output line, the loop becomes narrower. At resonance, the loop shrinks to a sharp rising S-curve with saturation. With this setting, the device can still be used as a switch if it is driven to full range. Alternately, it can be used to amplify the modulation present in the beam; however, the device is still passive so that the amplified amplitude is limited to the extent of the input power only.

Two types of nonlinear optical elements can be used to build optical bistable devices: dispersive nonlinear elements and absorptive nonlinear elements. In the former, the refractive index n is a function of light intensity. In the latter, the absorption coefficient a is a function of optical intensity. Feedback mechanisms are provided either externally or internally (Fig. 12.10).

Quantum-Well Absorptive Bistable Devices

Bistable devices can be built using multiple quantum wells with nonlinear absorptive materials [10]. MQW heterostructures are grown epitaxilly with alternating thin lay-

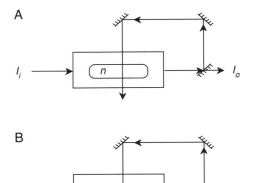

Figure 12.10   Dispersive and absorptive optical bistable devices. (A) A light beam fed back from the output passes through a dispersive medium to change its refractive index and establish a bistable operation. (B) A light beam fed back from the output passes through a dissipative medium to change its absorption coefficient to establish bistability. After H. S. Hinton [10].

ers of semiconductor materials of different bandgap energies, such as GaAs and AlGaAs shown in Fig. 12.11A. Layers as thin as 10 nm are grown in a cell of total thickness 1 μm. As the bandgap energy of AlGaAs is greater than that of GaAs, quantum potential wells are formed that confine the electrons to the GaAs layer. To use this cell as an optical switch, a reverse electrical potential from an external source is applied to the cell through a large resistance. The absorption coefficient of the material is a nonlinear function of the applied voltage. Absorption increases when the voltage across the MQW cell is decreased. In operation, the laser wavelength is chosen so that the photocurrent generated by the input light power increases the voltage drop in the series resistor which causes the voltage across the cell to drop, thereby increasing the absorption. This also serves as the internal feedback mechanism for bistability.

The bistable hysteresis loop can be traced as before by assigning increasing and then decreasing input powers and recording the outputs (Fig. 12.11B). Note that the cyclic hysteresis loop seems backward compared with that discussed in the previous section. This device is also called a self-electroptic effect device (SEED). The symmetric SEED (S-SEED) consists of two MQW $p-i-n$ diodes connected in series and is reverse biased as shown in Fig. 12.12A. Multiple quantum wells are built into the intrinsic region. When the diodes are connected in this fashion they become complementary. When one of the diodes is on, that is, reflecting light, the other is off, that is, absorbing light. The switching of the state of the diodes is determined by the power of the light beam directed at each diode, not the absolute intensity of the input beam. If the optical bistable loop is centered around the point where the two inputs

A

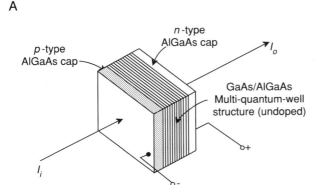

B

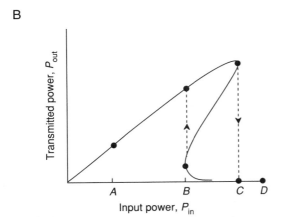

Figure 12.11   Multiple-quantum-well absorptive optical bistable devices. (A) This structure is a multiple-quantum well (MQW) using GaAs and GaAlAs semiconductor materials. It is a *p–i–n* diode with a MQW built into the intrinsic region. (B) The bistable states are shown as a hysteresis loop. Note that the bistable sense seems inverted.

are equal, the device will change its state when the power ratio exceeds 1.3 or is less than 0.7. A S-SEED 2-module can be built that serves as the basic building block of logic devices. Large arrays of S-SEED pairs can be fabricated readily by batch processing [11, 12].

The operating characteristics of a series-connected S-SEED are shown in Fig. 12.12A. To build a 2-module with an S-SEED, enabling lights can be injected with the signal lights as shown in Fig. 12.12B. The signal can be advanced step by step in a flip-flop manner as in a digital computer. To date, this is perhaps the best way to build an optical flip-flop circuit. It is small and can be integrated and mass produced. But there remain many problems, for example, the extra power supply and

A

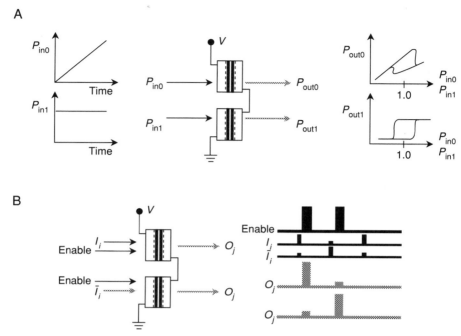

Figure 12.12 Pair of S-SEEDs in series with reverse bias. (A) Operating characteristics. (B) S-SEED-based 2-module. Enable light pulses advances the digits. Reprinted with permission from H. S. Hinton [10].

the power dissipation in a large number of devices when integrated. Research on SEEDs is growing, with respect to both telecommunications and optical computers.

Other Optical Switch Schemes

Use of a holographic technique to switch optical signals is another possibility. Holograms can be written optically in planar form to connect input and output fibers in two-dimensional arrays. Thick holograms can form efficient gratings to redirect a single optical beam [13]; however, because of the engineering problems encountered in constructing these complex switches, this technique does not look promising in the near future. The energy requirements for this process are excessive.

Physical implementation of a waveguide crossbar switch can be achieved with the crossbar configuration in Fig. 12.2C. Each of the $N \times N$ points formed by the intersection of N incoming channels and N outgoing channels is a cross point where the channels are connected. To change the connections from the ith to the jth channel, it

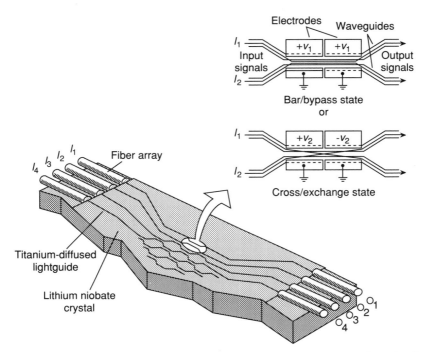

Figure 12.13 A dilated Benes network of waveguide directional couplers. Directional couplers are arranged to form an optical switch to route signals in communication networks. The switch has low insertion loss and little crosstalk by special arrangement. The device is also integratable. After S. K. Korothy and R. C. Alferness. *Optical Fiber Telecommunication, II.* (S. E. Miller and I. P. Kaminous, Eds.), © 1988 Academic Press.

is necessary only to switch the i, j cross point in the array from the crossover state to the straight-through state. The basic structure consists of a number of waveguide directional couplers deposited on $LiNbO_3$ substrate. Each coupler is equipped with electrodes on which voltages are applied to effect the change in state as described under Electrooptically Controlled Switches. An 8×8 version of this structure is shown in Fig. 12.13 [14.15]. The couplers are joined in a pattern known as the dilated Benes network. The Benes rearrangeable network limits loss to a logarithmic instead of a linear increase with switch size. Crosstalk is minimized because any coupler can have only one of its inputs active at any instant. Directional couplers can be integrated to form a large switching network, but both the couplers and their bending radii limit the maximum size of the integrated array (size 32×32).

A dissipative nonlinear medium, for which the absorption coefficient a or the reflectance r is a function of the optical intensity, can also be used to make optical switches. For example, an array of liquid crystal dots are arranged on a planar display.

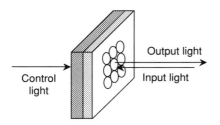

Figure 12.14   A liquid crystal spatial light array used as switches. Reprinted with permission from B. E. A. Saleh and M. C. Teich, *Fundamentals of Photonics.* © 1991 John Wiley & Sons.

The reflectance of these dots is controlled by a light source on the back of the array. Different points on the liquid crystal surface have different reflectances and act as independent switches controlled by the input light (Fig. 12.14).

Summary

In this chapter, we have described optical switches, many of which have already appeared on the market; however, the architecture of switching has been left untouched. Even time division multiplexing and wavelength division multiplexing were not reviewed.

Optical switches may be applied to both communications and computers. Optical computers are beyond the scope of this book, so we have confined our discussion of optical switches to applications in communications, or the routing of signals by optical means.

Mechanically controlled optical switches are the simplest to implement. They are insensitive to wavelength or polarization. But they are physically bulky, very slow in operation, imprecisely timed, and certainly not suitable for large-scale fabrication.

Directional waveguide couplers exchange optical energies along the coupling length of a four-port guide. For a coupling length corresponding to one-half of a wavelength, switching from one port to other, or vice verse, is accomplished. The device is wavelength sensitive. Although the insertion loss may be low, and the response is broad, its large dimensions (in millimeters to centimeters) and low response time (in milliseconds) make it unattractive.

Electrooptically controlled switches offer low insertion loss (below 1 dB) and subnanosecond switching time. But the required electrodes and external potential source supply add to the dimensions of the device and load the driving circuit capacitively. Crosstalk between outputs is high.

Holographic optical switches make use of two-dimensional planar arrays of optically written holograms to switch input fibers to output fibers through free space. At present, the energy requirements of the photorefractive materials for the hologram are prohibitively large.

Reactive and absorptive bistable devices have been proposed to activate optical switching. But the main interest in bistable devices is in developing optical computers, not in telecommunication switching.

Perhaps the most promising device thus developed in optical switching is the self-electrooptic device (SEED). It is a hybrid switch, using the change in the absorption property of a material embedded in the intrinsic layers of a multiple-quantum-well cell to effect light transmission. Two different types of SEEDs have been developed: the S-SEED (symmetric SEED) and the F-SEED (field effect SEED). The strength of SEEDs is that large arrays of small cells with uniform characteristics (125 × 256 or 32K array) can be fabricated by batch processing, and integration with other electronic devices is possible to produce gain and to improve the speed of response. The weaknesses of SEED, at present, are that the optical energy required to effect switching is about 1 pJ and the light source must contribute equally among all S-SEEDs in the circuit. The switching energy is actually smaller than that required for most electronic devices. With time, these problems may be solved.

References

1. E. Nussbaum, Overview of switching needs for 1990–2000 plus. Presented at IEEE/OAS Topical Meeting on Photonic Switching, Incline Village, Nevada, March 18–20, 1987.
2. P. W. Smith and W. J. Tomlinson, Bistable optical devices promise subpicosecond switching. *IEEE Spectrum* **18**, 26–33 (1981).
3. A. M. Hill, One-sided re-arrangeable networks. *IEEE J. Lightwave Technol.* **LT-4**, 785–789 (1986).
4. R. C. Alferness, L. L. Buhl, S. K. Korotky, and R. S. Tucker, High-speed delta–beta directional coupler switches. Presented at IEEE/OAS Topical Meeting on Photonic Switching, Incline, Nevada, March 18–20, 1987.
5. G. A. Bogert, E. J. Murphy, and R. T. Ku, Low cross-talk Ti:LiNbO$_3$ optical switch with permanently attached polarization maintaining fiber array. *J. Lightwave Technol.* **LT-4**, 1542–1545 (1986).
6. P. J. Duthie, M. J. Whale, and I. Bennion, New architecture for large integrated optical switch array. Presented at *IEEE/OSA* Top. Meet. on Photonic Switching, Incline Village, NV. Mar. 28–20, 1987.
7. E. K. Torok, J. A. Krawczak, G. L. Nelson, B. S. Fritz, W. A. Harvey, and F. G. Hewitt, Photonic switching with stripe domains. In *Photonic Switching* (T. K. Gustafson and P. W. Smith, Eds.), Springer-Verlag, Berlin, 1988.
8. A. Korpel, Acousto-optics—A review of fundamentals. *Proc. IEEE* **69**, 48–53 (1981).
9. M. Shiraski, F. Wada, H. Takamastsu, H. Nakajima, and K. Asama, Magnetoptical 2 × 2 switch for single-mode fibers. *Appl. Opt.* **23**, 3271–3276 (1984).
10. H. S. Hinton, Free-space digital optics and photonic switching. *Photonics Spectra,* pp. 123–128 (Dec. 1991).
11. D. A. B. Miller, Quantum well self electro-optic effect devices. *Opt. Quantum Electron.* **22**, S61–S89 (1990).
12. A. L. Lentine *et al.*, Integrated SEED photonic switching nodes, multiplexers, demultiplexers, and shift registers. In *Proceedings of 1991 Topical Meeting on Photonic Switching, Salt Lake City, March 1991,* pp. 60–66.

13. J. P. Huignard, Wave mixing in nonlinear photorefractive materials: Applications to dynamic beam switching and deflection. In *Proceedings of 1991 Topical Meeting on Photonic Switching, Salt Lake City, March 1991*.

14. H. S. Hinton, Switching to photonics. *IEEE Spectrum*, pp. 42–45 (Feb. 1992).

15. M. N. Islam, C. R. Menyak, C. J. Chen, and C. E. Soccolich, Chirp Mechanisms in soliton dragging logic gates. Opt. Lett., v. 16, pp. 214–216, Feb. 1991.

CHAPTER 13

Photonic Interconnections

Introduction

The interconnections between system components or between electronic devices have become a problem only recently. It was felt first in the telecommunications field, where the number of subscribers multiplied exponentially. Next was the microelectronic integration field, where there is an ever-increasing demand to integrate more components into shrinking real estate. We also doubt whether optical computers could ever be built without replacing the metallic interconnections with something that can accommodate high-density, high speed, and reliable interconnections with as little power dissipation as possible.

If interconnections would not to use the existing metallic wires, what else could they use? This chapter explores the possibility of using optics to resolve the problem of interconnections.

We can identify at least three different fields that would benefit from the development of new interconnection methods: the telecommunication field, including local area networks (LANs), broadband integrated service digital networks (B-ISDNs), and very large scale integration (VLSI) systems, and the optical computing system field. Although the common demands of higher density, higher speed, and higher power dissipation are the same, individual implementation may vary with the system. Each is a special subject and many specific publications are available. The purpose of this chapter is to introduce briefly the highlights of each case. Readers are advised to refer to the references for detailed study [1–3].

Optical Interconnections

As the success of using optical fibers to transmit signals via optic frequencies has been recognized, one begins to investigate how optics could be used to resolve the problem of interconnections.

In Chapter 12, we enumerated the advantages of using photonic switches rather than the mechanical and other types of logic switches now in use in the above-mentioned fields. We use a similar argument to explore the role of optics in interconnections.

The advantages of optical interconnections are several: (1) The speed of optical signal propagation does not depend on the number of components that handle the signal. The speed depends only on the speed of light in the medium in question. (2) Optical switches offer no capacitive loading effect to the circuit as in electronic and/or photonic switches. Freedom from the circuit loading effect is responsible for the greater flexibility of optical interconnection with respect to fan-in and fan-out capabilities. (3) Optical interconnections are immune to mutual interference effects. There is no electrical coupling between the high-frequency modulations of two proximate beams of light. (4) Free-space light beams can pass over or through other light beams in the same space without significant interaction. These properties of light simplify the signal routing without worrying about short-circuiting as in wire crossing. More advantages may become evident when individual systems are described.

Classification of Optical Interconnections

The different fields may need to develop different optical interconnections to realize the same advantages of high speed, wide bandwidth, and immunity to electromagnetic interference. Each has its own needs, and we address these individual cases separately.

Telecommunication Services

In communication systems in which a large number of substations are interconnected to form a network, the interconnections were formerly made by conducting wires or cables. One can visualize easily the large number of wires that occupy such telephone substations. As the number of subscribers increases, so does the number of interconnections. To minimize the number of connecting cables in the system, we divide the subscribers into groups, and use a few trunk cables to interconnect the groups, taking advantage of the fact that not all subscribers will use the trunk all the time. The most urgent problem is then how to route signals from one group to another through service trunks. Routing is designed to use the least number of trunks between groups to serve as many subscribers as possible. All this should be done at the highest speed possible and cover a wide band of frequencies. The routing scheme should also be designed to provide the flexibility of automatically finding a connection through the maze of networks to reach the desired destination within the shortest time.

Microintegrated Circuits

In microintegrated circuits, the demand is even more stringent. Although the number of components (gates and transistors) that can be put on a chip is increasing exponentially, the chip size limits the increase. The number of pins available for connections is not expected to increase freely. In fact, the limit has already been felt. For example, a recent design for a complex algorithm demanded 1000 input–output pins per chip. With the maximum space for 250 pins with electric connections now available, the algorithm would run far too slowly. Even if it were possible to fabricate such 1000-pin chips with electric connectors, the failure rate tied to that number of pins and the associated bonds would render the system insufficiently reliable. Thus, to limit the chip size to, say, a few square centimeters and the number of pins to fewer than 250, interconnection within the chip poses a real problem. As a way out, VLSI systems consisting of large numbers of MOS gates per chip ($>100,000$ gates) have been fabricated on multilevel design schemes. Within a chip, interconnections are still needed for different circuits. Between layers, interconnections are required to function. There is also the need to connect chip to chip, from IC to IC, as well as from board to board. The problem is that there is so little space and so many connections to be made. Electrical connections, even in a three-dimensional configuration, will never be able to perform the task. Most likely, optical interconnections or hybrid structures will be the only solution.

Optical Computing

Digital logic circuits have been used successfully in electronic digital computers with microelectronic integrated logic circuits. As the computer geometry grows smaller and denser, and as computations become faster and more complex, the capacity of electronic circuits is being outstripped. The connections simply cannot handle electrical signals swiftly and reliably. The power drain and the number of wires linking the circuit boards have become prohibitive. Optical interconnections are being sought.

How can optics help? Optical computing systems can have high space–bandwidth and time–bandwidth products; hence, many independent channels could be exploited for the demanding computations. Optical computing is capable of communicating many channels in parallel. Optical interconnections offer the combination of large conductor and large fan-out. *Fan-out* is the number of parallel loads driven simultaneously by one signal. Bandwidths of gigabits and fan-out up to 100 loads from a single fiber are possible. Optical signals can propagate through each other in separate channels with essentially no interference and can propagate in parallel channels without interference and crosstalk.

Even with the encouraging successes of optical communications and local area networks, many required optical components are not yet commercially available, particularly those with lower power consumption. At present, developers are using hybrid schemes to supplement the existing electronic connections with optical connectors at key points. It is hoped that eventually optical interconnections will be ushered into fully optical computing.

Methods to Implement Optical Interconnections

Conventional Optical Connections

Any switch with its control permanently affixed to the on-position can be considered a connection. Under this definition, all discussions in Chapter 12 apply to optical connections. One has only to judge the usefulness of the individual switch by its suitability to the assigned connection.

Conventional optical components—mirrors, lenses, prisms, beam splitters, and so on—can be used in numerous optical systems to establish optical connections. Simple interconnection maps can be created by using single optical components. Figure 13.1 shows some examples. A lens can fan-in, fan-out, magnify, and reduce as shown in Fig. 13.1A. A cylindrical lens can connect all points of each row in the input plane to a corresponding point in the output plane as shown in Fig. 13.1B. Two-dimensional planar arrays can be imaged by a lens over distances as short as a few millimeters and could compete on power efficiency grounds with electrical interconnects. A prism could be used to bend or shift the rays to establish an ordered interconnection as shown in Fig. 13.1C. Two prisms can be oriented to perform a perfect-shuffle interconnection map as shown in Fig. 13.1D. The perfect shuffle is an operation used in sorting algoriths and in the fast Fourier transform.

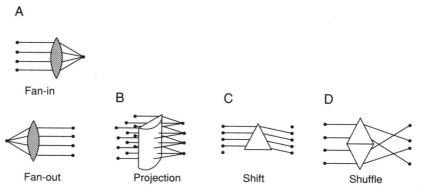

Figure 13.1   Examples of simple optical interconnections created by conventional optical components: (A) lens for fan-in or fan-out; (B) a cylindrical lens to connect all points of each row in the input plane to corresponding points in the output plane; (C) a prism for shifting connections to different directions; (D) using two prisms to perform a perfect shuffle.

Future extentive use of optical array interconnections within semiconductor wafers and between wafers in microintegration is anticipated.

Optical Fibers

The state of the art in existing optical interconnections is still limited to lower-level interconnections, such as within a chip, from chip to chip, and from package to package. We may use a medium such as fibers, free space, or waveguides to establish the connections.

Fiber interconnects can draw on the rich technological resources that already exist or that are in advanced development for other applications. New optical fibers with speeds up to 560 Mbit/s developed for long-distance communication and those developed for local area networks at 10 Mbit/s are available commercially.

Optical fibers have been used by Honeywell and others to connect board-to-board in a massively parallel computer with thousands of relatively simple microprocessors. Each optical fiber connects 512 input–output channels at more than 1 Gbit/s.

Optical Waveguides

Light can move through well-defined channels made of material as lightguides. Small optical waveguides, when placed close to each other and with the help of underlying electrodes, can function as a switch or connector. They can be integrated into packages to replace optical fibers for interconnections between chips and boards. Techniques have been developed for connecting VLSI chips with high-density optical waveguides on a silicon carrier.

Free-Space Light Interconnections

Free-space interconnection offers a large-scale connectivity that cannot be achieved in the more mature electronic technology. This includes the use of mirrors, lenses, prisms, and holograms. In microelectronic integrated circuits, even active optical connections using lasers and detectors can be included.

Free-space interconnects may be either focused or unfocused. Unfocused types can broadcast the signal over the entire chip, a simple way for many fan-out connections.

Holographic Interconnections

The use of diffraction gratings for holographic maps makes the interconnection very flexible. The period and orientation of the gratings can be varied by the spatial frequencies of the reference wave. Holography provides a method for drawing maps of interconnections in recording. Arbitrary interconnection maps may therefore be created by selecting the appropriate (grating) spatial frequencies at each point to record the hologram. Holographic interconnection devices are capable of establishing one-to-many or many-to-one interconnections. For complex interconnection maps, computer-generated holographic programs need to be developed.

Holographic Method

Holography involves the recording and reconstruction of optical waves. To record an optical image photographically, the emulsions used to make the transparencies are sensitive to optical intensity only. They do not contain the phase information. Phase information is important in reconstructing optical images in holography. The interference pattern produced by mixing the optical wave (object wave) with a reference wave for recording provides the phase information on a transparency. To reconstruct the image, the transparency is exposed to the original reference wave again. This process can be adapted for interconnections in optical communication systems.

Recording and Reconstruction of a Hologram

The information in the preceding section can be put into analytical terms by studying the interference patterns of two superimposed monochromatic waves of complex amplitudes $A_1(r)$ and $A_2(r)$, respectively. Let

$$A_1(r) = I_1^{1/2}(r) \exp(i\theta_1)$$
$$A_2(r) = I_2^{1/2}(r) \exp(i\theta_2)$$

where I_1 and I_2 are the intensities of A_1 and A_2, and θ_1 and θ_2 are their phase angles. $I_1 = A_1^2$ and $I_2 = A_2^2$. The resultant intensity can be expressed as

$$I = [A_1(r) + A_2(r)]*[A_1(r) + A_2(r)] \tag{13.1}$$
$$I = I_1 + I_2 + 2(I_1 I_2)^{1/2} \cos \theta$$

where $\theta = \theta_2 - \theta_1$. Note that the intensity now contains a term representing the interference effect that varies with the cosine of phase difference between the original waves. Thus, the phase information is contained in the last term of Eq. (13.1).

In holography, the original waves A_1 and A_2 are called the object wave reference wave, respectively. The interference pattern is represented by the third term in Eq. (13.1). All these data are recorded on the transparency as a hologram.

To reconstruct the object, the holographic transparency is illuminated with the reference wave. The result is a wave with complex amplitude

$$A = A_2 I_2 + A_2 I_1 + A_1 I_2 + A_2^2 A_1^* \tag{13.2}$$

The third term in Eq. (13.2) represents the image field which is now multiplied by the intensity of the reference wave. If I_2 is uniform, this term is the desired reconstructed wave. It can be separated from the other terms readily. Ordinary lighting will not result in reconstruction, because the phases θ_1 and θ_2 are random quantities that are uniformly distributed between 0 and 2π, so that the average of $\cos \theta$ is zero. The last term is the conjugate of the object wave.

Volume Holography

If a relatively thick medium is used to record the hologram, we will see some interesting effects. The interference pattern between the object and reference waves is now a function of x, y and z:

$$I(x,\ y,\ z) = I_1 + I_2 + 2(i_1 I_2)^{1/2} \cos(\mathbf{k}_1 * \mathbf{r} - \mathbf{k}_2 * \mathbf{r}) \qquad (13.3)$$

where $\mathbf{k}_1$ and $\mathbf{k}_2$ are the wavevectors of the object and reference waves, respectively. Note that a sinusoidal pattern of period $\Lambda = 2\pi/(\mathbf{k}_1 - \mathbf{k}_2)$ is developed that is known as the pattern of the diffraction grating. The Bragg condition is satisfied such that $\sin(\theta/2) = \lambda/2\Lambda$. This grating is useful for making holographic maps for interconnections.

High-speed operation of optical computers and imaging processers can depend on holographic technique for interconnections. Efficient holographic interconnection systems can be designed for many applications. They have been proven to outperform state-of-the-art electronic interconnects in density, power, and speed. But the cost of installing such a system is still high. In addition, the system is sensitive to environmental changes and mechanical alignment.

Hybrid Systems

Any combination of the above-listed methods can also serve as an interconnect for a particular system. The choice depends on the merit in doing so, either to reduce power consumption, to ease implementation, or to be more cost effective.

In complex logical processing, an all-electronic implementation seems more cost effective (at present) than its all-optical counterpart except that these processors are communication limited. This is to say they have interconnection problems. Here, optical interconnections may be helpful. In some designs, the data processing by electonic logic circuits is limited to some "islands." Different islands are then interconnected by optical links. This is known as the "smart pixel" or optically interconnected electronic island approach. Very efficient and multidimensional operations have been developed.

Practical Implementation of Optical Interconnections

In the following sections, we intend to introduce practical implementations of optical interconnects in various fields to provide some feeling for these systems. The presentation is no means complete and exhaustive. Only sketches are given. In each case, one or two short examples are cited without going into much detail. References are given for further study.

Telecommunication Systems

Switches and Switching

Photonic interconnections are intimately related to photonic switches. The use of these switches to route messages is called *switching*. Although photonic switching can be applied to both optical computing and telecommunications, we limit our discussion to the latter application in this section. Optical connections in optical computors is a specialized topic that we can touch on only briefly in this chapter. Comprehensive coverage can be found in other publications [4].

Before we start the discussion, let us make a distinction between the two types of switching, the toggle type and the logic type. The former type performs only the on–off task, leaving the control management separately. The logic type derives its flow of information by a clocked scheme that logically leads to signal routing. In electronic switching, this makes little difference because of the predominance of electronic logic. But in optical switching, it is important as we will see later.

In chapter 12, we introduced photonic switches as discrete components. The discussion in this chapter centers on the switching schemes; however, after each scheme, the merit of adapting which type of switches is commented on.

Telecommunication Schemes

Crossbar and Multistage Networks

In telecommunication systems, we wish to connect many substations in a network such that any station can be connected to a free line without blocking. The simplest switching method is to connect N input lines to N output lines in an $N \times N$ crossbar with an open/close switch at each crossing. An $N \times N$ crossbar scheme is shown in Fig. 13.2. We borrow this scheme from the electronic switching technique and replace the electric paths and electronic switches with optical lines and switches located at each crossing point to form an optical $N \times N$ switch crossbar. To connect

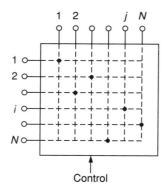

Figure 13.2   An $N \times N$ crossbar interconnection consisting of N inputs and N outputs arranged in crossbar form. At each cross point, an on/off switch operates to control the data flow.

input i to output at j, one merely has to close the switch at (i, j). One notes immediately that the number of discrete switches required is N^2. The path length also varies dramatically between the input and output ports. The longest path is that spanning the corners of the square diagonally. In optical terms, it means that the insertion loss will vary widely between paths. Obviously this simple scheme does not work efficiently even for a small station.

Advantage was taken of the fact that not all substations in a system will use the line continuously all the time, and a scheme for dividing the subscribers into groups according to location (or any other criteria). Then a much smaller number of trunk lines can be installed between the groups. A switching scheme is provided to let each substation within each group reach the trunk line only when it is needed. Successful electronic switching schemes are the CLOS network [5], and the BENES network [6], under which the number of discrete switches required to connect N inputs to N outputs in a nonblocking manner is greatly reduced. By modifying the electronic switches with optical switches and optical waveguides for optical signal carriers, a modern system is produced. Burke *et al.* improved the CLOS system by allowing stages with gain to be inserted in the optical path and by expanding the matrix to include a 128×128 lithium niobate switch for larger networks [7]. With the advant of erbium fiber amplifiers, implementaion of such switching systems is not far away.

Figure 13.3 is the schematic layout of a CLOS network showing twofold expansion in the center stages arising from the absolute nonblocking requirement. It is a three-stage $N \times M$ network composed of n input stages, m output stages, and $n + m - 1$ center stages, the number required to ensure a nonblocking operation. This scheme allows large matrices to be assembled from smaller buiding blocks, but it still requires complex interconnections. One advantage is that many dissimilar multistages networks may operate together. For example, the input and output stages might be time or wavelength switches, whereas the center stage might be a space switch. Gain stages may also be inserted in the optical path between chosen stages.

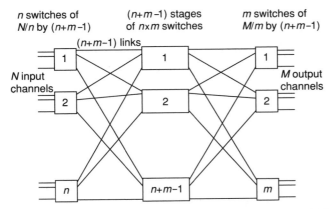

Figure 13.3 Schematic layout of CLOS network showing twofold expansion in center stages arising from absolute nonblocking requirement. After C. Clos [5], with permission.

A very attractive multistage switching network with ultralarge throughput switches known as STARLITE has been reported [8]. It offers self-routing capabilities with the result that packet-type data with suitable header addresses can worm their way through a complex network, setting up the switches and their own route as they proceed.

Guided-Wave Optical Switches

Optical switches using directional waveguide couplers have been described in Chapter 12. A typical layout for use of these devices as interconnections is shown in Fig. 13.4. A diffused Ti:LiNbO₃ four-port waveguide coupler consists of two input single-mode guides, two output guides, and an interaction region inbetween. The substrate is a lithium niobate crystal. A pair of electrodes over the interaction region controls the phase relationship of the waveguides and the two output guides. The switch works satisfactorily as a single element. Let us investigate the possibility of using a buildup of these directional couplers for a crossbar interconnection. The overall dimension of the resulting device are typically length 1 mm to 1 cm and width perhaps 10 to 20 μm. The electrodes that cover the interaction region are also long and requires 5 to 20 V to induce switching. It can be seen that the device exhibits large capacitance to the electrical driving circuit, thus limiting the switch reset time.

To connect these waveguides as a crossbar switch array, which may need 100 inputs and 100 outputs, or 10,000 units, the electrical drive would be immeasurable. At present, arrays about 16 × 16 in size and built on a single substrate represent the state-of-the-art along this line. Other architectures are being investigated [9, 10].

Recently, NEC of Japan demonstrated the feasibility of a 128 × 128 switching

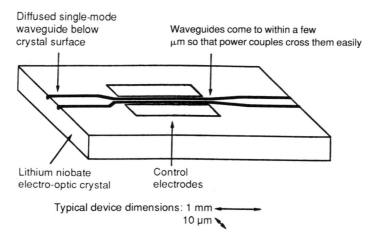

Diffused single-mode
waveguide below
crystal surface

Waveguides come to within a few
μm so that power couples cross them easily

Lithium niobate   Control
electro-optic crystal   electrodes

Typical device dimensions: 1 mm ⟵——⟶
10 μm ⟵

Figure 13.4 Typical guidewave directional coupler. On the substrate of lithium niobate, a single-mode directional waveguide coupler is deposited below the surface. The interaction region is narrowed to a few micrometers with a pair of control electrons deposited on top of the region. After R. V. Schmidt and R. C. Alferness, *Trans CAS,* **26** 1099–1128. © 1979, *IEEE.*

experiment, using lithium niobate 4 × 4 switching arrays as the basic building block to construct a larger switch [7]. Midwinter made an analysis of the demonstration to estimate the cost of implementing such a system and called it a "hero's" experiment [10]. The awesome complexity of the system, the need to drive thousands of electrical control signals for the pathway, and the required insertion of hundreds of semiconductor laser amplifiers to make up for the 48-dB matrix insertion loss are just too much for practical systems.

Although optical waveguide switches possess a wide bandwidth of about 4000 GHz per port, and therefore provide a huge throughput capability, the route reconfiguration time is set by the electric control system, which may be slow. Moreover, the implementation of this system is probably more complex than implementation of systems with digital electronic switches. It seems doubtful that these switching systems will be suited for switching byte-multiplexed traffic and asynchronous time multiplexing (ATM) systems.

Wavelength Division Multiplexing Switches

Optical fibers offer about 30-nm spectral bandwidth, which corresponds to about a 4000-GHz electrical spectrum. This brings a completely new dimension to switching through the wavelength domain. A generalized wavelength division multiplexing (WDM) network is shown in Fig. 13.5. It consists of N tunable lasers, each tuned to

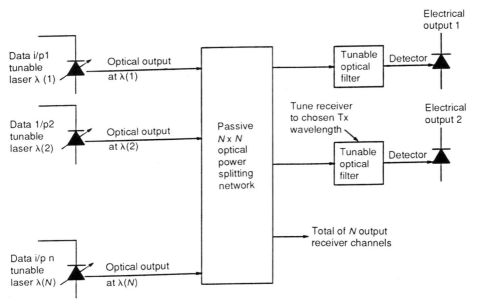

Figure 13.5 Generalized WDM network. Inputs from different tunable wavelength lasers are combined at a passive $N \times N$ optical power splitting network for processing. Outputs from this network are distributed through many tunable optical filters and lead to detectors of each channel as shown. After C. A. Brackett [11], with permission.

a wavelength and fed to a controller. At the output, an equal number of tunable optical filters and detectors are connected. The controller is a passive $N \times N$ optical power splitting network that dictates the transmission of signals to the receivers. The system is very flexible. Detailed analysis reveals that the complexity in component design is overwhelming [11].

Time Division Multiplexing Switches

Time division multiplexing (TDM) is commonly achieved in electronic switching by storing the signals in a buffer memory system and subsequently releasing them in timely order to be transmitted in proper time slots. Unfortunately, in optics, there is no such equivalent. One has to resolve the time delay by using recirculating delay lines of various (fiber) lengths. Note that the speed of light in an optical fiber corresponds to about 20 cm/ns so that an 8-bit delay at 1.2 Gbit/s corresponds to about 1.3 m of fiber. For 32-channel multiplexing, we need delay line lengths of 0, 1.3, 2.6, 5.2, 10.4, and 41.6 m that can be combined in each channel with a total of about 2.4 km of fiber. If the number of channels were increased to 256, the total delay fiber length would increase to 170 km. It can be seen how impractical this can be. This technique eliminates the possibility of total integration as was done in electronic switching.

Optical switches and the switching implementation of the devices described thus far remain a discrete component technology. Many functions that digital electronic data processing can do can be duplicated by optical means with the added advantage of wide bandwidth capability; however, it is difficult to see how the waveguide approach can be scaled to high-density switching schemes as in electronic switching technology. This may change in view of the recent invention of SEEDs, a planar multi-quantum-well optoelectronic device. *SEED* stands for self-electrooptic effect device [12]. A large array of SEEDs were proposed for switching [13, 14]. In Chapter 12, under Quantum-Well Absorptive Bistable Devices, the operating principle is briefly described.

Application of SEEDs to telecommunications is on the rise. At CLEO 1990, AT&T Bell Laboratories demonstrated a prototype multistage interconnection network using SEED elements in 32 $\times$ 64 arrays with free-space interconnections and crossover [15].

Very Large Scale Integrated Interconnections

In very large scale integrated (VLSI) circuits, interconnections occupy a large portion of the available chip area. Efforts to use optical interconnections increase as high-speed, high-density, microelectronic circuitry and the emergence of parallel processing rapidly progress.

Electronic interconnections are planar and cannot overlap or cross without insulation. Free-space optical interconnections can be three-dimensional and can pass through each other without mutual interference. This allows for a much greater density of interconnections. Optics offers greater flexibility for fan-out and fan-in inter-

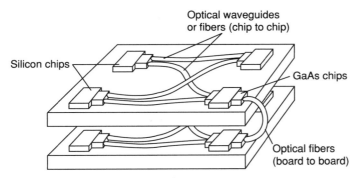

Figure 13.6 Optical fibers for chip-to-chip and board-to-board interconnections. After L. D. Hutcheson and P. Haugen [16], with permission.

connections as a result of the shorter delay time compared with electronic circuitry. The density of optical interconnections is not affected by the bandwidth of the data carried by each connection, and the power requirement is limited only by the sensitivity of the photodetectors.

Optical Fiber Interconnections

Optical fibers are used for interconnections between chips and within a single chip. Figure 13.6 shows one example [16]. Chips of different constructions, such as silicon and GaAs chips, are connected together within a board and between boards as shown in the example.

Waveguide Interconnections

Optical waveguides can be made as small as 1 to 10 μm in diameter and placed as close as 10 to 100 μm. They can be integrated into packages when the thickness and bending radius of the optical fiber seem cumbersome within chips on small, closely packed boards. A typical design by Honeywell's Physical Science Center is shown in Fig. 13.7. Thin-film optical waveguides, formed by depositing tantalum pentaoxide, aluminum oxide, or zinc oxide, form waveguides on a silicon carrier that transmit optical signals between silicon VLSI chips recessed into the carriers. Note that right-angle bends in the path will contain the optical signal, and signals can actually intersect without interfering with each other.

Free-Space and Holographic Interconnections

The unique property of the optical interconnection is that it can be done in free space, a feature not available in electronic systems. To appreciate the order of magnitude of the density of such interconnections, let us consider that we wish to connect as many

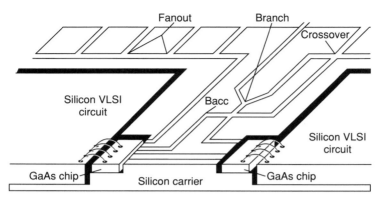

Figure 13.7 Waveguides used in VLSI interconnections between different types of substrates. The small GaAs transceiver chips convert the electric signals from the silicon chips into light signals. After L. D. Hutcheson and P. Haugen [16], with permission.

as 1000×1000 independent points/mm^2 in the object plane to a corresponding 1000×1000/mm^2 in the image plane. To implement this electrically, a million non-intersecting and insulated conductors would be required per square millimeter. To do it optically, only one microlens could complete the connection.

Free-space interconnections can be established with conventional optical components or by holographic images.

Optical interconnections may be implemented within a microelectronic chip by using a source, a detector, and a mirror or holographic reflector. The source, either an LED or laser, converts an electronic signal to an optical signal, which is directed to a photodiode and converted back to the electronic signal for further processing. The reflector or the routing device could be either a mirror or a hologram (Fig. 13.8). The hologram has the added advantage that it can be changed or programmed to suit the required connections.

A number of difficulties are encountered in this approach. The required light source and detector are usually made with GaAs semiconductors, whereas silicon is used exclusively for logic gates. These semiconductor materials cannot be integrated well because they have lattice-matching problems as well as different thermal coefficients. The accuracy of beam location poses another problem. The beam sharpness and the direction of the beam are sensitive to the temperature of the device in operation, thus making reliable operation difficult.

Optical Computing

It is logical to assume that optical computers would use optical interconnections all the way; however, optical computers are still in their infancy. Analog optical computers have been in operation since the 1950s, when the mathematical transform properties of lenses were exploited [17], but these devices lack the accuracy required

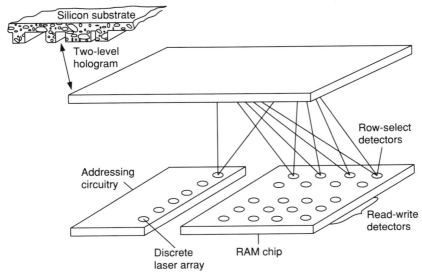

Figure 13.8 Holographic interconnections provide dynamically reconfigurable connections between chips.

by most applications. To follow digital logic to design optical computers by changing electronic components into optical components we lack workable components comparable to their electronic counterparts. For example, consider even the basic optical digital logic circuit. S-SEEDs are still in a developing stage. We do not discuss the interconnections for optical computers here.

On January 29, 1990, Alan Huang of AT&T Bell Laboratories unveiled an optical processor that, in theory, could perform all the logical operations necessary for digital computing. The processor used infrared laser beams to turn on and off a million times a second and process 32 channels of information at once. But practical realization of this computer may still be a decade away.

High-Level Electrooptic Integration

The importance of the development of the self-electrooptic effect device (SEED) should be emphasized here. A SEED has a PIN diode structure placed over an epitaxially grown dielectric reflector stack as shown in Fig. 13.9. This drawing is more detailed than that in Chapter 12. The intrinsic region of the diode consists of 50 to 100 multiple-quantum-well layers each about 10 nm thick so that the total structure is typically a few micrometers thick. When voltage is applied across the layers, the optical transmission of the quantum wells changes strongly, acting as an optical modulator that can modulate the incident light to a depth better than 20 dB with about

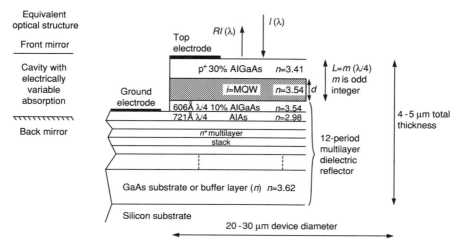

Figure 13.9 Structure of a SEED, a self-electrooptic effect device that is bistable and suitable for fabrication in very large arrays. After F. Grindle and J. E. Midwinter, *Electron. Lett.* **27**, 2327–2329. © 1991, *IEE*.

a 5-V drive. The insertion loss is very small, about 3 dB. The same device, operating in reverse bias, works as a very efficient photodetector.

SEEDs can be operated as optically triggered logic devices. When a SEED is connected to an electric bus through a chosen resistive load, or when two symmetric SEEDs are connected in series to a bias voltage, the device shows strong switching behavior as a function of the incident optical power as shown in Fig. 13.10. The hysteresis loop in input/output characteristics is clear. This opens the possibility of optically activated logical devices. The option of using two symmetric SEEDs is attractive with respect to large-scale integration in switching. The discussion in Chapter 12 under Quantum-Well Bistable Absorptive Devices is extended here to emphasize its importance.

As indroduced in Chapter 12, operation of the S-SEED depends on the optical absorption coefficient of the device exhibiting nonlinear behavior on application of voltage. As the voltage across the device (SEED) is dependent on the optical intensity I, so the light absorbed by the material creates charge carriers that alter its conductance. When two SEEDs (MQW diodes) are connected in tandem, they become complementary in that when one of the diodes is on, the other is off. The diode that is off absorbs energy (as if it were a resistor); the one that is on transmits. Increasing voltage (or power, as they both carry the same current) on a given diode decreases absorption and, hence, photocurrent. In the language of electrical engineering, a negative resistance results that gives rise to bistability. Thus for certain input powers, there are two possible output powers, or the device is bistable. An advantage of the S-SEEDs, which we repeat, is that changing states is a function of the ratio of the two input powers and not of the absolute intensity of the input beam. That is, the

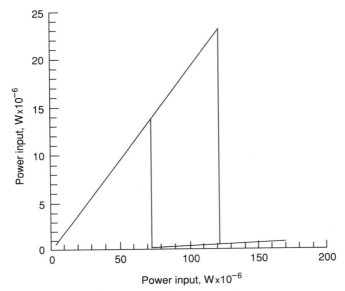

Figure 13.10 The bistable switching property of a SEED showing the hysteresis loop in input/output characteristics. After F. Grindle and J. E. Midwinter, *Electron. Lett.* **27**, 2327–2329. © 1991, *IEE.*

optical bistable loop is centered on the point where the two inputs are equal. It can be seen that the device will remain in its current state until the ratio exceeds 1.3 or 0.7. This is called ratio switching and can result in a better noise figure, the diode close to the switching point does not have to be biased to achieve signal gain.

A symmetric SEED (S-SEED) can be used as a base 2-module switching mode by adding an enable signal to both inputs and transmitting the state of the device to the next stage. (see Fig. 12.12). This forms the basic building block for large-scale switching networks. The possibility of using these schemes for interconnections is under investigation.

The great advantage of SEEDs is that they occupy very small space, typically $10 \times 20 \ \mu m$. An S-SEED is composed of two MQW pin diodes that are interconnected. Being planar in design, S-SEEDs can be fabricated in very large arrays with high yield. They can also be integrated with other devices on the same substrate; however, although SEEDs are small by optical standards (as compared with waveguides) they remain large by electronic standards. Just how seriously this will affects the expectation of replacing electronic logic remains to be seen.

A S-SEED implementation of multistage interconnections has been demonstrated by McCormick *et al.* [15].

Optical interconnections in microelectronics constitute a substantial research subject at present. Rapid development has been reported [18, 19].

Photonics in Switching

There has been a significant amount of interest in applying the new photonic technology to telecommunication switching systems [20]. Although we do not have the time and space to discuss this important switching problem in this book, a few examples of the use of photonics technology in switching may peak the reader's interest into this subject.

The devices we describe here are those concerned with the use of free-space interconnections for future ultrahigh data flow processors. They apply to large telecommunication systems as well as to microintegration within the wafer and between the wafer and the outside world.

Two types of switching wiring patterns occur repeatedly, crossover and perfect shuffle, that cause much trouble in metallic wiring implementation in crowded spaces. Crossover and perfect-shuffle interconnection patterns are shown schematically in Fig. 13.11. Imagine that to connect two large 128×128 arrays with metallic wires in these patterns, some 16,000 insulated metal lines are required. It is impossible if the space is limited.

Can optics help to resolve this problem? The answer is yes. Free-space optical interconnections have been developed by AT&T Bell Laboratories. We describe two examples.

A Crossover Network Scheme

Figure 13.12 shows an optical crossover using a prism grating, mirrors, bulk lenses, and a beam splitter. Implementation of the crossover network is designed for free-space digital optics intended to connect an array of two-dimensional input signals to another two-dimensional array at the output. S-SEEDs are used at the output image plane. The device consists of a polarization beam splitter (PBS), four lenses (L_1-L_4), two quarter-wavelength plates (Q_1 and Q_2), a prism grating (PG), a mirror (M_1), and an input array. Although the device is very cumbersome, it is worth remembering that the single optical assembly is capable of providing crossover connections be-

A   B

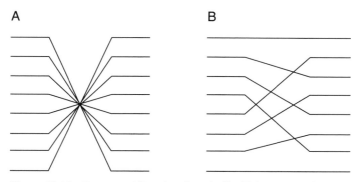

Figure 13.11   Crossover (A) and perfect-shuffle (B) interconnection patterns.

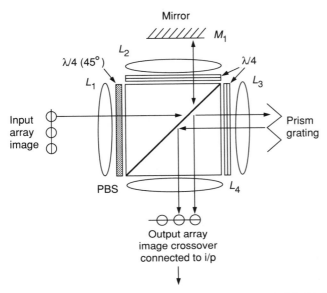

Figure 13.12 Optical crossover device developed by AT&T Bell Laboratories. After H. S. Hinton [20], with permission.

tween two arrays as large as 128×128, thus replacing some 16,000 insulating metallic lines.

This assembly is as follows: An input image enters the assembly from the left at (a) through L_1. Each pixel or spot is circularly polarized. At the PBS, the light is split into parallel and perpendicular components. The perpendicular component of the light is imaged onto the mirror M_1 through L_2. The light reflected from M_1 will have its polarization rotated by Q_1, allowing it to pass through the PBS where it can be imaged onto the output image plane through L_4. The parallel component of the input will pass the PBS, through L_3, and be imaged onto the prism grating (PG). The light being imaged onto the grating corresponds to the crossover patterns of the crossover interconnection topology. The light reflected from the prism grating will have its polarization rotated by the quarter-wavelength plated Q_2, allowing it to be redirected and focused through L_4 onto the output image plane. Thus, both straight-through and crossed interconnections have been achieved. The shift in space is related to the period of the prism grating.

A Clocked Space Multiplexed Beam Combination

By using a similar assembly, with a slight change in the location of the quarter-wavelength plate and mirror, one can construct a space-multiplexed beam combination (Fig. 13.13). Here, an additional quarter-wavelength plate has been inserted in front of lens L_4, making a total of three quarter-wavelength plates surrounding the

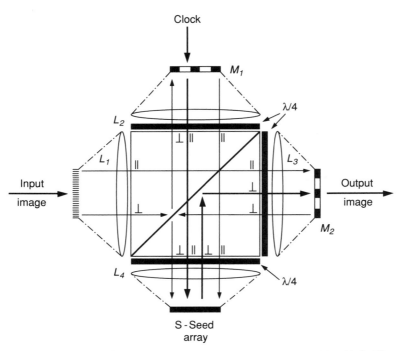

Figure 13.13 A clocked space multiplexed beam combination. After H. S. Hinton [20], with permission.

polarized beam splitter (PBS). The prism grating has being replaced by mirror M_2, and the output is now taken from M_2 as shown. M_1 and M_2 are small patterned mirrors designed for special purposes. At the bottom of the combiner is the S-SEED array. The device now accepts two input signals plus a clock or bias beam. Some constraints are noted: the beam spots must be small (<5 μm), forcing the entering signals to use the full aperture of an imaging lens, and the signals must not cohenently interfere at the device's optical window.

An input image, composed of two linear polarizations, enters the beam combination system on the left, the perpendicular component is directed upward through Q_1 and L_1 to reach the mask M_1, while the parallel component passes through the PBS. Both components travel through the quarter-wavelength plates, where linear polarization is converted to circular polarization. The light traveling upward is imaged onto patterned mirrors of mask M_1. This mask consists of an array of small mirrors that are located in the image field of L_2 so that light reflected from them will be imaged onto the upper half of the rectangular window of the S-SEED array. Therefore the image traveling upward reflects off M_1 and passes again a quarter-wavelength plate, which changes the polarization from circular to parallel, allowing the light to pass through the PBS onto the upper half of the S-SEED array.

The parallel component of the input image passes through the PBS and is imaged through Q_2 and L_2 onto the small patterned mirrors of mask M_2, reflected by the PBS, and then imaged onto the upper half of the S-SEED array. As these two inputs will have different orientations, they will not interfere.

The optical clock enters the beam combination unit by being imaged in the transparent region below the small mirrors of mask M_1. This collection of high-power spots will be imaged onto the bottom half of the S-SEED array. The reflected output signal of the S-SEED array, derived from the clock, will have its polarization rotated, allowing it to be reflected by the PBS and then imaged onto the transparent region below the small mirrors of mask M_2. This output image can then be collected and used by an optical interconnection or another beam combination unit.

Elegant two-dimensional perfect-shuffle interconnections have also been built using bulk optics [21].

Summary

In this chapter we have briefly discussed the implementation of optical interconnections to relieve the congestion in large communication terminals and/or very large scale integration (VLSI) chips in microelectronics.

Optical interconnections offer the combination of large bandwidth, large fan-out, and high throughput. Bandwidths of gigabits and fan-outs up to 100 loads from a single fiber are feasible. Optical interconnections may consume less power, are immune to crosstalk and interference, and lend themselves to multiplexing and switching. The number of optical fiber links in the system can be much lower than that in equivalent electric connections. In general, even in the early stage of development, optical interconnections offer superior performance compared with electronics.

Difficulties arise in meeting the stringent package and fabrication demands; in aligning with components within the tolerances allowed; and in preventing the chip from heating up, which would cause thermal expansion and move the optoelectronic devices out of alignment.

Let us comment on how optical interconnections overcome the problems facing both the telecommunication and microintegration electronic industries with present-day technology. One obvious fact is that for low intensities, optical rays may cross each other without interference, thus enabling us to increase the density of interconnections. The possibility of parallelism should count as a big advantage for optical interconnections. Other advantages over conducting wires include the ability to provide lower energy per connector, lower on-chip power dissipation per chip, lower crosstalk between connections, and lower interconnection skew. The possibility of using hybrid interconnections to take advantage of both types of interconnections, electronic and optical, is under study.

For optical computing, we still look forward to all-optical components, including optical bistable devices and interconnections.

References

1. Special issue on optical interconnections. *Appl. Opt.* (Information Proceedings) **29**, No. 8 (1990).
2. Special issue on optical interconnectins and networks. *SPIE* **1281** (1990).
3. Special issue on optical interconnections in the computer environment, vol. **1178**, (1990).
4. Optical computing, a special issue. *Proc. IEEE* **72**, pp. 753–974 (1984); A. D. McAulay, *Optical Computer Architectures,* Wiley, New York, 1991; Advances in optical information processing IV. *SPIE* **1296** (1990).
5. C. Clos, A study of non-blocking switching networks. *Bell Syst. Tech. J.* **32**, 407–424 (1953).
6. V. E. Benes, Optical rearrangeble multistage connecting networks. *Bell Syst. Tech. J.* **43**, 1641–1656 (1964).
7. C. Burke, M. Funuwara, M. Yamaguchi, H. Nishimoto, and H. Honmou, Studies of a 128 line space division switch using lithium niobate switch matrices and optical amplifiers. In *Topical Meeting on Photonic Switching, Salt Lake City, Utah, March 6–8, 1991,* Optical Society of America, Washington, DC, 1991.
8. A. Huang and S. Knauer, Starlite, a wideband digital switch. In *Proceedings of IEEE Global Telecommunications Conference, Atlanta, Georgia,* pp. 121–125, IEEE, New York, 1984.
9. R. A. Spanke, Architectures for large nonblocking optical space systems. *IEEE J. Quantum Electron.* **QE-22**, 964–967, 1986; R. A. Spanke, Architectures for guided-wave optical space switching systems. *IEEE Commun.* **25**, 42–47 (1987).
10. J. E. Midwinter, Photonics in switching: The next 25 years of optical communications. An inaugural address delivered before the IEEE convention on October 10, 1990. *IEEE Proc. J.* **139**, 1–12 (1992).
11. C. A. Brackett, Capacity of multi-wavelength optical-star packet switches and implications for packet length. In *Topical Meeting on Photonic Switching, Salt Lake City, Utah, March 6–8, 1991,* Optical Society of America, Washington, DC, 1991.
12. D. A. B. Miller, Quantum well self-electro-optic effect devices. *Opt. Quantum Electron.* **22**, S61–S98 (1990).
13. T. J. Cloonan, M. J. Herron, F. A. P. Tooley, G.ˇ. Richards, F. B. McCormick, E. Kerbis, J. L. Brubaker, and A. L. Lentine, An all-optical implementation of a 3D crossover switching network. *IEEE Photonic Technol. Lett.* **2**, 438 (1990).
14. L. M. F. Chirovsky, Large arrays of symmetric self electro-optic effect devices. In *Proceedings of 1991 Toptical Meeting on Photonic Switching, Salt Lake City, Utah, March 6–8, 1991,* pp. 56–59, Optical Society of America, Washington, DC, 1991.
15. F. B. McCormick, F. A. P. Tooley, T. J. Cloonan, J. L. Brubaker, A. L. Lentine, S. J. Hinterlong, and M. J. Herron, A digital free space photonic switching network demonstration using S-SEEDs. In *CLEO 1990, Technical Digest Series,* Vol. 7, Post Deadline Paper CDPD-1, Optical Society of America, Washington, DC, 1990; C. Burke, M. Fujiwara, M. Yamaguche, H. Nishimoto, and H. Honmoou, Studies of a 128 line space division switch using lithium niobate switch matrices and optical amplifiers. In *Topical Meeting on Photonic Switching, Salt Lake City, Utah, March 6–8, 1991,* Optical Society of America, Washington, DC, 1991.
16. L. D. Hutcheson and P. Haugen, Optical interconnects replace hardwire. *IEEE Spectum,* pp. 30–35 (Mar. 1987).
17. L. J. Cutrona, E. N. Leith, C. J. Palermo, and L. J. Porecello, Optical data processing and filtering system. *Trans. Inform. Theory* **IT-6**, 386–400 (1960).
18. R. L. Freeman, Telecommunication system engineering. In *Digital Switching & Networks,* Wiley, New York, 1989.
19. Special issue on optical interconnects. *Appl. Opt. Inform. Process.* **29**, No. 8 (1990).
20. H. S. Hinton, Photonic switching fabrics. *IEEE Commun. Mag.,* pp. 71–89 (Apr. 1990).
21. C. Wu and T. Feng, Universality of the shuffle-exchange network. *IEEE Trans. Comput.* **C-30**, 324–332 (1981).

CHAPTER 14

Photonic Image Processing

Introduction

Signal processing is a technology that extends, improves, and automates the detection of intelligent signals carried by electromagnetic waves. In this chapter, the signal we encounter is the optical image; thus we call the technology *photonic image processing*.

A common form of optical image detection is by the use of optical lenses and photographic films. Optical cameras consisting of a few lenses are used to catch the image on films as in photography. The film is then chemically developed to reveal the image. To extend the wavelength range to include infrared through the visible and into the X-ray regions, special films for each range are developed. Processing of these optical images is simple and inexpensive. The disadvantage is that films need to be developed; thus it is not a real-time processing system.

Many image processing methods are available. A general image processing system is represented schematically in the block diagram in Fig. 14.1. Light-based images are captured by optics and a sensor array; stored in a buffer to be processed in any of a number of ways, including digitizing, intensifying, filtering, using computers if necessary; and finally displayed on a screen.

Signal comes in two forms, analog and digital; each needs its own distinct processing technique. Typical processing techniques, analog and digital, are discussed under each category. The final image is focused on a film for storage or on a screen of a cathode-ray oscillograph for display.

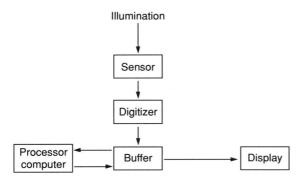

Figure 14.1   General layout of an image processing system. In a typical image processing system, the object image is captured by optics and a sensor array and is stored in a buffer from which further processing may be carried out. Signals may have to be digitized for digital processing. In the processor, image information may be enhanced or filtered as required. The final product is displayed on film or on scope. In digital processing, images are converted back to analog form before display.

A Simple Optical Imaging System

Photography is perhaps the simplest, most effective method to process an image. It is analog processing. Under the illumination of a visible light source, the objective image is focused through a positive lens on a film sensitized with a chemical deposit that changes its response with light intensity. In this scheme, only one positive lens enclosed in a box is needed. The only control is a variable aperture to adjust the opening of the lens. To view the image, a frost glass sheet is placed at the image plane. Elementary geometrical optics [1] shows that an object at distance d_1 from the lens will produce an image on the frost glass plate at d_2 on the other side of the lens that has a focal length f. If we assume the object is in the object plane, a sharp and inverted image will be presented at the image plane as shown in Fig. 14.2. Photographic film can be used to record the image. The system is focused so that from each point on the object plane, the emitted paraxial ray will reach a single point in the image plane. The lens property, defined by the focal length f, satisfies the relation $1/f = 1/d_1 + 1/d_2$. Note that the image is inverted, and the focal length is less than d_1 or d_2. If the object is not exactly in focus, that is, the image plane does not locate exactly at d_2, then the image of a point object will be blurred into a fuzzy spot; the image becomes fuzzy. The size of the blurred spot, assumed circular in shape, or the radius of the blurred spot is an inverse measure of the resolving power of the lens. It is a measure of the quality of the image. If it is assumed that the focusing error is ϵ, and

$$\epsilon = \frac{1}{d_1} + \frac{1}{d_2} - \frac{1}{f} \tag{14.1}$$

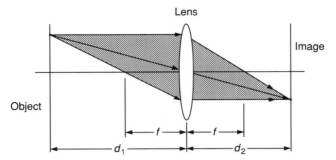

Figure 14.2 A single lens is the simplest image processing system. A lens with focal length f is placed between the object plane at d_1 and the image plane at d_2. When the system is focused, each point from the object plane reaches a corresponding point in the image plane.

then the radius of the blurred spot $r_b = \epsilon d_2 D/2$, where D is the diameter of a circular aperture of the lens. A small aperture corresponds to an increased depth of focus.

This simple optical image system fits the pattern described in Fig. 14.1. But no other elements have been added to control the quality of the image. There is no room for improvement. The quality of the image is determined by the quality of the lens. This simple system does not require any sensor or detector, because the human eye acts as both the sensor and detector of visible light. Slight modification of the system design will, however, allow many controls to be added for quality improvement. First let us consider a two-lens system. By adding a second lens to the system as shown in Fig. 14.3, using identical lenses, the image appearing at the image plane is not

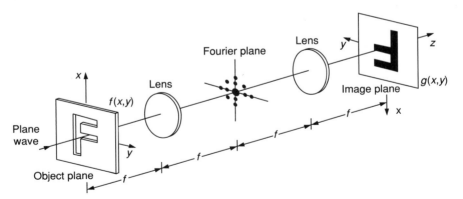

Figure 14.3 A 4-f two-lens system to introduce Fourier transform operations. The first lens performs Fourier transformation of the object which is shown at the Fourier plane. The second lens performs the inverse Fourier transformation to recover the image. After B. E. A. Saleh and M. C. Teich, *Fundamentals of Photonics.* 1990. Reprinted by permission. © 1991 John Wiley & Sons.

magnified. The interesting feature revealed by this addition is that we have introduced a Fourier plane between the lenses. A Fourier plane is a plane at which the Fourier transform of a real function, $f(x, y)$, would appear after transformation into $F(v_x, v_y)$ in the frequency domain.

A simple lens can perform a mathematically complex operation on light passing through it. This operation is called Fourier transformation [2]. If the complex amplitude function of a plane wave traveling in the z direction in the object plane is represented by $f(x, y)$, where the variables x and y represent the spatial coordinates in a plane, according to Fourier theory, it may similarly be written as a superposition of harmonic functions of x and y of the form $F(v_x, v_y)\exp\{-i2\pi(v_x x + v_y y)\}$, where $F(v_x, v_y)$ is the complex amplitude and v_x and v_y are the spatial frequencies in the x and y directions, respectively. The transformation of $f(x, y)$ into $F(v_x, v_y)$ functions is called the *Fourier transform*. The plane that contains this function, $F(v_x, v_y)$, is called the *Fourier plane*, where v_x and v_y are now the new coordinates. $F(v_x, v_y)$ lies in a frequency domain. Furthermore, the inverse Fourier transform of $F(v_x, v_y)$ becomes the complex amplitude of the image function $g(x, y)$ shown in the image plane. Examples of Fourier transforms of some known functions are given in the next section. Here we present the important properties of the Fourier transform in an example: The Fourier transform of a constant function is a single value. Thus, a uniformly illuminated light in the object plane has a mathematically equivalent single bright point in the Fourier plane. For any real lens of finite diameter, the transformation is a single point but becomes alternately bright and dark concentric rings. Larger lenses will make smaller ring diameters in the Fourier transform plane. Return now to our example shown in Fig. 14.3. The first lens transforms the object function $f(x, y)$ into its Fourier equivalent $F(v_x, v_y)$ in the Fourier plane. A second lens inverse-transforms it back into spatial function $g(x, y)$ in the image plane. The inverse transform reconstructs the original function and becomes a perfect replica of the object on the image plane. Note that the coordinate system of the image plane has been reversed.

Let us now reveal the usefulness of the Fourier transform. In this Fourier plane, the transform of combined signal and noise appears at different locations. If one can block the noise pattern in the Fourier plane by spatial filters, the noise can be removed and the reformulated image becomes crisp and sharp. Thus, the quality of the image can be upgraded. This is accomplished by merely inserting a spatial filter between the lenses as shown in Fig. 14.4. The original signal is shown in Fig. 14.4A. The Fourier transform of the original function is shown in Fig. 14.4B. Note that the signal and noise spectra are separated because of their frequency difference. Figure 14.4C shows the passband of a low-pass filter. Insertion of this filter into the Fourier plane (to be described later) produces the clarified picture shown in Fig. 14.4D, where the noise has been removed.

Other processes, such as convolution, correlation, scaling, and frequency translation, can be provided with appropriate elements.

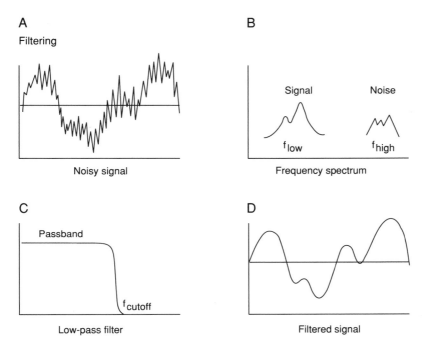

Figure 14.4 Effect of filters: (A) The original noisy signal; (B) fourier transform of the function representing the noisy signal; note the separation of the spectra as a result of the differences in their frequency contents; (C) characteristic of a low-pass filter. (D) When the filter is used, the clarified signal reappears. After H. S. Hecht, *High Technology,* p. 56. © 1983 *High Technology.*

The Fourier Transform

Let us start with the Fourier transform of one variable t in the time domain. A complex-valued function $f(t)$ may be decomposed as a superposition integral of harmonic functions of different frequencies and complex amplitudes as

$$f(t) = \int_{-\infty}^{\infty} F(\nu)\exp(i2\pi\nu t)d\nu \tag{14.2}$$

(We have avoided using the factor $1/2\pi$ in front of the integrating sign by replacing ω with $\nu = \omega/2\pi$). The component with frequency ν has a complex amplitude $F(\nu)$ given by

$$F(\nu) = \int_{-\infty}^{\infty} f(t)\exp(-i2\pi\nu t)dt \tag{14.3}$$

$F(\nu)$ is called the Fourier transform of $f(t)$, and $f(t)$ is the inverse Fourier transform of $F(\nu)$. Equations (14.2) and (14.3) are called a Fourier transform pair. Both func-

TABLE 14.1
Selected Functions and Their Fourier Transforms

Function		$f(t)$	$F(\nu)$	
Uniform		1	$\delta(\nu)$	
Impulse		$\delta(t)$	1	
Rectangular		$\text{rect}(t)$	$\text{sinc}(\nu)$	
Exponential[a]		$\exp(-|t|)$	$\dfrac{2}{1+(2\pi\nu)^2}$	
Gaussian		$\exp(-\pi t^2)$	$\exp(-\pi\nu^2)$	
Chirp[b]		$\exp(j\pi t^2)$	$e^{j\pi/4}\exp(-j\pi\nu 2)$	
Sum of $M = 2S + 1$ impulses		$\sum\limits_{n=-s}^{s}\delta(t-n)$	$\dfrac{\sin(M\pi\nu)}{\sin(\pi\nu)}$	
Infinite sum of impulses		$\sum\limits_{n=-\infty}^{\infty}\delta(t-n)$	$\sum\limits_{n=-\infty}^{\infty}\delta(\nu-n)$	

[a]The double-sided exponential function is shown. The Fourier transform of the single-sided exponential, $f(t) = \exp(-t)$ with $t \geq 0$, is $F(\nu) = 1/[1 + j2\pi\nu]$. Its magnitude is $1/[1 + 2\pi\nu)^2]^{1/2}$.
[b]The functions $\cos(\pi t)^2$ and $\cos(\pi\nu^2)$ are shown.

tions represent the signal with $f(t)$ in the time domain and $F(v)$ in frequency domain. The square of the absolute value of $f(t)$ represents the signal power. The square of the absolute value of $F(v)$ represents the spectral density. Table 14.1 shows some selected functions and their Fourier transforms.

Properties of the Fourier Transform

A full discussion of the properties of the Fourier transform can be found in any standard textbook on the subject. Here we mention two theorems, the convolution and correlation theorems, because of their importance in future applications to optical problems.

Convolution of the two functions $f_1(t)$ and $f_2(t)$ in the time domain is equivalent to multiplication of $F_1(v)$ and $F_2(v)$ in the frequency domain. Thus, if $f_1(t)$ and $f_2(t)$ are Fourier transformable into $F_1(v)$ and $F_2(v)$, respectively, then

$$F(v) = F_1(v)F_2(v) \qquad (14.4)$$

and

$$f(t) = \int_{-\infty}^{\infty} f_1(\tau)f_2(t - \tau)d\tau \qquad (14.5)$$

is the inverse transform. The $t - \tau$ represents a time delay.

The correlation between two complex functions $f_1(t)$ and $f_2(t)$ is defined as

$$f(t) = \int_{-\infty}^{\infty} f_1 * (\tau)f_2(t + \tau)d\tau, \qquad (14.6)$$

and

$$F(v) = F_1 * (v)F_2(v) \qquad (14.7)$$

Correlation is comparison of a known signal with another signal that may or may not be similar to it. In autocorrelation a signal is compared with itself as shown in Fig. 14.5. Here, a stored signal is compared with the signal being examined for similarities. The result is shown in Fig. 14.5C: a high peak indicates close resemblance of the two signals. Cross-correlation compares two signals from different origins. Figure 14.6 shows the sequence from no correlation (in Fig. 14.6C) to strong positive correlation (in Fig. 14.6F).

The signal energy integrated throughout the time domain is related to the spectral density integrated over the spectral density as

$$\int_{-\infty}^{\infty} f(t)^2dt = \int F(v)^2dv \qquad (14.8)$$

This is Parevel's theorem.

A

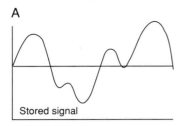

Stored signal

B

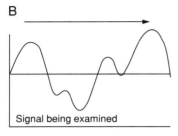

Signal being examined

C

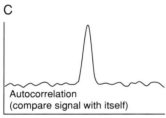

Autocorrelation
(compare signal with itself)

Figure 14.5 Autocorrelation. The peak in the output signal indicates the correlation or match. After H. S. Hecht, *High Technology,* p. 56. © 1983 *High Technology.*

All these properties of the Fourier transform pair can be used to improve the quality of image processing. Examples are cited in later sections.

Two-Dimensional Fourier Transform

Fourier transform works equally well for two-dimensional space. The Fourier transform pair becomes

$$f(x, y) = \int\!\!\int_{-\infty}^{\infty} F(v_x, v_y)\exp\{-i2\pi(v_x x + v_y y\}_x dv_y \qquad (14.9)$$

$$F(v_x, v_y) = \int\!\!\int_{-\infty}^{\infty} f(x, y)\exp\{i2\pi(v_x + v_y)\}dx\ dy \qquad (14.10)$$

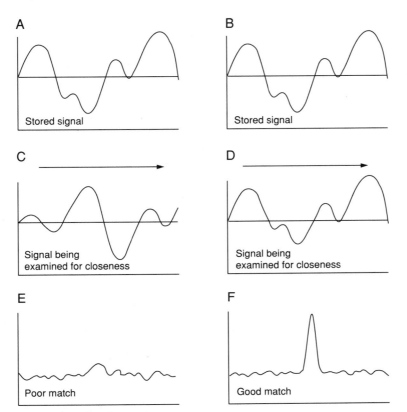

Figure 14.6 Cross-correlation. Signals A and C are compared and show little correlation as shown in E. Signals D and B are compared and show strong correlation as in F. After H. S. Hecht, *High Technology*, p. 56. © 1983 *High Technology*.

Most of the properties of the Fourier transform for one variable can be generalized for the two dimensional case. An example of a circularly symmetric function is cited here. The Fourier transform of a circularly symmetric function is also circulary symmetric. i.e., for the function

$$f(x, y) = 1, \text{ for } (x^2 + y^2)^{1/2} \leq 1 \quad (14.11)$$
$$f(x, y) = 0, \text{ otherwise,} \quad (14.12)$$

the Fourier transform is

$$F(\nu_x, \nu_y) = J_1 \frac{(2\pi\nu_s)}{\nu_p}, \ \nu_p = (\nu_x^2 + \nu_y^2)^{1/2} \quad (14.13)$$

where J_1 is the Bessel function of order 1.

Spatial Filters

Image quality can be improved by performing operations in the Fourier plane between the two lenses. The first element we wish to introduce to a processor is to modify the image by introducing a filter in the Fourier plane.

1. Figure 14.7 comprises a series of pictures showing how a low-pass filter, a high-pass filter, and a vertical-pass filter can be used to change image quality. In Fig. 14.7A, the spatial filter is a mask with a circular hole of diameter D. The intention of the mask is to cut off the high-frequency components of the original signal function $f(x, y)$. For a circularly symmetric hole in the mask with an aperture D, it passes spatial frequencies that are smaller than ν_s and blocks higher frequencies, where $D/2 = \nu_s \lambda f$ and $\nu_s = (\nu_x^2 + \nu_y^2)^{1/2}$.

2. In Fig. 14.7B, a high-pass filter is introduced. It consists of a transparent plate

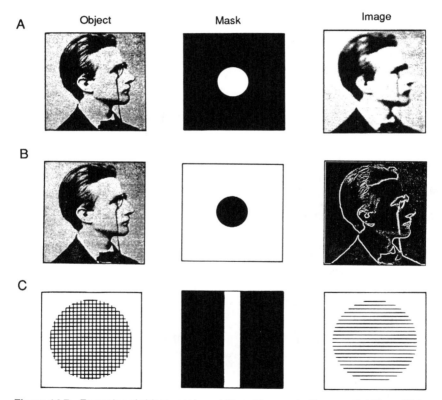

Figure 14.7 Examples of object, mask, and filtered images for three spatial filters: (A) low-pass filter, (B) high-pass filter, (C) vertical-pass filter. Black indicates that transmittance is zero; white, that the transmittance is unity. After B. E. A. Salek and M. C. Teich, *Fundamentals of Photonics*. Reprinted by permission. © 1991 John Wiley & Sons.

with an opaque central dot of diameter D. The filter output is high at regions of high rate of change and small at regions of smooth or slow variation of the object. The filter then enhances the edges of the image as shown. One sees the outline of the picture.

3. In Fig. 14.7C, a vertical-pass filter is used in the transform plane. The vertical-pass filter blocks horizontal frequencies and transmits vertical frequencies. Thus, only variations in the x direction are transmitted. The resultant image is shown on the right-hand side. If the filter is a mask consisting of a vertical slit of width D, the highest transmitted frequency is $v_y = (D/2)/\lambda f$.

Sensors

The effectiveness of any imaging system is a function of many factors that have different impacts depending on the applications. A sensor, however, remains an important element responsible for system performance irrespective of the application. A brief discussion of common sensors frequently used in many applications and their properties follows.

One type of sensor that we omit here is the thermal sensor. Many thermal sensing devices have been in use for many years.

Film

The importance of using photographic films to catch images has been recognized for a long time. Film offers high spatial resolution in the visible region, reasonable wavelength sensitivity and dynamic range, and moderate cost. It does not, however, offer real-time imaging. Also, the sensitivity, resolution, and dynamic range of the film are mutually dependent so that compromises must be made among these properties.

Charge-Coupled Devices

Among the many sensors available for image detection, we favor the charge-coupled device (CCD) [3]. We limit our discussion to this type of sensor.

The Principle of Operation

The charge-coupled device is a silicon integrated circuit in which information is carried by charge packets stored under a set of clocked electrodes. Metallic electrodes are deposited on an oxide layer or other insulator on top of the semiconductor substrate, and are arranged to form a closely spaced array. By application of appropriate voltages to the electrodes, depletion regions are created in the underlying semiconductor. The result is a series of potential wells in which charges can be stored under the selected potential well. Charges can also be shifted from one well to another by

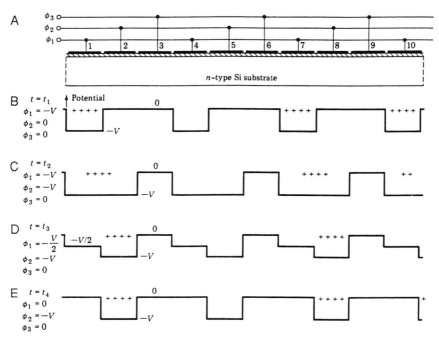

Figure 14.8 Illustrating charge transfer in a charge-coupled device. (A) Every third electrode is at the same potential and three phase voltages ϕ_1, ϕ_2, and ϕ_3 are applied. (B–E) The potential profile variations during one shift interval. The potential energy for positive charge is proportional to the potential; hence, these curves also represent the potential energy wells for holes. After J. Milliman, *Microtectonics*. © 1979 McGraw-Hill.

properly varying the voltage on the adjacent electrodes. Figure 14.8 depicts a three-phase CCD together with the voltages applied to the electrodes. These electrodes are arranged in groups. Within a group of three electrodes (gates), each gate is connected to a separate clock. An imaging element, the pixel, is defined by these gates. Incident light carrying information in the form of intensity variation creates electron–hole pairs, and the electrons collected in the potential wells are defined by the gates (electrodes). When one of these clocks is turned on, that is, at $t = t_1$ let $\phi_1 = -V$, a local potential well is formed beneath that gate in every pixel (shown in Fig. 14.8B). The pixels detect and convert photons into electrons which are stored in these wells. By the application of sequentially clocked pulses to the gates, as in Figs. 14.8C, D, and E, the charge is transported to the next gate. The charge can be transported, line by line, to the readout register located at the end of the array. The stored electrons are then read out in a serial manner to form an electronic replica of the scene imaged on the device. The number of electrons in a pixel is proportional to the number of photons incident on that particular pixel.

Charge-coupled devices usually operate as parallel registers. An image projected on a two-dimensional sensor array produces a charge pattern on the parallel sensor

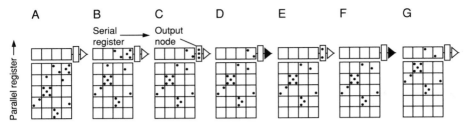

Figure 14.9 Sequential charge transfer between wells in a CCD array enables the electronic image to be read out in the detection process. After K. Prettyjohns [4], with permission.

array (Fig. 14.9A). The different numbers of dots per square indicate the charges in these squares. The serial readout register is shown at the top of each stage. In succeeding stages, 2 through 7, charges are shifted up the parallel register one row at a time and are also sequentially shifted to the right of the register. The charges at the output node are collected for signal processing.

Charge-coupled devices have many advantages over other types of sensors: (1) they are solid-state and require low power consumption; (2) they can be integrated into high resolution of matrices as large as 4096 × 4096 pixels; (3) they respond to a broad range of wavelengths; (4) they have very low readout noise (1–6 electrons rms); (5) they have a linear response, high geometric fidelity, and large dynamic range (greater than 18 bits); and (6) its quantum efficiency is extremely high as can be seen in the plot of quantum efficiency against photon energy (Fig. 14.10). As can be seen from this plot, CCDs are capable of responding in the ultraviolet, visible, and near-infrared regions. In addition, CCDs may be made sensitive to X rays.

Charge-coupled devices have been developed and established as excellent detectors of visible radiation. The goal of fabricating a single-chip, all-solid-state, fully TV-compatible image sensor and video camera has been achieved. Devices for scientific and medical applications are progressing rapidly.

Charge-Coupled Devices for Spectroscopy

Until recently, charge-coupled devices have been designed mostly for visual imaging applications. Rapid progress has been made to develop improved CCDs to meet the special demands of spectroscopy.

For spectroscopic applications, CCDs must be capable of operating at a low light level and must have a high signal-to-noise ratio, low readout noise, a large aspect ratio, and a wide wavelength range. We attempt to compare present-day CCDs with earlier products with respect to these properties [4].

1. *Wavelength range.* For spectroscopy, an ideal CCD array would cover a wavelength range from 300 to 1000 nm. Before 1986, the range attainable was 300 to 850 nm. In 1992, a wavelength range from 120 to 1000 nm was reported [4].

2. *Dark noise.* System noise ultimately determines the minimum optical power a

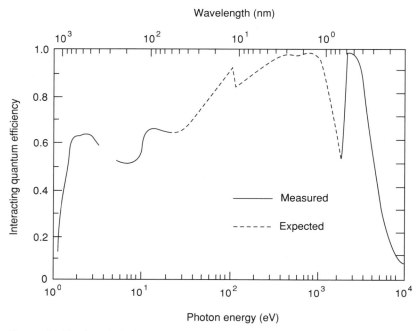

Figure 14.10 A typical charge-coupled device exhibits high quantum efficiency over a wide wavelength range. From D. Heidtman, *LFW,* p. 96, Dec. 1991. © 1991 PennWell Publishing Co.

CCD can detect. The dark noise that arises from statistical variations in the dark current resulting from thermally generated carriers per pixel contributes most to this noise. Spectroscopists often operate CCDs with liquid nitrogen Dewars to reduce thermally generated dark current. Before 1986, CCDs operated at liquid nitrogen temperature used to have a dark current of about 10^{-2}/pixel/s. In 1992, this number was reduced to less than 3×10^{-5}/pixel/s. This reduction was made possible by adopting a multipinned-phase device architecture which causes most of the thermally generated electrons to recombine with holes before they can enter a potential well [5].

3. *Readout noise.* Readout noise is the noise associated with each individual readout event. In very low light level applications, such as high-resolution Raman spectroscopy studies, the readout noise of the CCD array may be system limiting. For a given signal-to-noise ratio, less readout noise means that less time is required to accumulate spectral data. In recent years, there has been improvement in readout noise by at least one order of magnitude, from 10 to 100 units to about 4 units in 1992.

4. *Aspect ratio.* For spectroscopy, one needs to have as many pixels as possible and arrange these along the long axis in the dispersive direction, across the entire flat field of view of the spectrograph. The spectral resolution of the CCD (in pixels per millimeter) determines the wavelength resolution of the spectrometer. Early CCDs were designed for rectangular TV/video format, which has a typical aspect ratio of

520 × 312 pixels for imaging applications. They come with dimensions of about 1.5 × 1.0 and 1 × 1 cm. The limited number of pixels along the dispersion axis may result in poor resolution. New CCD designs for spectroscopy have an aspect ratio in the range 3 : 1 to 6 : 1 and dimensions typically of 1024 × 256 pixels. Spectral resolution of 25 μm in the dispersion axis has been reported.

Many designers also supply softwares package for CCD systems so that data manipulation for desired analysis can be done by simple keystroke commands. The price of the system has also decreased drastically.

A 160 × 244-element PtSi/Si Schottky barrier diode focal-plane CCD array has also been used for imaging in the reflected, short-wave 1- to 3-μm infrared band or 3- to 5-μm medium-wave infrared band. PtSi imagers require cooling to 77 K (liquid nitrogen temperature) [6]. Back-side illumination can be used to increase quantum efficiency. PtSi imager and camera system manufacturers using even larger format arrays of sizes are commercially available, for example, 640 × 486 (by Kadak), 512 × 488 (by Loral Fairchild), 640 × 480 (by Sarnoff, formerly RCA).

Acoustooptic Image Processors

The interaction between sound energy and light energy has been used to process and display radio signals for many years [7]. Use of acoustooptic devices in real-time signal convolution and correlation has increased dramatically because of improvements in device characteristics and implementation techniques. Acoustooptic processors can process many signals in parallel, thus speeding up the processing. We devote this section to developments in this field.

The Fundamental Principle

The fundamental principle of acoustooptic interaction was discussed in Chapter 2. Brillouin predicted that moving acoustic wavefronts cause the incident light to be diffracted and their frequency Doppler shifted. The basic Doppler shift frequency equals the sound frequency; however, its importance in practical applications was realized only after the invention of the laser in the 1960s.

With the aid of computers, Acoustooptics has been developed into a major image processing technique in the scientific and medical fields.

Light waves and acoustic waves can be brought to interact with each other on a common ground, known as the Bragg cell [8]. The phenomenon is closely analogous to Bragg's X-ray diffraction. So, the new phenomenon is called the acoustic Bragg diffraction.

Also introduced in Chapter 2 were the two Bragg devices: the bulk Bragg cell and the surface-acoustic-wave (SAW) cell. Interaction takes place deep within the bulk cell. One end of the cell is attached on a transducer; the other end terminates with an absorber to reduce reflections. To build a two-arm interferometer, two bulk cells are required. A narrow-width Bragg cell is also of the bulk type but it has a narrow width. It can be made comparable to the acoustic wavelength. Narrow Bragg cells are useful in building switches and many other devices.

In the SAW cell, the interaction takes place on the surface. Two acoustic transducers, one at each end, can function on the same SAW cell simultaneously. Two oppositely directed acoustic waves are thus generated in the same cell to modulate the signal and reference beam, respectively. Refer to Fig. 2.6 for the types of Bragg cells.

Acoustooptic Image Processing

Power spectrum analyzers use a Bragg cell to deflect a light beam at an angle proportional to the frequency of an input microwave signal as dictated by Eq. (2.28) [9]. Laser light is used with a spatial filter, and is collimated by a set of lenses and directed to the Bragg cell. The microwave signal to be analyzed is introduced to the Bragg cell through a transducer, which converts the microwave into an acoustic wave. That wave modulates the light beam, breaking it up into separate beams for every component of the microwave signal and deflecting each beam in proportion to its frequency. A lens focuses these beams onto a bank of photodetectors. The detectors convert each frequency component into electric current to indicate the original frequency composition. Figure 14.11 is a schematic diagram of an acoustooptic spectrum analyzer.

Acoustooptic Spectrum Analyzer

A good spectrum analyzer requires a bandwidth of about 60 to 100 MHz, with good resolution and pulse response that can detect pulses as short as 100 ns. A large dynamic range on the order of 50 to 60 dB will help the system to probe a broad range for information. Interferometric spectrum analyzers may be required to accommodate a great number of inputs.

Acoustooptic Convolution and Correlation

Correlators are used to determine similarities and differences among microwave signals. A correlator integrates the product of mathematical functions representing the signals to be compared and comes up with a value that denotes the degree of similarity.

An advanced one-dimensional acoustooptic correlator using a SAW Bragg cell is shown schematically in Fig. 14.12. Two transducers, one for the signal and one for a reference signal (signals 1 and 2), are mounted on two ends of a SAW transducer. A laser beam is split into two beams and directed to interact with the two acoustic waves in the SAW cell. The interaction of the laser light with the acoustic waves traveling in opposite directions will diffract the beams differently, causing both upshifts and downshifts in frequencies. The intensity of the emerging beam is pro-

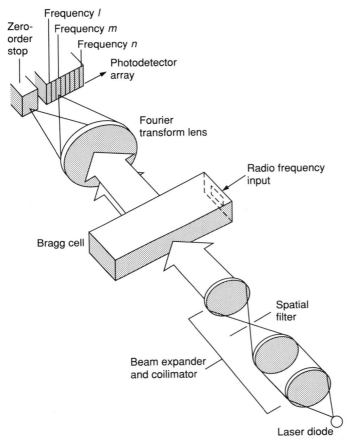

Figure 14.11 Acoustooptic spectrum analyzer. A bulk Bragg cell is illustrated here. Radio frequency containing many component frequencies to be analyzed is fed into a transducer, which converts it to acoustic frequencies. The laser beam interacts with the different frequency components diffracting them at different angles; they can be separated and displayed on the detector array. After A. P. Gontzonlis, *Digital electronics meets its match. Spectrum* **25**, 22. © 1988 *IEEE.*

portional to the product of the time-varying intensities of the two microwave signals, and the location of the beam's peak on the photodetector array is proportional to the difference between the arrival times of the two signals. The two diffracted and modulated beams that result follow essentially the same path to the photodetectors, where they are heterodyned. The linear detector array integrates the product over time, producing an electrical output proportional to the correlation between the microwave signals.

Acoustooptic correlators can handle up to 3000 delays across up to 200 MHz with up to 60 dB in dynamic range. If the same operations were done digitally, the com-

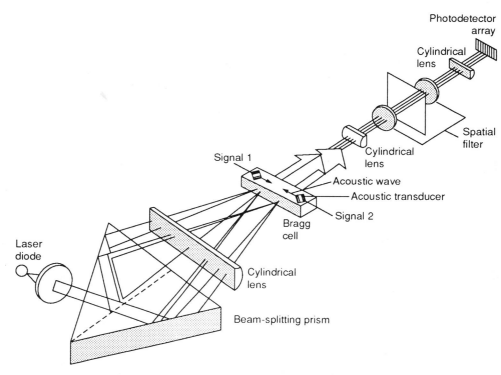

Figure 14.12 Acoustooptic correlator. A one-dimensional acoustooptic correlator uses a surface-acoustic-wave Bragg cell. Signals 1 (reference) and 2 (signal) are applied to two end transducers as shown. A split laser beam interacts with the signals separately. The beams are heterodyned and processed to obtain correlation. After A. P. Gontzonlis, *Digital electronics meets its match. Spectrum* **25**, 22. © 1988 *IEEE.*

puter would have to do about 300,000 complex multiplications and additions. An acoustooptic correlator is simple to operate and requires much less equipment. It is therefore preferred. With a slight modification, the system can be used for convolution analysis.

Holographic Imaging

Holography is the technique of producing images by wavefront recording and reconstruction. Gabor in 1948 was the first to recognize the concept of holography [10]. Practical implementation was possible after the emergence of lasers [11].

A hologram is constructed by recording on a photosensitive film the interference pattern of the phase and amplitude of coherent light from two laser beams, one a reference light from the source and the other, light reflected or scattered from an object. The film is developed chemically and the hologram can be kept for further processing. To view the hologram, the film is exposed to one of the coherent light sources again. The image of the original object is then revealed.

Phase information is vital information to be recorded with the amplitude in holography. The interference with a reference wave provides this information. Fortunately, both photographic films and photodetectors are responsive to the intensity of the total wave, which includes the interference term that carries the phase information.

To reconstruct the original image, the reference wave is again introduced. For easy separation of the original image from the reference wave, the two waves are usually directed to the hologram at different angles. The reference wave is usually a monochromatic source. If only a white light source is available, reconstruction of the original image can still be accomplished provided that a region of illumination that is spatially coherent were available.

Optical holography has many uses in industry to produce three-dimensional, highly realistic images unmatched by other photographic methods. It can check for flaws in aircraft structure, boiler welding, and other objects. It can also be used to measure velocities of moving particles. Holograms can be Fourier transformed to achieve spatial filtering and other applications.

Many books and articles have been written about the theory and applications of holography. Readers may refer to these materials for further exploration [12–14].

A new application, use of the holographic imaging method to capture light in flight, has been reported [15]. We remember the time when Edgerton's stroboscopic light captured the image of a splashing milk drop in film as spectacular and useful.

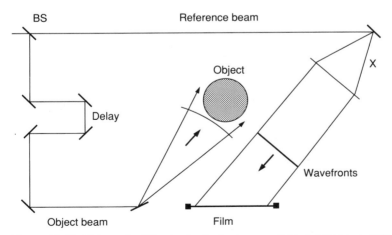

Figure 14.13 Holography. Ultrashort optical pulses are divided at the beam splitter (BS) to form an object beam path and the reference beam. X is a beam expander. The delay line is adjusted to ensure that the light scattered from the object reaches the film in synchronism with the reference pulse. After J. A. Valdmanis and N. H. Abramson [15], with permission.

In freezing the time frame at the microsecond level, many mechanical problems were solved using this technique. What can one say about the presently available techniques that freeze time at the picosecond and femtosecond (10^{-15} s) levels? In 100 femtoseconds (fs), a photon travels 30 μm. Although methods to capture single snapshot images are available, a holographic image could record a temporal continuum of three-dimensional, frameless images, each having a temporal resolution of 100 fs. The scientific value of these images can be anticipated. Studying how light propagates through a variety of media can reveal vast information about the physical world. A light-in-flight holographic experimental setup is shown in Fig. 14.13.

Only a slight modification of the conventional setup is required to perform this experiment. Here a mode-locked short coherent pulsed laser is used. In the object wave path, a delay line is added and adjusted to ensure that the light scattered from the object reaches the holographic plate simultaneously with the light from the reference wave. To capture the object in motion or other nonrepetitive events, video cameras can be used.

Digital Signal Processing

Most electrical signals have waveforms that are analog; that is, their waveforms vary continuously over time. Analog signal processing is a well-developed technology. Amplifiers and filters are more commonly built as analog devices. They work directly on the signal; however, the design of analog processing system is extremely component sensitive. Frequent readjustment for component aging and drift is required. There is very little that can be done in analog processing to improve the quality of the output.

Signals can also be processed digitally; we first sample the signal at regular intervals and then present the data in numerical form and process the data in computers. Finally, we convert the numerical values back to analog form for viewing. In digital processing, a lot can be done to improve the quality of the image. Image enhancement, which removes "snow" from a TV picture, increasing the contrast of a picture to render greater detail, and image restoration to fill in more details can all be done in digital processing [16, 17].

Analog-to-Digital Conversion

As signals come mostly in analog form, it becomes necessary to digitize the analog signals before processing digitally. This process is called *analog-to-digital conversion* indicated by D in Fig. 14.1, the original signal is captured by a scanning sensor and translated into a series of numbers that can then be manipulated at high speed by a computer in the same manner as any other numbers. The conversion is done by sampling and digitizing. The signal is sampled at regular intervals and a numerical value is assigned to each sampled unit. These values are then stored as information in serial form in a register waiting for processing.

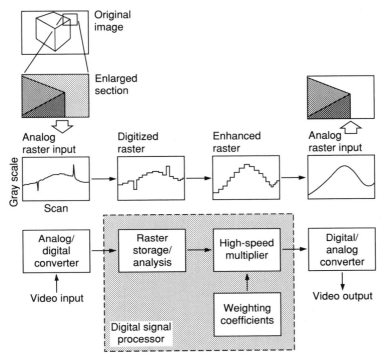

Figure 14.14 Image enhancement block diagram. In the processor, the signal is digitized by an analog-to-digital converter and stored in the digital signal processor. Each picture element in the scan is compared with its neighbor so that aberrations caused by noise can be removed and weighting coefficients can be applied to increase the range of the gray-scale values. Finally the enhanced set of elements is converted to analog form for viewing. The result is more detailed picture. After R. Cook, *High Technology*, pp. 25–30. © 1985 *High Technology*.

Image Processing

Image processing is performed by running a series of programs prewritten for the specific purpose. The computer hardware consists of a digital signal processing (DSP) chip which can be programmed to do the necessary steps for processing on computers. For example, for background noise reduction, the program may require that data from many frames of signal be averaged and then subtracted from the present data. Or each picture element of the scan is compared with its neighbor so that aberrations caused by noise can be removed and weighting coefficients can be applied to increase the range of gray-scale values. Figure 14.14 is a block diagram of this setup. When the processing is done, the enhanced set of elements is converted to analog form, resulting in a more detailed picture with reduced noise.

Poor optics, focus, or acquisitive electronics will often blur an image by attenuating higher spatial frequencies. In digital processing, one can selectively change the spatial frequency structure and reconstruct the image.

Image Recognition

Digital signal processing chips are remarkably versatile. The same chip can be used for different purposes by simply developing a suitable program. For example, in image recognition, digital images of many types of enemy airplanes can be stored in the memory, waiting to be compared with the image of an airplane in flight, and then performing correlation tests can be performed with each of the stored images to identify the type.

Adaptive Filtering

A filter circuit that can be adapted in real time to shifting circuit conditions is called an *adaptive filter.* It is very useful in long-distance voice transmission. In this type of communication circuit, echo cancellation between repeaters, which change constantly with the line characteristics, becomes an important task in maintaining perfect

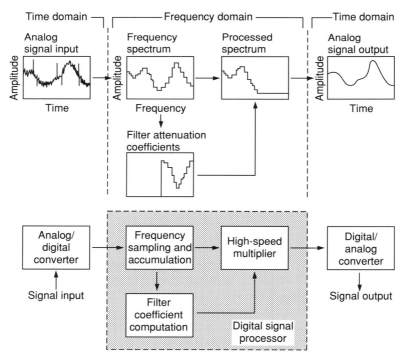

Figure 14.15 Adaptive filter block diagram. The echo pattern is identified in the digitized format in the Fourier transformed pattern. An inverse pattern is formed by reversing the filter coefficient and is added to the spectrum to cancel the noise component. After R. Cook, *High Technology.* © 1985 *High Technology.*

communication. Fast response of the equipment is a must. Digital processing is the only process fast enough to handle the huge number of computations needed to adjust the filtering property.

An adaptive filter processor of this kind is illustrated in Fig. 14.15. The analog signal in the time domain is sampled and converted into digital form as before. In the processor, a frequency spectral analysis is performed using Fourier transform technique. The frequency component responsible for echo is identified (usually located apart from that of the signal because of the different frequencies involved). A negative version of the echo pattern is synthesized and applied to the signal to cancel out the echo. The block diagram indicates each step schematically.

Image Measurement

Image processing systems are quite accurate at measuring and counting for scientific exploration and laboratory experiments. An image analysis system extends human vision by providing accurate image measurement capabilities and the possibility of automating the entire analysis process.

Image processing has its shortcomings. Automatic machine processing can do only what it does best, for example, improving and measuring images and automating some vision tasks. But for general-purpose selection, for example, to select areas of interest for examination, human eye can usually do it easier. It takes an image processing unit considerable work, both in programming and in computation, to approximate this ability poorly.

Other Optical Imaging Systems

In addition to the aforementioned systems, there are other possibilities of optical processors. A few are mentioned here.

Wide-Angle Fiberoptic Image Arrays

Coherent fiberoptic bundles have been used to convey pattern information from the input to the output end, to be viewed directly by the human eye, or to be put in immediate contact with a focal plane imager such as a CCD image array. These devices have no built-in magnification. Although a bundle of glass fibers can be coherently expanded at the output end to achieve true magnification, the process is subject to aberrations. Moslehi *et al.* developed a compound eye vision system by adding a non-imaging optic coupler to enhance the input or the output [18]. For example, in a concentrator, one end of the coupler of the fiber bundle (of 100 fibers) terminates on a CCD sensor. The other end is loosely molded into an expander fitted into a 10 × 10 array distributed on a portion of a spherical surface such that each fiber is pointing in a different direction, like the compound eye of an insect. The view imaged on this device is thus concentrated onto the CCD sensor ready to be processed.

Graded-Index Rod Imager

The graded-index (GRIN) rod is a special fiber of large diameter specially designed so that its radial index of refraction is a function of the radius, or its radial index of refraction is graded [19]. The unique property useful for imaging processing is that for a half-pitched rod length, the rod acts as an inverted mirror of the image at the input with no magnification. Arrays of these rods have been used extensively in modern copying machines. They simplify the optical lens design, making the copies smaller and less expensive.

Bar Code Reader

Everybody is aware of the bar code readers employed in most supermarkets and department stores that read the prices of items when they are passed by a lighted window. They read so accurately and fast and presents customers with a detailed list of the items they brought, the price per piece, and the total price. This well-received addition to markets, department stores, and manufacturers has many potential applications. Libraries and factories are already using these devices to keep inventories. Bar code readers may find use in any system that needs simple computation and recordkeeping.

The bar code reader is an optical image processor. The bar code is a collection of short bars of varying width arranged in a sequence intended to identify an item for sale or for storage. It contains the necessary data on the item, such as type, name, unit price, and total price. The preprinted bar-coded tag is pasted on the item for sale in the inventory. When the item is brought to the counter, it is passed by a bar code reader. The reader is a laser beam that illuminates the bar code. The reflected light is detected by photodetectors which then send a signal to process the sale through a computer. The computer manipulates simple multiplications and additions as required. The computer results are printed on a sales slip to be given to the customer and to keep records. The device is so simple, fast, accurate, and convenient to use that it is welcomed in almost all applications.

The original bar code reader, the so-called "point-of-sale," was first introduced in Europe. It was a full-size optical processor consisting of a laser light source, a set of optical lenses, rotating mirrors, and a detector. Modern devices make use of a laser beam and rotating hologram or hologram scanner, reducing the size to merely 5 cm in height.

A hologram scanner is a device using holographic optical elements to move a finely focused beam of light in a systematic pattern over a surface to sense or write out an image. A single rotating disk contains several holograms, each with a different diffraction characteristic. The optical system scans the laser beam either horizontally, vertically, or diagonally. The output light is divided into multiple beams by a set of mirrors and then directed to the bar-coded item. The moving beams may strike the bar-coded tag in any direction and any angle. When one of the beams strikes the bar code, processing is initiated. The multiple-beam technique removes the uncertainty of the beam direction and the need to hold the item against the light source in a fixed position. Light scattered back from the bar code is rediffracted through the hologram and converged by an optical lens onto a photodetector. The image of the bar-coded tag that coincides with the written hologram on the disk is selected and displayed on the photodetector for registration. Fujitsu's hologram bar code reader is shown in Fig. 14.16. A laser (He–Ne) light is directed through a series of holographic facets mounted on a spinning disk. The holograms focus the beam, and the rotation of the disk causes it to scan. A sequence of mirrors then directs the beam up through a scan plate. The holograms also reconverge the light scattered back from the bar code label onto the photodetector. This system has advantages over the old optical system: (1) The hologram replaces the complex lens-and-mirror combination of the old reader, thus reducing the overall size of the device. (2) The mirrors are set up so that multiple rays are aimed at the bar code label simultaneously and continuously, increasing the probability that the beam will read the bar code and making the reader insensitive to the location of the label. (3) The use of multiple beams ensures that at least one of them will be reflected, detected, and recorded [20].

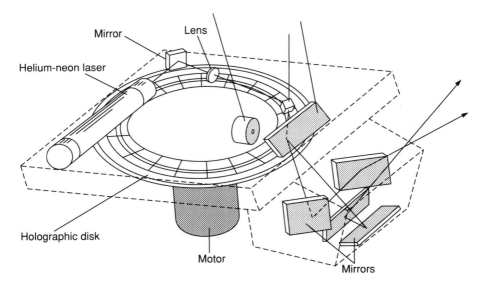

Figure 14.16 Fujitsu's bar code reader. Laser light is directed through a series of hologram facets mounted on a spinning disk. The holograms focus the beam, and the rotation of the disk causes it to scan. A sequence of mirrors directs the beam up through a scan plate. The holograms also reconverge the light scattered back from the bar code label onto the photodetector. After T. Inagaki [20], with permission.

Bimodal Imaging System

We have seen analog optical imaging systems at work. For example, the synthetic aperture radar used an analog device to process the scanning radar signal returned by of the terrain in 1953 [16]. The resolution was very high; however, a large volume of data were needed to process the returning signals numerically. With the digital technology of the 1950s, this was prohibitively difficult. To overcome this difficulty, much faster optical processing was introduced. Holographic films were taken of the radar signals modulated by a local oscillator. Then, when the film was illuminated by a coherent light source, the Fresnel zones created a map of radar reflectivity in the image plane. Despite remarkable speeds, however, such systems lacked the accuracy of digital processors.

H. John Caulfield of Alabama A&M University suggested a bimodal scheme as reported by Shimazu [21]. This model incorporates both analog and digital processing. An analog processor is used to solve a problem closely related to the problem ultimately to be solved. A digital processor then calculates errors in the same region to a high accuracy. The problem is reposed to the analog processor as a problem of error reduction. Thus, iterative improvements in accuracy are gained. In this manner, the advantages of both analog and digital processors are used to the full extent without investing too heavily in either system. Would this be a nice way to solve all image processing problem? Only time will tell.

Summary

We began this chapter by describing traditional optical image processing using optical lenses in the simplest way. Then, by introducing the Fourier transformation technique, we demonstrated that the quality of an image can be improved through Fourier plane manipulation. The simplicity and the elegance of this method have led to many applications.

Image processing can be handled in either the analog or the digital format. The analog format is simple and direct. It can also handle parallel operations, which helps to speed up the processing. Its disadvantages include the lack of exactness, difficulty in maintaining equipment in check with the changing operating conditions, and little room for improvement. On the other hand, digital processing is essentially a serial operation. An analog-to-digital converter is used at the input and then the processed information is converted back to the analog format again at the output.

We also covered the sensors used for to gather input information. Among these are charge-coupled devices, which have become popular not only in visual and video cameras, but in scientific and medical applications.

The use of acoustooptic devices for image processing is on the rise in both medicine and science. The contributions of acoustooptic devices in practical weaponry should never be neglected.

Ultrasonic image processing, the use of ultrasonic frequencies above 20 kHz to

30 MHz (and higher) in medical image applications, often called echo processing, has risen in popularity. It has become a speciality in itself. Readers are referred to another publication on this topic [22].

At the end of this chapter, we mentioned a few odds and ends of devices in the field of optical processing. The ideals are simple, but the practical applications and the future development of these devices should never be overlooked.

So many innovative image processing devices are already on the market and many more will be invented soon. Our coverage represents only a very small fraction of the applications. It serves only to stimulate our readers to look forward to future developments.

References

1. J. R. Meyer-Arent, *Introduction to Classical and Modern Optics,* Prentice-Hall, 3rd ed., Englewood Cliffs, NJ, 1989.
2. E. Kamen, *Introduction to Signal and Systems,* 2nd ed., Macmillan, New York, 1990.
3. Special issue on charge transfer devices, *IEEE Trans. Electron. Devices,* **ED-23,** No. 2, 71–126 (1976).
4. K. Prettyjohns, Improved CCDs meet special demands of spectroscopy. *Laser Focus World* **28,** No. 10, 127–136 (1992).
5. C. A. Murray and S. B. Dierker, Use of an unintensified charge-coupled device detector for low light-level Raman spectroscopy. *J. Opt. Soc. Am.* **3,** 2151 (1986).
6. B. Y. Tsaur, C. K. Chen, and J. P. Mattia, "PtSi Schottky-barrier focal plane arrays for multispctral imaging in ultraviolet, visible and infrared spectral bands. *IEEE Electron. Device Lett.* **11,** 162–164 (1990).
7. F. Okolicsany, The wave-slot, an optical television system. *Wireless Eng.* **14,** 527–536 (1937).
8. L. Brillouin, Diffusion de la lumiere et des rayons X par un corps transparent homogene. *Ann Phys. (Paris)* **17,** 88–122 (1922).
9. A special issue on acoustooptic signal processing. *Proc. IEEE* **69,** No. 1, 48–118 (1981).
10. D. Gabor, *Light and Information,* pp. 109–153, 1961.
11. E. N. Leith and J. Upatnicks, Photography by laser. *Sci. Am.* **212,** No. 6, 24–35 (1965).
12. J. B. DeVelis and G. O. Reynolds, *Theory and Applications of Holography,* Addison–Wesley, Reading, MA, 1967.
13. N. Abramson, *The Making and Evaluation of Hologram.* Academic Press, New York, 1981.
14. G. Saxby, *Practical Holography,* Prentice-Hall, Englewood Cliffs, NJ, 1989.
15. J. A. Valdmanis and N. H. Abramson, Holographic imaging captures light in flight. *Laser Focus World,* pp. 111–117 (Feb. 1991).
16. L. J. Cutrona, E. N. Leith, C. J. Palermo, and L. J. Porecello, Optical data processing and filtering systems. *IRE Trans. Inform. Theory* **IT-6,** 386–400 (1960).
17. G. C. Goodwin, R. H. Middleton, and H. V. Poor, High-speed digital signal processing and control. *Proc. IEEE* **80,** 240–259 (1992).
18. B. Moslehi, W. Parkyn, P. Harvey, D. Zaleta, and P. Soltan, Wide-angle fibre optic imaging arrays for vision and display system applications. *Electron. Lett.* **27,** 1022–1024 (1991).
19. J. D. Rees and W. Lame, Some radiometric properties of gradient-index fiber lenses. *Appl. Opt.* **19,** 1065 (1980).
20. T. Inagaki, Hologram lenses lead to compact scanners. *IEEE Spectrum* **26,** No. 3, 39–43 (1989).
21. M. Shimazu, Optical computing coming of age. *Photonic Spectra,* pp. 66–76 (Nov. 1992).
22. G. S. Kino, *Acoustic Waves: Device, Imaging, and Analog Signal Processing,* Prentice-Hall, New York, 1987.

Index

Aberration compensation, phase conjugation, 209–210
Absorption, photon–atom interactions, 14
Absorption coefficient
  light–light interactions, 14–15
  semiconductor photodetectors, quantum efficiency, 128–129
Acoustic Bragg diffraction, image processing, 310
Acoustooptically controlled switches, photonic switch design, 256–259
Acoustooptic devices
  image processors, 309–312
  modulator, photonic applications, 243–244
  operating principles, 309–310
  photonic applications, 241–244
  spectrum analyzer, 310–311
Active mode-locking devices, locking in fiber lasers, 72–73
Adaptive filtering, digital signal processing, 315–316
Adjustable delay line, solitons, 192
ADL, *see* Adjustable delay line
Alexandrite lasers, tunability, 94–95
Amplified spontaneous emission
  fiber laser
    background, 59
    operating principles, 60–61
  parametric optical amplifiers
    erbium-doped paramps, 168–169

    noise, 170–171
  soliton transmission systems, 190
Amplifiers, *see specific types*
Amplitude modulation, photonic applications, 237–238
Angular momentum, photon electromagnetic field, 12–13
APD, *see* Avalanche photodiode
Arclamps, solid-state laser pumping, 88
Argon fluoride lasers
  applications, 114–115
  commercial applications, 123–124
  excimer technology, 111–113
Argon ion laser, 106–107
  commercial applications, 120–122
  excimer technology, 111–112
ASE, *see* Amplified spontaneous emission
Aspect ratio
  image processing, charge-coupled devices, 309
Asymmetric stretch vibrational state, 107
Asynchronous time multiplexing systems, 283
ATM, *see* Asynchronous time multiplexing systems
Atomic bending bonds, 107
Atomic pumping, helium-neon laser, 104–105
Atomic structure
  laser technology, 103
  photon interactions with, 13–14
Attenuators
  graded-index rod lenses, 225–226